*Carsten Steger, Markus Ulrich,
and Christian Wiedemann*
**Machine Vision Algorithms and
Applications**

GMNT-TUKF-PJRL-NYMV

*Carsten Steger, Markus Ulrich,
and Christian Wiedemann*

Machine Vision Algorithms and Applications

WILEY-VCH Verlag GmbH & Co. KGaA

The Authors

Dr. Carsten Steger
MVTec Software GmbH
Machine Vision Technologies, München
authors@machine-vision-book.com

Dr. Markus Ulrich
MVTec Software GmbH
Machine Vision Technologies, München

Dr. Christian Wiedemann
MVTec GmbH
Machine Vision Technologies, München

1st Edition 2007
1st Reprint 2008
2nd Reprint 2012

All books published by Wiley-VCH are carefully produced. Nevertheless, authors, editors, and publisher do not warrant the information contained in these books, including this book, to be free of errors. Readers are advised to keep in mind that statements, data, illustrations, procedural details or other items may inadvertently be inaccurate.

Library of Congress Card No.:
applied for

British Library Cataloguing-in-Publication Data
A catalogue record for this book is available from the British Library.

Bibliographic information published by the Deutsche Nationalbibliothek
Die Deutsche Nationalbibliothek lists this publication in the Deutsche Nationalbibliografie; detailed bibliographic data are available on the Internet at http://dnb.d-nb.de.

© 2008 WILEY-VCH Verlag GmbH & Co. KGaA, Weinheim

All rights reserved (including those of translation into other languages). No part of this book may be reproduced in any form – by photoprinting, microfilm, or any other means – nor transmitted or translated into a machine language without written permission from the publishers. Registered names, trademarks, etc. used in this book, even when not specifically marked as such, are not to be considered unprotected by law.

Typesetting Uwe Krieg, Berlin
Printing Betz-Druck GmbH, Darmstadt
Bookbinding Litges & Dopf GmbH, Heppenheim

Printed in the Federal Republic of Germany
Printed on acid-free paper

ISBN: 978-3-527-40734-7

Contents

Preface *IX*

1   **Introduction** *1*

2   **Image Acquisition** *5*
2.1   Illumination *5*
2.1.1   Electromagnetic Radiation *5*
2.1.2   Types of Light Sources *7*
2.1.3   Interaction of Light and Matter *8*
2.1.4   Using the Spectral Composition of the Illumination *10*
2.1.5   Using the Directional Properties of the Illumination *12*
2.2   Lenses *17*
2.2.1   Pinhole Cameras *18*
2.2.2   Gaussian Optics *18*
2.2.3   Depth of Field *24*
2.2.4   Telecentric Lenses *27*
2.2.5   Lens Aberrations *30*
2.3   Cameras *35*
2.3.1   CCD Sensors *36*
2.3.2   CMOS Sensors *40*
2.3.3   Color Cameras *42*
2.3.4   Sensor Sizes *45*
2.3.5   Camera Performance *45*
2.4   Camera–Computer Interfaces *49*
2.4.1   Analog Video Signals *49*
2.4.2   Digital Video Signals: Camera Link *54*
2.4.3   Digital Video Signals: IEEE 1394 *55*
2.4.4   Digital Video Signals: USB 2.0 *57*
2.4.5   Digital Video Signals: Gigabit Ethernet *58*
2.4.6   Image Acquisition Modes *61*

3	**Machine Vision Algorithms**	**65**
3.1	Fundamental Data Structures	65
3.1.1	Images	65
3.1.2	Regions	66
3.1.3	Subpixel-Precise Contours	69
3.2	Image Enhancement	70
3.2.1	Gray Value Transformations	70
3.2.2	Radiometric Calibration	73
3.2.3	Image Smoothing	78
3.2.4	Fourier Transform	89
3.3	Geometric Transformations	94
3.3.1	Affine Transformations	94
3.3.2	Projective Transformations	95
3.3.3	Image Transformations	96
3.3.4	Polar Transformations	100
3.4	Image Segmentation	102
3.4.1	Thresholding	102
3.4.2	Extraction of Connected Components	111
3.4.3	Subpixel-Precise Thresholding	113
3.5	Feature Extraction	115
3.5.1	Region Features	116
3.5.2	Gray Value Features	120
3.5.3	Contour Features	124
3.6	Morphology	125
3.6.1	Region Morphology	125
3.6.2	Gray Value Morphology	140
3.7	Edge Extraction	144
3.7.1	Definition of Edges in 1D and 2D	145
3.7.2	1D Edge Extraction	149
3.7.3	2D Edge Extraction	155
3.7.4	Accuracy of Edges	162
3.8	Segmentation and Fitting of Geometric Primitives	169
3.8.1	Fitting Lines	170
3.8.2	Fitting Circles	174
3.8.3	Fitting Ellipses	175
3.8.4	Segmentation of Contours into Lines, Circles, and Ellipses	177
3.9	Camera Calibration	180
3.9.1	Camera Models for Area Scan Cameras	181
3.9.2	Camera Model for Line Scan Cameras	185
3.9.3	Calibration Process	188
3.9.4	World Coordinates from Single Images	193
3.9.5	Accuracy of the Camera Parameters	196

3.10	Stereo Reconstruction	*198*
3.10.1	Stereo Geometry	*199*
3.10.2	Stereo Matching	*207*
3.11	Template Matching	*211*
3.11.1	Gray-Value-Based Template Matching	*212*
3.11.2	Matching Using Image Pyramids	*217*
3.11.3	Subpixel-Accurate Gray-Value-Based Matching	*221*
3.11.4	Template Matching with Rotations and Scalings	*222*
3.11.5	Robust Template Matching	*222*
3.12	Optical Character Recognition	*241*
3.12.1	Character Segmentation	*241*
3.12.2	Feature Extraction	*243*
3.12.3	Classification	*246*
4	**Machine Vision Applications**	*263*
4.1	Wafer Dicing	*263*
4.2	Reading of Serial Numbers	*269*
4.3	Inspection of Saw Blades	*276*
4.4	Print Inspection	*281*
4.5	Inspection of Ball Grid Arrays (BGA)	*285*
4.6	Surface Inspection	*293*
4.7	Measuring of Spark Plugs	*301*
4.8	Molding Flash Detection	*307*
4.9	Inspection of Punched Sheets	*316*
4.10	3D Plane Reconstruction with Stereo	*320*
4.11	Pose Verification of Resistors	*328*
4.12	Classification of Non-Woven Fabrics	*334*

References *339*

Index *347*

Preface

The machine vision industry has enjoyed a growth rate well above the industry average for many years. Machine vision systems currently form an integral part of many machines and production lines. Furthermore, machine vision systems are continuously deployed in new application fields, in part because computers become faster all the time and thus enable applications to be solved that were out of reach just a few years ago.

Despite its importance, there are few books that describe in sufficient detail the technology that is important for machine vision. While there are numerous books on image processing and computer vision, very few of them describe the hardware components that are used in machine vision systems to acquire images (illuminations, lenses, cameras, and camera–computer interfaces). Furthermore, these books often only describe the theory, but not its use in real-world applications. Machine vision books, on the other hand, often do not describe the relevant theory in sufficient detail. Therefore, we feel that a book that provides a thorough theoretical foundation of all the machine vision components and machine vision algorithms, and that gives non-trivial practical examples of how they can be used in real applications, is highly overdue.

The applications we present in this book are based on the machine vision software HALCON, developed by MVTec Software GmbH. To enable you to get a hands-on experience with the machine vision algorithms and applications that we discuss, this book contains a registration code that enables you to download, free of charge, a student version of HALCON as well as all the applications we discuss. For details, please visit www.machine-vision-book.com.

While the focus of this book is on machine vision applications, we would like to emphasize that the principles we will present can also be used in other application fields, e.g., photogrammetry or medical image processing.

We have tried to make this book accessible to students as well as practitioners (OEMs, system integrators, and end-users) of machine vision. The text requires only very little mathematical background. We assume that the reader has a basic knowledge of linear algebra (in particular, linear transformations between vector spaces expressed in matrix algebra) and calculus (in particular, sums and differentiation and integration of one- and two-dimensional functions).

Machine Vision Algorithms and Applications. Carsten Steger, Markus Ulrich, and Christian Wiedemann
Copyright © 2008 WILEY-VCH Verlag GmbH & Co. KGaA, Weinheim
ISBN: 978-3-527-40734-7

This book is based on a lecture and lab course entitled "Machine vision algorithms" that Carsten Steger has held annually since 1999 at the Department of Informatics of Technische Universität München. Parts of the material have also been used by Markus Ulrich in a lecture entitled "Close-range photogrammetry" held annually since 2005 at the Institute of Photogrammetry and Cartography of Technische Universität München. These lectures typically draw an audience from various disciplines, e.g., computer science, photogrammetry, mechanical engineering, mathematics, and physics, which serves to emphasize the interdisciplinary nature of machine vision.

We would like to express out gratitude to several of our colleagues who have helped us in the writing of this book. Wolfgang Eckstein, Juan Pablo de la Cruz Gutiérrez, and Jens Heyder designed or wrote several of the application examples in Chapter 4. Many thanks also go to Gerhard Blahusch, Alexa Zierl, and Christoph Zierl for proofreading the manuscript. Finally, we would like to express our gratitude to Andreas Thoß and Ulrike Werner of Wiley-VCH for having the confidence that we would be able to write this book during the time HALCON 8.0 was completed.

We invite you to send us suggestions on how to improve this book. You can reach us at authors@machine-vision-book.com.

München, May 2007   *Carsten Steger, Markus Ulrich, Christian Wiedemann*

1
Introduction

Machine vision is one of the key technologies in manufacturing because of increasing demands on the documentation of quality and the traceability of products. It is concerned with engineering systems, such as machines or production lines, that can perform quality inspections in order to remove defective products from production or that control machines in other ways, e.g., by guiding a robot during the assembly of a product.

Some of the common tasks that must be solved in machine vision systems are as follows [1]:

- Object identification is used to discern different kinds of objects, e.g., to control the flow of material or to decide which inspections to perform. This can be based on special identification symbols, e.g., character strings or bar codes, or on specific characteristics of the objects themselves, such as their shape.

- Position detection is used, for example, to control a robot that assembles a product by mounting the components of the product at the correct positions, e.g., in a pick-and-place machine that places electronic components onto a printed circuit board (PCB). Position detection can be performed in two or three dimensions, depending on the requirements of the application.

- Completeness checking is typically performed after a certain stage of the assembly of a product has been completed, e.g., after the components have been placed onto a PCB, to ensure that the product has been assembled correctly, i.e., that the right components are at the right place.

- Shape and dimensional inspection is used to check the geometric parameters of a product to ensure that they lie within the required tolerances. This can be used during the production process but also after a product has been in use for some time to ensure that the product still meets the requirements despite wear and tear.

- Surface inspection is used to check the surface of a finished product for imperfections such as scratches, indentations, protrusions, etc.

Figure 1.1 displays an example of a typical machine vision system. The object (1) is transported mechanically, e.g., on a conveyor belt. In machine vision applications,

Machine Vision Algorithms and Applications. Carsten Steger, Markus Ulrich, and Christian Wiedemann
Copyright © 2008 WILEY-VCH Verlag GmbH & Co. KGaA, Weinheim
ISBN: 978-3-527-40734-7

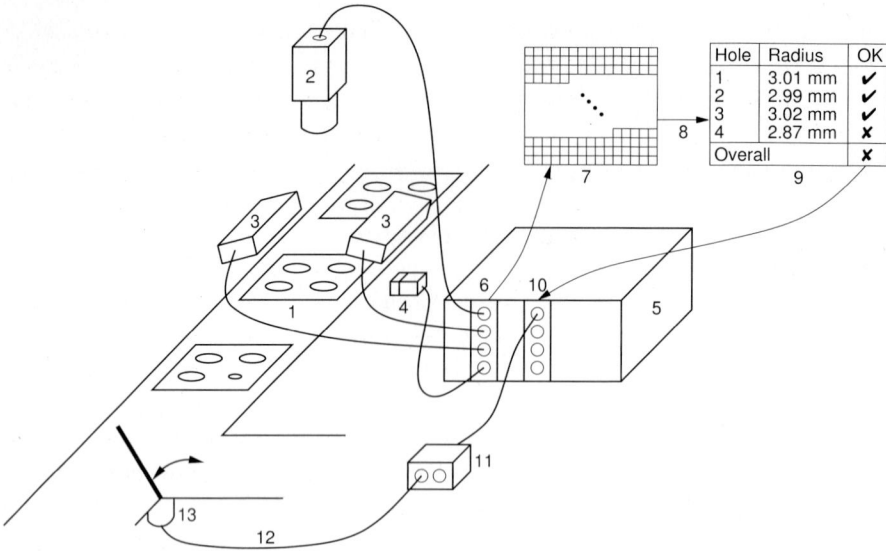

Fig. 1.1 The components of a typical machine vision system. An image of the object to be inspected (1) is acquired by a camera (2). The object is illuminated by the illumination (3). A photoelectric sensor (4) triggers the image acquisition. A computer (5) acquires the image through a camera–computer interface (6), in this case a frame grabber. The photoelectric sensor is connected to the frame grabber. The frame grabber triggers the strobe illumination. A device driver assembles the image (7) in the memory of the computer. The machine vision software (8) inspects the objects and returns an evaluation of the objects (9). The result of the evaluation is communicated to a programmable logic controller (PLC) (11) via a digital input/output (I/O) interface (10). The PLC controls an actuator (13) through a fieldbus interface (12). The actuator, e.g., an electric motor, moves a diverter that is used to remove defective objects from the production line.

we would often like to image the object in a defined position. This requires mechanical handling of the object and often also a trigger that triggers the image acquisition, e.g., a photoelectric sensor (4). The object is illuminated by a suitably chosen or specially designed illumination (3). Often, screens (not shown) are used to prevent ambient light from falling onto the object and thereby lowering the image quality. The object is imaged with a camera (2) that uses a lens that has been suitably selected or specially designed for the application. The camera delivers the image to a computer (5) through a camera–computer interface (6), e.g., a frame grabber. The device driver of the camera–computer interface assembles the image (7) in the memory of the computer. If the image is acquired through a frame grabber, the illumination may be controlled by the frame grabber, e.g., through strobe signals. If the camera–computer interface is not a frame grabber but a standard interface, such as IEEE 1394, USB 2.0, or Ethernet, the trigger will typically be connected to the camera and illumination directly or through a programmable logic controller (PLC). The computer can be a standard industrial PC or a specially designed computer that is directly built into the camera. The latter configuration is often called a smart camera. The computer may use a standard processor, a digital signal processor (DSP), a field-programmable gate array (FPGA), or a combination of the above. The machine vision software (8) inspects the objects and returns an evaluation of the objects (9). The result of the evaluation is communicated to a controller (11), e.g., a PLC or a distributed control system (DCS). Often, this communication is performed by digital input/output (I/O) interfaces (10). The PLC, in turn, typically controls an actuator (13) through a communication interface (12), e.g., a fieldbus or serial interface. The actuator, e.g., an electric motor, then moves a diverter that is used to remove defective objects from the production line.

As can be seen from the large number of components involved, machine vision is inherently multidisciplinary. A team that develops a machine vision system will require expertise in mechanical engineering, electrical engineering, optical engineering, and software engineering.

To maintain the focus of this book, we have made a conscious decision to focus on the aspects of a machine vision system that are pertinent to the system until the relevant information has been extracted from the image. Therefore, we will forgo a discussion of the communication components of a machine vision system that are used after the machine vision software has determined its evaluation. For more information, please consult [2, 3, 4].

In this book, we will try to give you a solid background on everything that is required to extract the relevant information from images in a machine vision system. We include the information that we wish someone had taught us when we started working in the field. In particular, we mention several idiosyncrasies of the hardware components that are highly relevant in applications, which we had to learn the hard way.

The hardware components that are required to obtain high-quality images are described in Chapter 2: illuminations, lenses, cameras, and camera–computer interfaces. We hope that, after reading this chapter, you will be able to make informed decisions about which components and setups to use in your application.

Chapter 3 discusses the most important algorithms that are commonly used in machine vision applications. It is our goal to provide you with a solid theoretical foundation that helps you in designing and developing a solution for your particular machine vision task.

To emphasize the engineering aspect of machine vision, Chapter 4 contains a wealth of examples and exercises that show how the machine vision algorithms discussed in Chapter 3 can be combined in non-trivial ways to solve typical machine vision applications.

2
Image Acquisition

In this chapter, we will take a look at the hardware components that are involved in obtaining an image of the scene we want to analyze with the algorithms presented in Chapter 3. Illumination makes the essential features of an object visible. Lenses produce a sharp image on the sensor. The sensor converts the image into an analog or digital video signal. Finally, camera–computer interfaces (analog or digital frame grabbers, bus systems like IEEE 1394 or USB 2.0, or network interfaces like Ethernet) accept the video signal and convert it into an image in the computer's memory.

2.1
Illumination

The goal of illumination in machine vision is to make the important features of the object visible and suppress undesired features of the object. To do so, we have to consider how the light interacts with the object. One important aspect is the spectral composition of the light and the object. We can use, for example, monochromatic light on colored objects to enhance the contrast of the desired object features. Furthermore, the direction from which we illuminate the objects can be used to enhance the visibility of features. We will examine these aspects in this section.

2.1.1
Electromagnetic Radiation

Light is electromagnetic radiation of a certain range of wavelengths, as shown in Table 2.1. The range of wavelengths visible for humans is 380–780 nm. Light with shorter wavelengths is called ultraviolet (UV) light. Electromagnetic radiation with even shorter wavelengths consists of X-rays and gamma rays. Light with longer wavelengths than the visible range is called infrared (IR) light. Electromagnetic radiation with even longer wavelengths consists of microwaves and radio waves.

Monochromatic light is characterized by its wavelength λ. If light is composed of a range of wavelengths, it is often compared to the spectrum of light emitted by a black body. A black body is an object that absorbs all electromagnetic radiation that falls onto it and thus serves as an ideal source of purely thermal radiation. Therefore,

Machine Vision Algorithms and Applications. Carsten Steger, Markus Ulrich, and Christian Wiedemann
Copyright © 2008 WILEY-VCH Verlag GmbH & Co. KGaA, Weinheim
ISBN: 978-3-527-40734-7

Tab. 2.1 The light spectrum. The names of the ranges for IR and UV radiation correspond to the German standard DIN 5031 [5]. The names of the colors are due to [6].

Range	Name	Abbreviation	Wavelength λ
Ultraviolet	Vacuum UV	UV-C	100 nm–200 nm
	Far UV		200 nm–280 nm
	Middle UV	UV-B	280 nm–315 nm
	Near UV	UV-A	315 nm–380 nm
Visible	Blue-purple		380 nm–430 nm
	Blue		430 nm–480 nm
	Green-blue		480 nm–490 nm
	Blue-green		490 nm–510 nm
	Green		510 nm–530 nm
	Yellow-green		530 nm–570 nm
	Yellow		570 nm–580 nm
	Orange		580 nm–600 nm
	Red		600 nm–720 nm
	Red-purple		720 nm–780 nm
Infrared	Near IR	IR-A	780 nm–1.4 µm
		IR-B	1.4 µm–3 µm
	Middle IR	IR-C	3 µm–50 µm
	Far IR		50 µm–1 mm

the light spectrum of a black body is directly related to its temperature. The spectral radiance of a black body is given by Planck's law [7, 8]:

$$I(\lambda, T) = \frac{2hc^2}{\lambda^5} \frac{1}{e^{hc/\lambda kT} - 1} \tag{2.1}$$

Here, $c = 2.997\,924\,58 \times 10^8 \text{ m s}^{-1}$ is the speed of light, $h = 6.626\,0693 \times 10^{-34}$ J s is the Planck constant, and $k = 1.380\,6505 \times 10^{-23}$ J K^{-1} is the Boltzmann constant. The spectral radiance is the energy radiated per unit wavelength by an infinitesimal patch of the black body into an infinitesimal solid angle of space. Hence, its unit is W sr^{-1} m^{-2} nm^{-1}.

Figure 2.1 displays the spectral radiance for different temperatures T. It can be seen that black bodies at 300 K radiate primarily in the middle and far IR range. This is the radiation range that is perceived as heat. Therefore, this range of wavelengths is also called thermal IR. The radiation of an object at 1000 K just starts to enter the visible range. This is the red glow that can be seen first when objects are heated. For $T = 3000$ K, the spectrum is that of an incandescent lamp (see Section 2.1.2). Note that it has a strong red component. The spectrum for $T = 6500$ K is used to represent average daylight. It defines the spectral composition of white light. The spectrum for $T = 10000$ K produces light with a strong blue component.

Because of the correspondence of the spectra with the temperature of the black body, the spectra also define so-called color temperatures.

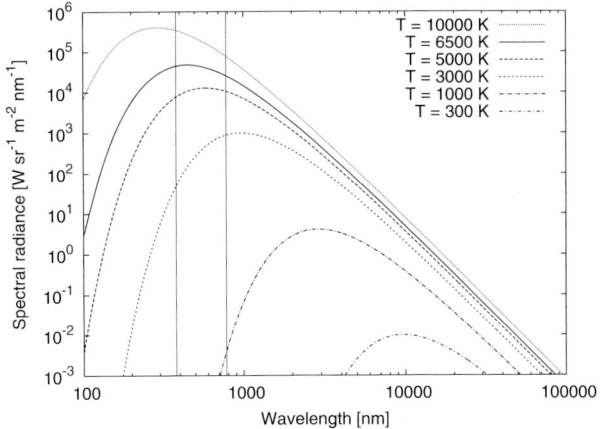

Fig. 2.1 Spectral radiance emitted by black bodies of different temperatures. The vertical lines denote the visible range of the spectrum.

2.1.2 Types of Light Sources

Before we take a look at how to use light in machine vision, we will discuss the types of light sources that are commonly used in machine vision.

Incandescent lamps create light by sending an electrical current through a thin filament, typically made of tungsten. The current heats the filament and causes it to emit thermal radiation. The heat in the filament is so high that the radiation is in the visible range of the electromagnetic spectrum. The filament is contained in a glass envelope that contains either a vacuum or a halogen gas, such as iodine or bromine, which prevents oxidation of the filament. Filling the envelope with a halogen gas has the advantage that the lifetime of the lamp is increased significantly compared to using a vacuum. The advantage of incandescent lamps is that they are relatively bright and create a continuous spectrum with a color temperature of 3000–3400 K. Furthermore, they can be operated with low voltage. One of their disadvantages is that they produce a large amount of heat: only about 5% of the power is converted to light; the rest is emitted as heat. Other disadvantages are short lifetimes and the inability to use them as flashes. Furthermore, they age quickly, i.e., their brightness decreases significantly over time.

Xenon lamps consist of a sealed glass envelope filled with xenon gas, which is ionized by electricity, producing a very bright white light with a color temperature of 5500–12 000 K. They are commonly divided into continuous-output short- and long-arc lamps as well as flash lamps. Xenon lamps can produce extremely bright flashes at a rate of more than 200 flashes per second. Each flash can be extremely short, e.g., 1–20 µs for short-arc lamps. One of their disadvantages is that they require a sophisticated and expensive power supply. Furthermore, they exhibit aging after several million flashes.

Like xenon lamps, fluorescent lamps are gas-discharge lamps that use electricity to excite mercury vapor in a noble gas, e.g., argon or neon, causing ultraviolet light to be emitted. This UV light causes a phosphor salt coated onto the inside of the tube that contains the gas to fluoresce, producing visible light. Different coatings can be chosen, resulting in different spectral distributions of the visible light, which are equivalent to color temperatures of 3000–6000 K. Fluorescent lamps are driven by alternating current. This results in a flickering of the lamp with the same frequency as the current. For machine vision, high-frequency alternating currents of 22 kHz or more must be used to avoid spurious brightness changes in the images. The main advantages of fluorescent lamps are that they are inexpensive and can illuminate large areas. Some of their disadvantages are a short lifetime, rapid aging, and an uneven spectral distribution with sharp peaks for certain frequencies. Furthermore, they cannot be used as flashes.

A light-emitting diode (LED) is a semiconductor device that produces narrow-spectrum (i.e., quasi-monochromatic) light through electroluminescence: the diode emits light in response to an electric current that passes through it. The color of the emitted light depends on the composition and condition of the used semiconductor material. The possible range of colors comprises infrared, visible, and near ultraviolet light. White LEDs can also be produced. They internally emit blue light, which is converted to white light by a coating with a yellow phosphor on the semiconductor. One advantage of LEDs is their longevity: lifetimes larger than 100 000 hours are not uncommon. Furthermore, they can be used as flashes with fast reaction times and almost no aging. Since they use direct current, their brightness can be controlled easily. In addition, they use comparatively little power and produce little heat. The main disadvantage of LEDs is that their performance depends on the ambient temperature of the environment in which they operate. The higher the ambient temperature, the lower the performance of the LED and the shorter its lifetime. However, since LEDs have so many practical advantages, they are currently the primary illumination technology used in machine vision applications.

2.1.3
Interaction of Light and Matter

Light can interact with objects in various ways, as shown in Figure 2.2.

Reflection occurs at the interfaces between different media. The microstructure of the object (essentially the roughness of its surface) determines how much of the light is reflected diffusely and how much specularly. Diffuse reflection scatters the reflected light more or less evenly in all directions. For specular reflection, the incoming and reflected light ray lie in a single plane. Furthermore, their angles with respect to the surface normal are identical. Hence, the macrostructure (shape) of the object determines the direction into which the light is reflected specularly. In practice, however, specular reflection is never perfect (as it is for a mirror). Instead, specular reflection

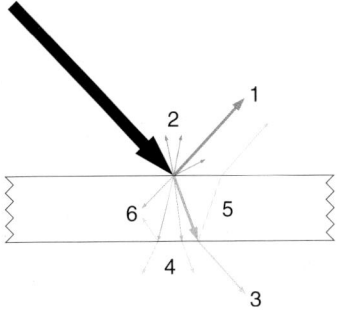

Fig. 2.2 The interaction of light with the object. The light that falls onto the object is visualized by the black arrow. (1) Specular reflection. (2) Diffuse reflection. (3) Direct transmission. (4) Diffuse transmission. (5) Backside reflection. (6) Absorption.

causes a lobe of intense reflection for certain viewing angles, depending on the angle of the incident light, as shown in Figure 2.3. The width of the side lobe is determined by the microstructure of the surface.

Reflection at metal and dielectric surfaces, e.g., glass or plastics, causes light to become partially polarized. Polarization occurs for diffuse as well as specular reflection. In practice, however, polarization caused by specular reflection dominates.

The fraction of light reflected by the surface is given by the bidirectional reflectance distribution function (BRDF). The BRDF is a function of the direction of the incoming light, the viewing direction, and the wavelength of the light. If the BRDF is integrated over both directions, the reflectivity of the surface is obtained. It depends only on the wavelength.

Note that reflection also occurs at the interface between two transparent media. This may cause backside reflections, which can lead to double images.

Transmission occurs when the light rays pass through the object. Here, the light rays are refracted, i.e., they change their direction at the interface between the different media. This is discussed in more detail in Section 2.2.2. Depending on the internal and surface structure of the object, the transmission can be direct or diffuse. The fraction of light that passes through the object is called transmittance. Like the reflectivity, it depends, among other factors, on the wavelength of the light.

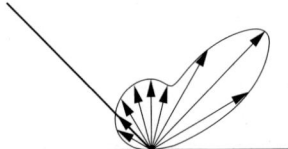

Fig. 2.3 Light distribution caused by the combination of diffuse and specular reflection.

Finally, absorption occurs if the incident light is converted into heat within the object. All light that is neither reflected nor transmitted is absorbed. If we denote the light that falls onto the object by I, the reflected light by R, the transmitted light by T, and the absorbed light by A, the law of conservation of energy dictates that $I = R + T + A$. In general, dark objects absorb a significant amount of light.

All of the above quantities except specular reflection depend on the wavelength of the light that falls onto the object. Wavelength-dependent diffuse reflection and absorption give opaque objects their characteristic color. Likewise, wavelength-dependent transmission gives transparent objects their characteristic color.

Finally, it should be noted that real objects are often more complex than the simple model described above. For example, an object may consist of several layers of different materials. The top layer may be transparent to some wavelengths and reflect others. Further layers may reflect parts of the light that has passed through the layers above. Therefore, finding out a suitable illumination for real objects often requires a significant amount of experimentation.

2.1.4
Using the Spectral Composition of the Illumination

As mentioned in the previous section, colored objects reflect certain portions of the light spectrum, while absorbing other portions. This can often be used to enhance the visibility of certain features by employing an illumination source that uses a range of wavelengths that is reflected by the objects that should be visible and is absorbed by the objects that should be suppressed. For example, if a red object on a green background should be enhanced, a red illumination can be used. The red object will appear bright, while the green object will appear dark.

Figure 2.4 illustrates this principle by showing a printed circuit board (PCB) illuminated with white, red, green, and blue light. While white light produces a relatively good average contrast, the contrast of certain features can be enhanced significantly by using colored light. For example, the contrast of the large component in the lower left part of the image is significantly better under red illumination because the compo-

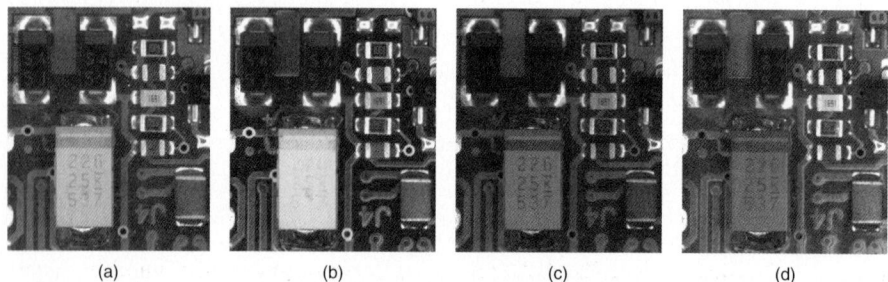

Fig. 2.4 A PCB illuminated with (a) white, (b) red, (c) green, and (d) blue light.

nent itself is light orange, while the print on the component is dark orange. Therefore, the component itself can be segmented more easily with red illumination. Note, however, that red illumination reduces the contrast of the print on the component, which is significantly better under green illumination. Red illumination also enhances the contrast of the copper-colored plated through holes. Blue illumination, on the other hand, maximizes the contrast of the light blue resistor in the center of the row of five small components.

Since charge-coupled device (CCD) and complementary metal–oxide–semiconductor (CMOS) sensors are sensitive to infrared light (see Section 2.3.3), infrared light can often also be used to enhance the visibility of certain features, as shown in Figure 2.5. For example, the tracks can be made visible easily with infrared light since the matt green solder resist that covers the tracks and makes them hard to detect in visible light is transparent to infrared light.

Fig. 2.5 A PCB illuminated with infrared light.

All of the above effects can also be achieved through the use of white light and color filters. However, since a lot of efficiency is wasted if white light is created only to be filtered out later, it is almost always preferable to use colored illumination right away.

Nevertheless, there are filters that are useful for machine vision. As we have seen above, CCD and CMOS sensors are sensitive to infrared light. Therefore, an infrared cut filter is often useful to avoid unexpected brightness or color changes in the image. On the other hand, if the object is illuminated with infrared light, an infrared pass filter, i.e., a filter that suppresses the visible spectrum and lets only infrared light pass, is often helpful.

Another useful filter is a polarizing filter. As mentioned in Section 2.1.3, light becomes partially polarized through reflection at metal and dielectric surfaces. To suppress this light, we can mount a polarizing filter in front of the camera and turn it in such a way that the polarized light is suppressed. Since unpolarized light is only partially polarized through reflection, an even better suppression of specular reflections can be achieved if the light that falls onto the object is already polarized. This principle is shown in Figure 2.6. The polarizing filters in front of the illumination and camera are called polarizer and analyzer, respectively.

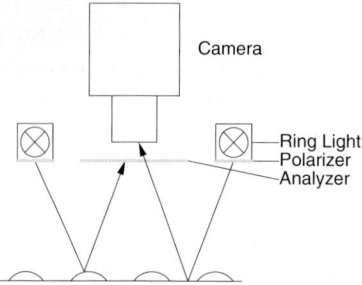

Fig. 2.6 The principle of a polarizer and analyzer. The polarizer and analyzer are polarizing filters that are mounted in front of the illumination and camera, respectively. If the analyzer is turned by 90° with respect to the polarizer, light polarized by specular reflection is suppressed by the analyzer.

The effect of using polarizing filters is shown in Figure 2.7. Figure 2.7(a) shows a PCB illuminated with a directed ring light. This causes specular reflections on the board, solder, and metallic parts of the components. Using a polarizer and analyzer almost completely suppresses the specular reflections, as shown in Figure 2.7(b).

Fig. 2.7 (a) A PCB illuminated with a directed ring light. Note the specular reflections on the board, solder, and metallic parts of the components. (b) A PCB illuminated with a directed ring light using a polarizer and analyzer. Note that the specular reflections have been suppressed almost completely.

2.1.5
Using the Directional Properties of the Illumination

While the spectral composition of light can often be used advantageously, most often the directional properties of the illumination are used to enhance the visibility of the essential features in machine vision.

By directional properties, we mean two different effects. On the one hand, the light source may be diffuse or directed. In the first case, the light source emits the light more or less evenly in all directions. In the second case, the light source emits the light in a very narrow range of directions. In the limiting case, the light source emits only parallel light rays in a single direction. This is called telecentric illumination. Telecentric illuminations use the same principles as telecentric lenses (see Section 2.2.4).

On the other hand, the placement of the light source with respect to the object and camera is important. Here, we can discern different aspects. If the light source is on the same side of the object as the camera, we speak of front light. This is also often referred to as incident light. However, since incident light often means the light that falls onto an object, we will use front light throughout this book. If the light source is on the opposite side of the object as the camera, we speak of back light. This is sometimes also called transmitted light, especially if images of transparent objects are acquired. If the light source is placed at an angle to the object so that most of the light is reflected to the camera, we speak of bright-field illumination. Finally, if the light is placed in such a way that most of the light is reflected away from the camera, and only light of certain parts of the object is reflected to the camera, we speak of dark-field illumination.

All of the above criteria are more or less orthogonal to each other and can be combined in various ways. We will discuss the most commonly used combinations below.

Diffuse bright-field front light illuminations can be built in several ways, as shown in Figure 2.8. LED panels or ring lights with a diffuser in front of the lights have the advantage that they are easy to construct. The light distribution, however, is not

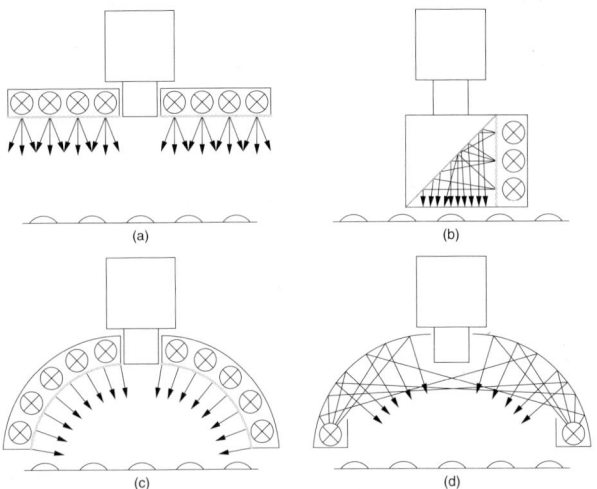

Fig. 2.8 Different diffuse bright-field front light constructions.
(a) A LED panel or ring light with a diffuser in front of the lights. (b) A coaxial diffuse light with a semi-transparent diagonal mirror and a diffuser in front of the lights. (c) A dome light with a a diffuser in front of the lights. (d) A dome light with an LED ring light and the dome itself acting as the diffuser.

perfectly homogeneous, unless the light panel or ring is much larger than the object to be imaged. Furthermore, no light comes from the direction in which the camera looks through the illumination. Coaxial diffuse lights are regular diffuse light sources for which the light is reflected onto the object through a semi-transparent mirror, through which the camera also acquires the images. Since there is no hole through which the camera must look, the light distribution is more uniform. However, the semi-transparent mirror can cause ghost images. Dome lights try to achieve a perfectly homogeneous light distribution either by using a dome-shaped light panel with a diffuser in front of the lights or by mounting a ring light into a diffusely reflecting dome. In the latter case, multiple internal reflections cause the light to show no preferred direction. Like the LED panels or ring lights, the dome lights have the slight disadvantage that no light comes from the direction of the hole through which the camera looks at the object.

Diffuse bright-field front light illuminations are typically used to prevent shadows and to reduce or prevent specular reflections. They can also be used to look through transparent covers of objects, e.g., the transparent plastic on blister packs (see Figure 2.9).

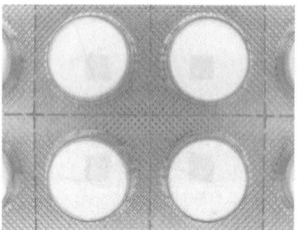

Fig. 2.9 Blister pack illuminated with a dome light of the type shown in Figure 2.8(d).

Directed bright-field front light illumination comes in two basic varieties, as shown in Figure 2.10. One way to construct them is to use tilted ring lights. This is typically used to create shadows in cavities or around the objects of interest. One disadvantage

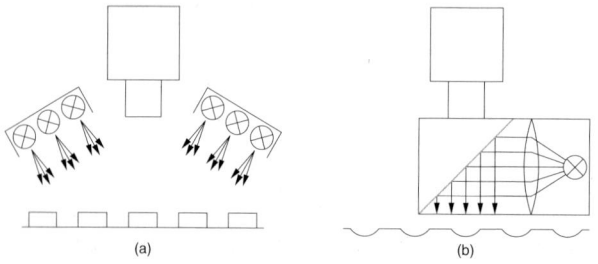

Fig. 2.10 Different directed bright-field front light constructions. (a) A focused ring light. (b) A coaxial telecentric illumination with a semi-transparent diagonal mirror.

of this type of illumination is that the light distribution is typically uneven. A second method is to use a coaxial telecentric illumination to image specular objects. Here, the principle is that object parts that are parallel to the image plane reflect the light to the camera, while all other object parts reflect the light away from the camera. This requires a telecentric lens to work. Furthermore, the object must be aligned very well to ensure that the light is reflected to the camera. If this is not ensured mechanically, the object may appear completely dark.

Directed dark-field front light illumination is typically constructed as an LED ring light, as shown in Figure 2.11. The ring light is mounted at a very small angle to the object's surface. This is used to highlight indentations and protrusions of the object. Hence, the visibility of structures like scratches, texture, or engraved characters can be enhanced. An example of the last two applications is shown in Figure 2.12.

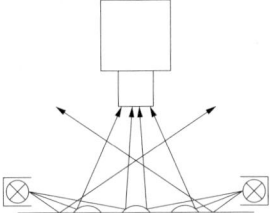

Fig. 2.11 Directed dark-field front light illumination. This type of illumination is typically constructed as an LED ring light.

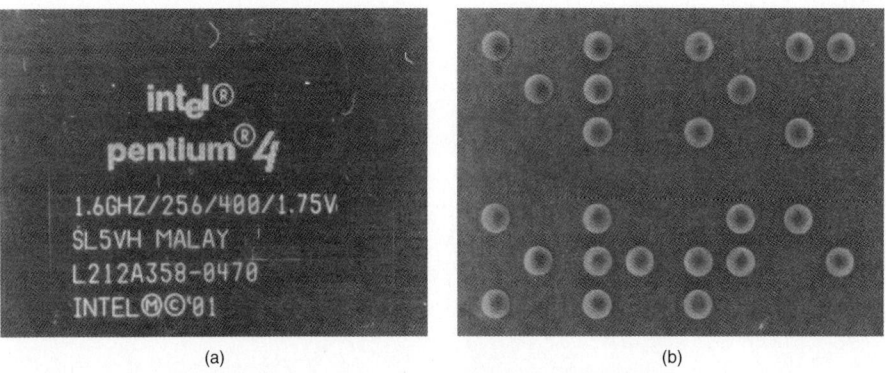

Fig. 2.12 (a) An engraved serial number on a CPU and (b) Braille print on a pharmaceutical package highlighted with directed dark-field front light illumination. Note that the characters as well as the scratches on the CPU are highlighted.

Diffuse bright-field back light illumination consists of a light source, often made from LED panels or fluorescent lights, and a diffuser in front of the lights. The light source is positioned behind the object, as shown in Figure 2.13(a). Back light illumination only shows the silhouette of opaque objects. Hence, it can be used whenever the essential information can be derived from the object's contours. For transparent objects, back light illumination can be used in some cases to make the inner parts of the objects visible because it avoids the reflections that would be caused by front light illumination. One drawback of diffuse bright-field back light illumination is shown in Figure 2.13(b). Because the illumination is diffuse, for objects with a large depth some parts of the object that lie on the camera side of the object can be illuminated. Therefore, diffuse back light illumination is typically only used for objects that have a small depth.

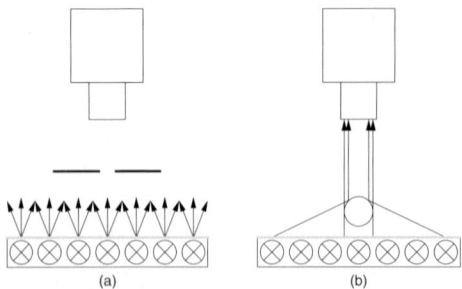

Fig. 2.13 (a) Diffuse bright-field back light illumination. This type of illumination is often constructed with LED panels or fluorescent lights with a diffuser in front of the lights. (b) Reflections on the camera side of the object can occur for objects with a large depth.

Figure 2.14 displays two kinds of objects that are usually illuminated with a diffuse bright-field back light illumination: a flat metal workpiece and a filament in a light bulb.

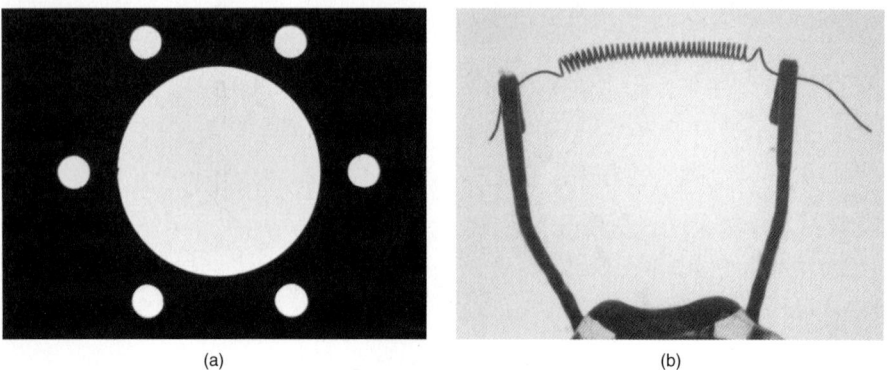

Fig. 2.14 (a) Metal workpiece and (b) filament in a light bulb illuminated with a diffuse bright-field back light illumination.

Directed bright-field back light illuminations are usually constructed with telecentric illuminations, as shown in Figure 2.15, to prevent the problem that the camera side of the object is also illuminated. For a perspective lens, the image of the telecentric light would be a small spot in the image because the light rays are parallel. Therefore, telecentric back light illuminations must be used with telecentric lenses. The illumination must be carefully aligned with the lens. This type of illumination produces very sharp edges at the silhouette of the object. Furthermore, because telecentric lenses are used, there are no perspective distortions in the image. Therefore, this kind of illumination is frequently used in measurement applications.

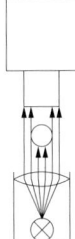

Fig. 2.15 A telecentric bright-field back light illumination.

Figure 2.16 compares the effects of using diffuse and telecentric bright-field back light illuminations on part of a spark plug. In both cases, a telecentric lens was used. Figure 2.16(a) clearly shows the reflections on the camera side of the object that occur with the diffuse illumination. These reflections do not occur for the telecentric illumination, as can be seen from Figure 2.16(b).

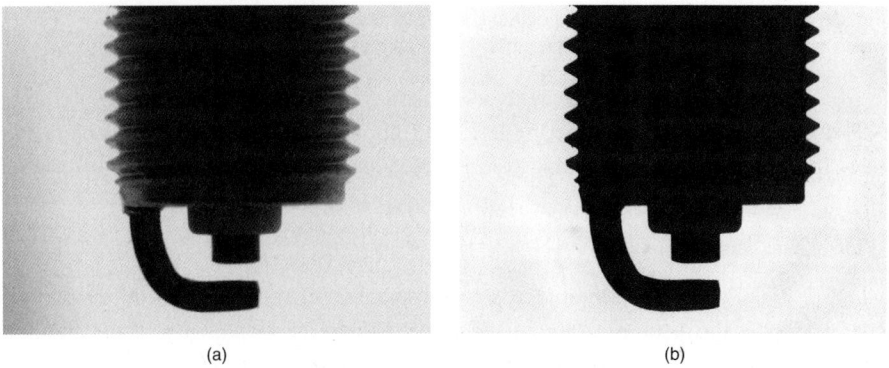

Fig. 2.16 A spark plug illuminated with (a) diffuse and (b) telecentric back light illumination. Note the reflections on the camera side of the object that occur with the diffuse illumination.

2.2 Lenses

A lens is an optical device through which light is focused in order to form an image inside of a camera, in our case on a digital sensor. The purpose of the lens is to create a sharp image in which fine details can be resolved. In this section, we will take a look

2.2.1
Pinhole Cameras

If we neglect the wave nature of light, we can treat light as rays that propagate in straight lines in a homogeneous medium. Consequently, a model for the image created by a camera is given by the pinhole camera, as shown in Figure 2.17. Here, the object on the left side is imaged on the image plane on the right side. The image plane is one of the faces of a box into which a pinhole has been made on the opposite side to the image plane. The pinhole acts as the projection center. The pinhole camera produces an upside-down image of the object.

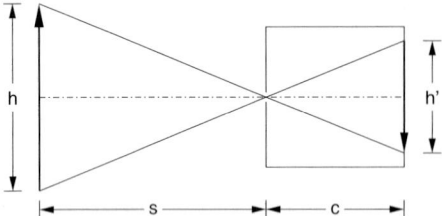

Fig. 2.17 The pinhole camera. The object is imaged in a box into which a pinhole has been made. The pinhole acts as the projection center.

From the similar triangles on the left and right sides of the projection center, it is clear that the height h' of the image of the object is given by

$$h' = h\frac{c}{s} \tag{2.2}$$

where h is the height of the object, s is the distance of the object to the projection center, and c is the distance of the image plane to the projection center. The distance c is called the camera constant or the principal distance. Equation (2.2) shows that, if we increase the principal distance c, the size h' of the image of the object also increases. On the other hand, if we increase the object distance s, then h' decreases.

2.2.2
Gaussian Optics

For the purposes of measuring objects in world coordinates through camera calibration (see Section 3.9), the pinhole camera model would almost be sufficient. This simple model does not reflect reality, however. Because of the small pinhole, very little light would actually pass to the image plane, and we would have to use very long exposure times to obtain an adequate image. Therefore, real cameras use lenses to collect light in the image. Lenses are typically formed from a piece of shaped glass or plastic. Depending on their shape, lenses can cause the light rays to converge or to diverge.

Lenses are based on the principle of refraction. A light ray traveling in a certain medium will travel at a speed v, slower than the speed of light in vacuum c. The ratio $n = c/v$ is called the refractive index of the medium. The refractive index of air at the standard conditions for temperature and pressure is 1.000 2926, which is typically approximated by 1. Different kinds of glasses have refractive indices roughly between 1.48 and 1.62.

When a light ray meets the border between two media with different refractive indices n_1 and n_2 at an angle of α_1 with respect to the normal to the boundary, it is split into a reflected ray and a refracted ray. For the discussion of lenses, we are interested only in the refracted ray. It will travel through the second medium at an angle of α_2 to the normal, as shown in Figure 2.18. The relation of the two angles is given by the law of refraction [9, 10]:

$$n_1 \sin \alpha_1 = n_2 \sin \alpha_2 \qquad (2.3)$$

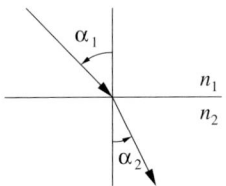

Fig. 2.18 The principle of refraction: the light ray entering from the top is refracted at the border between the two media with refractive indices n_1 and n_2.

The refractive index n actually depends on the wavelength λ of the light: $n = n(\lambda)$. Therefore, if white light, which is a mixture of different wavelengths, is refracted, it is split up into different colors. This effect is called dispersion.

As can be seen from Eq. (2.3), the law of refraction is nonlinear. Therefore, it is obvious that imaging through a lens is, in contrast to the pinhole model, a nonlinear process. In particular, this means that in general light rays emanating from a point (a so-called homocentric pencil) will not converge to a single point after passing through a lens. For small angles α to the normal of the surface, however, we may replace $\sin \alpha$ with α [10]. With this paraxial approximation, the law of refraction becomes linear again:

$$n_1 \alpha_1 = n_2 \alpha_2 \qquad (2.4)$$

The paraxial approximation leads to Gaussian optics, in which a homocentric pencil will converge again to a single point after passing through a lens that consists of spherical surfaces. Hence, Gaussian optics is the ideal for all optical systems. All deviations from Gaussian optics are called aberrations. The goal of lens design is to construct a lens for which Gaussian optics hold for angles α that are large enough that they are useful in practice.

Let us now consider what happens with light rays that pass through a lens. For our purposes, a lens can be considered as two adjacent refracting centered spherical surfaces with a homogeneous medium between them. Furthermore, we will assume that the medium outside the lens is identical on both sides of the lens. Lenses have a finite thickness. Hence, the model we are going to discuss, shown in Figure 2.19, is called the thick lens model [9, 10]. For the discussion, it should be noted that the light rays travel from left to right. All horizontal distances are measured in the direction of the light. Consequently, all distances in front of the lens are negative. Furthermore, all upward distances are positive, while all downward distances are negative.

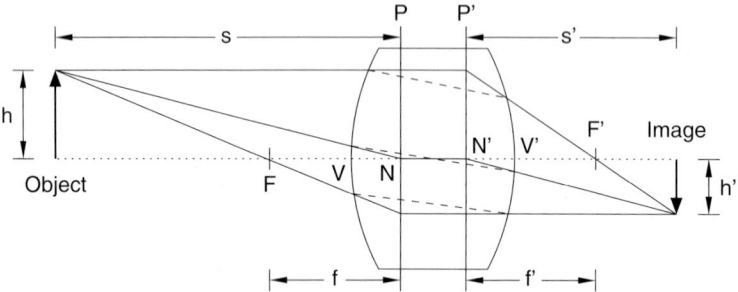

Fig. 2.19 The geometry of a thick lens.

The object in front of the lens is projected to its image behind the lens. The lens has two focal points F and F', at which rays parallel to the optical axis that enter the lens from the opposite side of the respective focal point converge. The principal planes P and P' are given by the intersection of parallel rays that enter the lens from one side with the corresponding rays that converge to the focal point on the opposite side. The focal points F and F' have a distance of f and f' from P and P', respectively. Since the medium is the same on both sides of the lens, we have $f = -f'$, and f' is the focal length of the lens. The object is at a distance of s from P (the object distance), while its image is at a distance of s' from P' (the image distance). The optical axis, shown as a dotted line in Figure 2.19, is the axis of symmetry of the two spherical surfaces of the lens. The surface vertices V and V' are given by the intersection of the lens surfaces with the optical axis. The nodal points N and N' have a special property that is described below. Since the medium is the same on both sides of the lens, the nodal points are given by the intersection of the principal planes with the optical axis. If the media were different, the nodal points would not lie in the principal planes.

With the above definitions, the laws of imaging with a thick lens can be stated as follows:

- A ray parallel to the optical axis before entering the lens passes through F'.
- A ray that passes through F leaves the lens parallel to the optical axis.

- A ray that passes through N passes through N' and does not change its angle with the optical axis.

As can be seen from Figure 2.19, all three rays converge at a single point. Since the imaging geometry is completely defined by F, F', N, and N', they are called the cardinal elements of the lens. Note that, for all object points that lie in a plane parallel to P and P', the corresponding image points also lie in a plane parallel to P and P'. This plane is called the image plane.

Like for the pinhole camera, we can use similar triangles to determine the essential relationships between the object and its image. For example, it is easy to see that $h/s = h'/s'$. Consequently, similar to Eq. (2.2) we have

$$h' = h \frac{s'}{s} \tag{2.5}$$

By introducing the magnification factor $\beta = h'/h$, we also see that $\beta = s'/s$. By using the two similar triangles above and below the optical axis on each side of the lens having F and F' as one of its vertices and the optical axis as one of its sides, we obtain (using the sign conventions mentioned above): $h'/h = f/(f-s)$ and $h'/h = (f'-s')/f'$. Hence, with $f = -f'$ we have

$$\frac{1}{s'} - \frac{1}{s} = \frac{1}{f'} \tag{2.6}$$

This equation is very interesting. It tells us where the light rays will intersect, i.e., where the image will be in focus, if the object distance s is varied. For example, if the object is brought closer to the lens, i.e., the absolute value of s is decreased, the image distance s' must be increased. Likewise, if the object distance is increased, the image distance must be decreased. Therefore, focusing corresponds to changing the image distance. It is interesting to look at the limiting cases. If the object moves to infinity, all the light rays are parallel, and consequently $s' = f'$. On the other hand, if we move the object to F, the image plane would have to be at infinity. Bringing the object even closer to the lens, we see that the rays will diverge on the image side. In Eq. (2.6), the sign of s' will change. Consequently, the image will be a virtual image on the object side of the lens, as shown in Figure 2.20. This is the principle that is used in a magnifying glass.

From $\beta = h'/h = f/(f-s) = f'/(f'+s)$, we can also see that, for constant object distance s, the magnification ratio β increases if the focal length f' increases.

Real lens systems are more complex than the thick lens we have discussed so far. To minimize aberrations, they typically consist of multiple lenses that are centered on their common optical axis. An example of a real lens is shown in Figure 2.21. Despite its complexity, a system of lenses can still be regarded as a thick lens, and can consequently be described by its cardinal elements. Figure 2.21 shows the position of the focal points F and F' and the nodal points N and N' (and thus the position of the

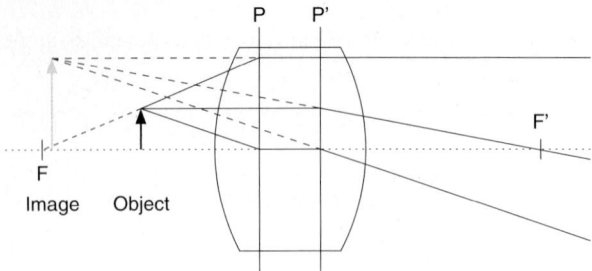

Fig. 2.20 The virtual image created by an object that is closer to the lens than the focal point F.

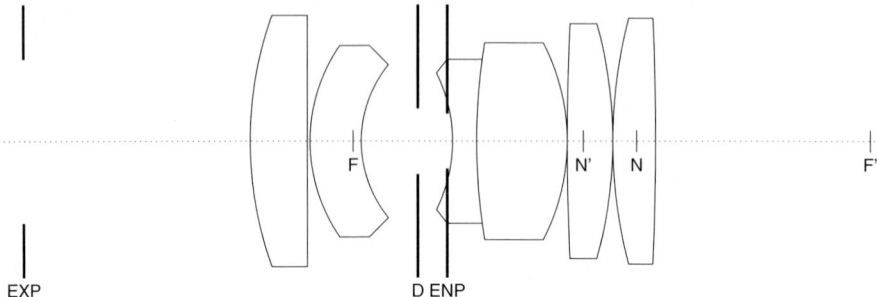

Fig. 2.21 The cardinal elements of a system of lenses. D is the diaphragm; ENP is the entrance pupil; EXP is the exit pupil.

principal planes). Note that for this lens the object side focal point F lies within the second lens of the system. Also note that N′ lies in front of N.

Real lenses have a finite extent. Furthermore, to control the amount of light that falls on the image plane, lenses typically have an iris diaphragm (also called iris) within the lens system. The size of the opening of the iris diaphragm can typically be adjusted with a ring on the lens barrel. The iris diaphragm of the lens in Figure 2.21 is denoted by D. In addition, other elements of the lens system, such as the lens barrel, may limit the amount of light that falls on the image plane. Collectively, these elements are called stops. The stop that limits the amount of light the most is called the aperture stop of the lens system [11]. Note that the aperture stop is not necessarily the smallest stop in the lens system because lenses in front or behind the stop may magnify or shrink the apparent size of the stop as the light rays traverse the lens. Consequently, a relatively large stop may be the aperture stop for the lens system.

Based on the aperture stop, we can define two important virtual apertures in the lens system [9, 11]. The entrance pupil defines the area at the entrance of the lens system that can accept light. The entrance pupil is the (typically virtual) image of the aperture stop in the optics that come before it. Rays that pass through the entrance pupil are able to enter the lens system and pass through it to the exit. Similarly, the exit pupil is the (typically virtual) image of the aperture stop in the optics that follow it. Only rays that pass through this virtual aperture can exit the system. The entrance and exit pupils of the lens system in Figure 2.21 are denoted by ENP and EXP, respectively.

We can single out a very important light ray from the pencil that passes through the lens system: the principal or chief ray. It passes through the center of the aperture stop. Virtually, it also passes through the center of the entrance and exit pupils. Figure 2.22 shows the actual path of the principal ray as a thick solid line. The virtual path of the principal ray through the centers of the entrance pupil Q and the exit pupil Q' is shown as a thick dashed line. In this particular lens, the actual path of the principal ray passes very close to Q. Figure 2.22 also shows the path of the rays that touch the edges of the entrance and exit pupils as thin lines. As for the principal ray, the actual paths are shown as solid lines, while the virtual paths to the edges of the pupils are shown as dashed lines. These rays determine the light cone that enters and exits the lens system, and consequently the amount of light that falls onto the image plane.

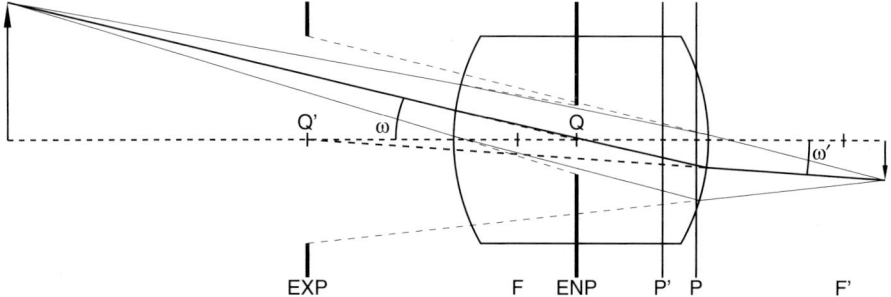

Fig. 2.22 The principal ray of a system of lenses.

Another important characteristic of the lens is the pupil magnification $\beta_P = d_{EXP}/d_{ENP}$, i.e., the ratio of the diameters of the exit and entrance pupils. It can be shown [10] that it also relates the object and image side field angles to each other:

$$\beta_P = \frac{\tan \omega}{\tan \omega'} \tag{2.7}$$

Note that ω' in general differs from ω, as shown in Figure 2.22. The angles are identical only if $\beta_P = 1$.

With the above, we can now discuss how the pinhole model is related to Gaussian optics. We can see that the light rays in the pinhole model correspond to the principal rays in Gaussian optics. In the pinhole model, there is a single projection center, whereas in Gaussian optics there are two projection centers: one in the entrance pupil for the object side rays and one in the exit pupil for the image side rays. Furthermore, for Gaussian optics, in general $\omega' \neq \omega$, whereas $\omega' = \omega$ for the pinhole camera. To reconcile these differences, we must ensure that the object and image side field angles are identical. In particular, the object side field angle ω must remain the same since it is determined by the geometry of the objects in the scene. Furthermore, we must create a single projection center. Because ω must remain constant, this must be done by (virtually) moving the projection center in the exit pupil to the projection center in the entrance pupil. As shown in Figure 2.23, to perform these modifications, we must virtually shift the image plane to a point at a distance c from the projection center Q in the entrance pupil, while keeping the image size constant. As described in Section 2.2.1, c is called the camera constant or the principal distance. This creates a new image side field angle ω''. From $\tan \omega = \tan \omega''$, we obtain $h/s = h'/c$, i.e.,

$$c = \frac{h'}{h} s = \beta s = f' \left(1 - \frac{\beta}{\beta_p}\right) \tag{2.8}$$

The last equation is derived in [10]. The principal distance c is the quantity that can be determined through camera calibration (see Section 3.9). Note that it can differ substantially from the focal length f' (in our example by almost 10%). Also note that c depends on the object distance s. Because of Eq. (2.6), c also depends on the image distance s'. Consequently, if the camera is focused to a different depth it must be recalibrated.

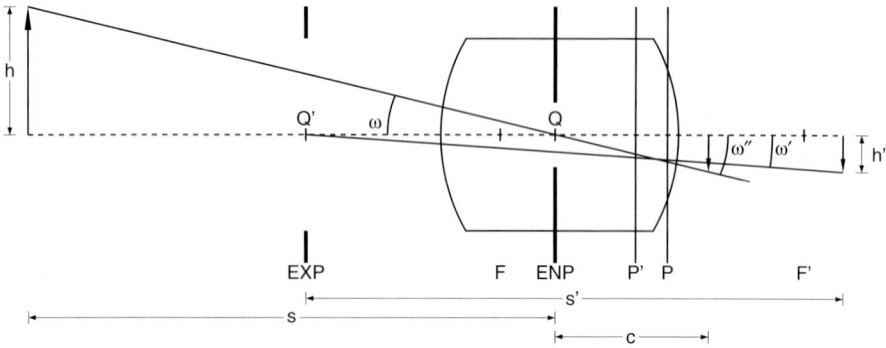

Fig. 2.23 The relation of the pinhole camera and Gaussian optics.

2.2.3
Depth of Field

Up to now, we have focused our discussion on the case in which all light rays converge again to a single point. As shown by Eq. (2.6), for a certain object distance s, the image plane (IP) must lie at the corresponding image distance s'. Hence, only objects lying in a plane parallel to the image plane, i.e., perpendicular to the optical axis, will be in focus. Objects not lying in this focusing plane (FP) will appear blurred. For objects lying farther from the camera, e.g., the object at the distance s_f in Figure 2.24, the light rays will intersect in front of IP at an image distance of s'_f. Likewise, the light rays of objects lying closer to the camera, e.g., at the distance s_n, will intersect behind IP at an image distance of s'_n. In both cases, the points of the object will be imaged as a circle of confusion having a diameter of d'_f and d'_n, respectively.

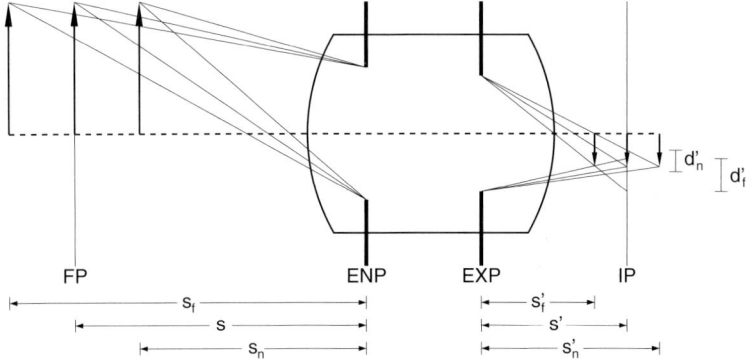

Fig. 2.24 Depth of field of a lens. The image of an object at the focus distance s is imaged in focus. The image of objects nearer or farther than the focus distance result in a blurred image.

Often, we want to acquire sharp images even though the object does not lie entirely within a plane parallel to the image plane. Since the sensors that are used to acquire the image have a finite pixel size, the image of a point will still appear in focus if the size of the circle of confusion is in the same order as the size of the pixel (between 5 and 10 μm for current sensors).

As we can see from Figure 2.24, the size of the blur circle depends on the extent of the ray pencil, and hence on the diameter of the entrance pupil. In particular, if the iris diaphragm is made smaller, and consequently the entrance pupil becomes smaller, the diameter of the circle of confusion will become smaller. Let us denote the permissible diameter of the circle of confusion by d'. If we set $d' = d'_f = d'_n$, we can calculate the image distances s_f and s_n of the far and near planes that will result in a diameter of the circle of confusion of d' [10]:

$$s_{f,n} = \frac{sf'^2}{f'^2 \pm Fd'(s + f'/\beta_p)} \qquad (2.9)$$

Here, F denotes the f-number of the lens, which is given by

$$F = f'/d_{\text{ENP}} \qquad (2.10)$$

The f-number can typically be adjusted with a ring on the lens barrel. The f-numbers are usually specified using the standard series of f-number markings [11] as powers of $\sqrt{2}$, namely: $f/1, f/1.4, f/2, f/2.8, f/4, f/5.6, f/8, f/11, f/16, f/22$, etc. A ratio of $\sqrt{2}$ is chosen because the energy E that falls onto the sensor is proportional to the exposure time t and proportional to the area A of the entrance pupil, i.e., inversely proportional to the square of the f-number:

$$E \sim \frac{t}{A} \sim \frac{t}{F^2} \qquad (2.11)$$

2 Image Acquisition

Hence, increasing the f-number by one step, e.g., from $f/4$ to $f/5.6$, halves the image brightness.

From Eq. (2.9), it follows that the depth of field Δs is given by

$$\Delta s = \frac{2sf'^2 Fd'(s+f'/\beta_p)}{f'^4 - F^2 d'^2 (s+f'/\beta_p)^2} \tag{2.12}$$

If we assume that s is large compared to f'/β_p, we can substitute $s+f'/\beta_p$ by s. If we additionally assume that f'^4 is large with respect to the rest of the denominator, we obtain

$$\Delta s \approx \frac{2s^2 Fd'}{f'^2} \tag{2.13}$$

Hence, we can see that the depth of field Δs is proportional to $1/f'^2$. Consequently, a reduction of f' by a factor of 2 will increase Δs by a factor of 4. On the other hand, reducing the f-number by a factor of 2 will lead to an increase of Δs by a factor of 2. Note that this means that the exposure time will need to be increased by a factor of 4 to obtain the same image brightness.

Figure 2.25 displays two images of a depth-of-field target taken with $F = 2$ and $F = 16$ with a lens with $f' = 12.5$ mm. The depth-of-field target is a set of scales and lines mounted at an angle of $45°$ on a metal block. As predicted from (2.13), the image with $F = 2$ has a much smaller depth of field than the image with $F = 16$.

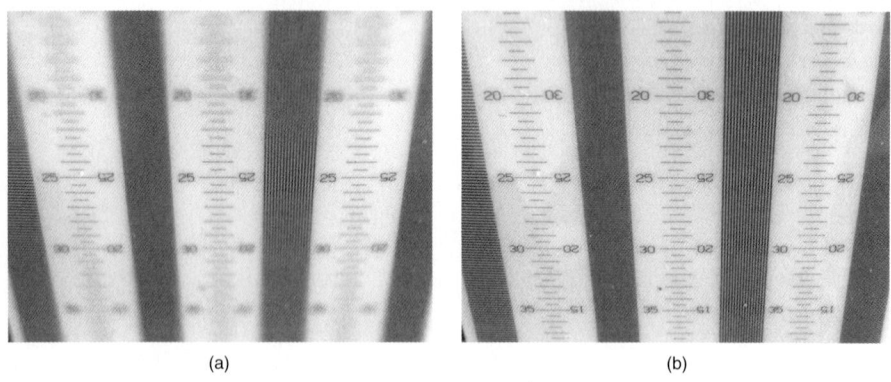

Fig. 2.25 Two images of a depth-of-field target taken with (a) $F = 2$ and (b) $F = 16$ with a lens with $f' = 12.5$ mm. Note the small depth of field in (a) and the large depth of field in (b). Also note the perspective distortion in the images.

It should be noted that it is generally impossible to increase the depth of field arbitrarily. If the aperture stop becomes very small, the wave nature of light will cause the light to be diffracted at the aperture stop. Because of diffraction, the image of a point in the focusing plane will be a smeared-out point in the image plane. This will limit the sharpness of the image. It can be shown [9, 10] that, for a circular aperture stop with an effective f-number of F_e, the image of a point will be given by the Airy disk:

$$I = \left(\frac{2J_1(v)}{v}\right)^2 \tag{2.14}$$

where $J_1(x)$ is the Bessel function of the first kind, $v = \pi\sqrt{x'^2 + y'^2}/(\lambda F_e)$ and $F_e = F(1 - \beta/\beta_p)$. Figure 2.26 shows the Airy disk in two-dimensional (2D) as well as a one-dimensional (1D) cross-section. The radius of the first minimum of Eq. (2.14) is given by $r' = 1.21967\,\lambda F_e$. This is usually taken as the radius of the diffraction disk. Note that r' increases as the effective f-number increases, i.e., as the aperture stop becomes smaller, and that r' increases as the wavelength λ increases. For $\lambda = 500$ nm (green light) and $F_e = 8$, r' will already be approximately 5 μm, i.e., the diameter of the diffraction disk will be in the order of 1–2 pixels. We should keep in mind that the Airy disk will occur for a perfectly focused system. Points lying in front of or behind the focusing plane will create even more complicated brightness distributions that have a radius larger than r' [9, 10].

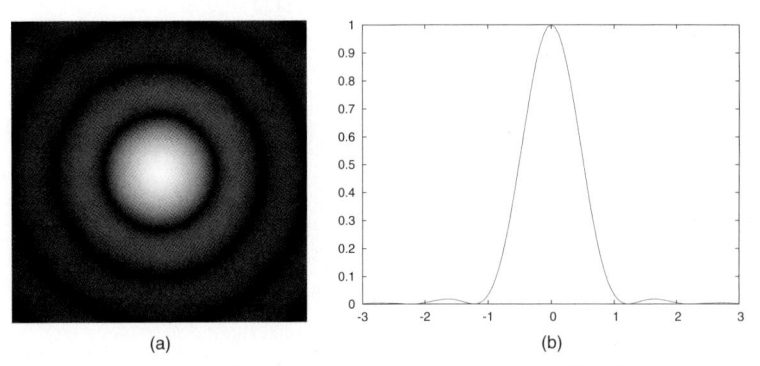

Fig. 2.26 Diffraction will cause a point to be imaged as an Airy disk. (a) 2D image of the Airy disk. To make the outer rings visible, a look-up table (LUT) with a gamma function with $\gamma = 0.4$ has been used. (b) 1D profile through the Airy disk.

2.2.4
Telecentric Lenses

The lens systems we have considered so far all perform a perspective projection of the world. Objects that are closer to the lens produce a larger image, which is obvious from Eqs. (2.2) and (2.8). Consequently, the image of objects not lying in a plane parallel to the image plane will exhibit perspective distortions. In many measurement applications, however, it is highly desirable to have an imaging system that performs a parallel projection because this eliminates the perspective distortions and removes occlusions of objects that occur because of the perspective distortions.

From a conceptual point of view, a parallel projection can be achieved by placing an infinitely small pinhole aperture stop at the image-side focal point F′ of a lens system. From the laws of imaging with a thick lens on page 20, we can deduce that the aperture stop only lets light rays parallel to the optical axis on the object side pass, as shown in Figure 2.27. Consequently, the lens must be at least as large as the object we would like to image.

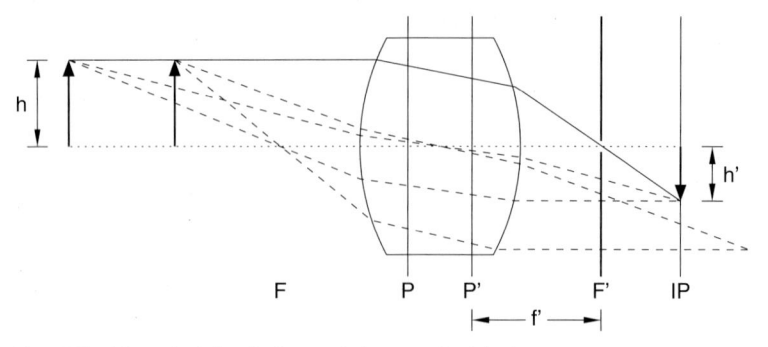

Fig. 2.27 The principle of telecentric lenses. A pinhole aperture stop is placed at the image-side focal point F′, letting only light rays parallel to the optical axis on the object side pass.

Like the pinhole camera, this construction is impracticable because too little light would reach the sensor. Therefore, the aperture stop must have a finite extent, as shown in Figure 2.28. Here, for simplicity the principal planes have been drawn at the same position (P = P′). As can be seen from the principal ray, objects at different object distances are imaged at the same position. Like for regular lenses, objects not in the focusing plane will create a circle of confusion.

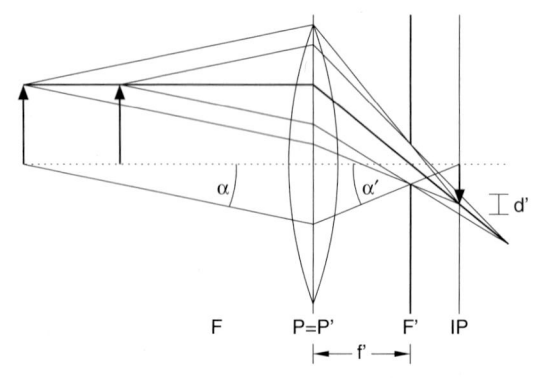

Fig. 2.28 An object-side telecentric lens. From the principal ray, drawn as a thick line, it can be seen that objects at different object distances are imaged at the same position. Like for regular lenses, objects not in the focusing plane will create a circle of confusion.

If we recall that the entrance pupil is the virtual image of the aperture stop in the optics that come before it, and that the aperture stop is located at the image-side focal point F′, we can see that the entrance pupil is located at $-\infty$ and has infinite extent. Since the center of the entrance pupil acts as the projection center, the projection center is infinitely far away, giving rise to the name telecentric perspective for this parallel projection. In particular, since the projection center is at infinity on the object side, such a lens system is called an object-side telecentric lens. Note that the exit pupil of this lens is the aperture stop.

In contrast to the simple pinhole telecentric lens discussed above, an object-side telecentric lens must be larger than the object by an amount that takes the size of the aperture stop into account, as shown by the rays that touch the edge of the aperture stop in Figure 2.28.

The depth of field of an object-side telecentric lens is given by [10]

$$\Delta s = -\frac{d'}{\beta \sin \alpha} = \frac{d'}{\beta^2 \sin \alpha'} \tag{2.15}$$

As before, d' is the permissible diameter of the circle of confusion (see Section 2.2.3), β is the magnification of the lens, given by $\beta = -\sin \alpha / \sin \alpha'$, and $A = \sin \alpha$ is the numerical aperture of the lens. It is related to the f-number by $F = 1/(2A)$. Likewise, $\sin \alpha'$ is the image-side numerical aperture.

Another kind of telecentric lens can be constructed by positioning a second lens system in such a way behind the aperture stop of an object-side telecentric lens that the image-side focal point F'_1 of the first lens system coincides with the object-side focal point F_2 of the second lens system (see Figure 2.29). From the laws of imaging with a thick lens on page 20, we can deduce that now also the light rays on the image side of the second lens system will be parallel to the optical axis. This construction will move the exit pupil to ∞. Therefore, these lenses are called bilateral telecentric lenses.

The magnification of a bilateral telecentric lens is given by $\beta = -f'_2/f'_1$ [10]. Therefore, the magnification is independent of the object position and of the position of the image plane, in contrast to object-side telecentric lenses, where the magnification is only constant for a fixed image plane.

The depth of field of a bilateral telecentric lens is given by [10]

$$\Delta s = -\frac{d'}{\beta \sin \alpha} \tag{2.16}$$

Figure 2.30 shows an image of a depth-of-field target taken with a telecentric lens with $\beta = 0.17$ and $F = 5.6$. Note that, in contrast to the perspective lens used in Figure 2.25, there are no perspective distortions in the image.

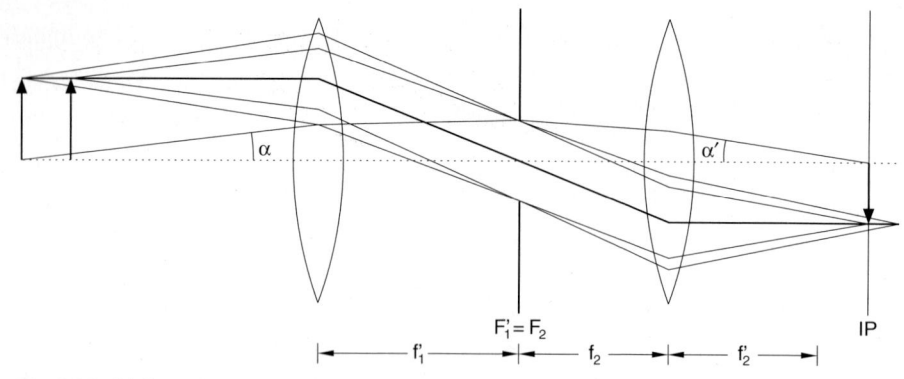

Fig. 2.29 A bilateral telecentric lens. The lens behind the aperture stop is positioned in such a way that $F'_1 = F_2$. From the principal ray, drawn as a thick line, it can be seen that objects at different object distances are imaged at the same position. Like for regular lenses, objects not in the focusing plane will create a circle of confusion.

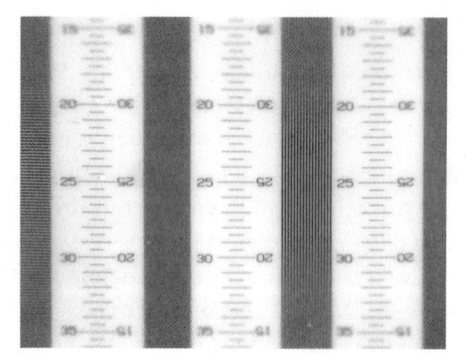

Fig. 2.30 An image of a depth-of-field target taken with a telecentric lens with $\beta = 0.17$ and $F = 5.6$. Note that there are no perspective distortions in the image.

2.2.5
Lens Aberrations

Up to now, we have assumed Gaussian optics, where a homocentric pencil will converge to a single point after passing through a lens. For real lenses, this is generally not the case. In this section, we will discuss the primary aberrations that can occur [9, 12].

Spherical aberration, shown in Figure 2.31(a), occurs because light rays that lie far from the optical axis will not intersect at the same point as light rays that lie close to the optical axis. This happens because of the increased refraction that occurs at the edge of a spherical lens. No matter where we place the image plane, there will always be a circle of confusion. The best we can do is to make the circle of confusion as small as possible. The spherical aberration can be reduced by setting the aperture stop to a larger f-number, i.e., by making the iris diaphragm smaller because this prevents the light rays far from the optical axis from passing through the lens. However, as described at the end of Section 2.2.3, diffraction will limit how small we can make the aperture stop. An alternative is to employ lenses that use non-spherical surfaces instead of spherical surfaces. Such lenses are called aspherical lenses.

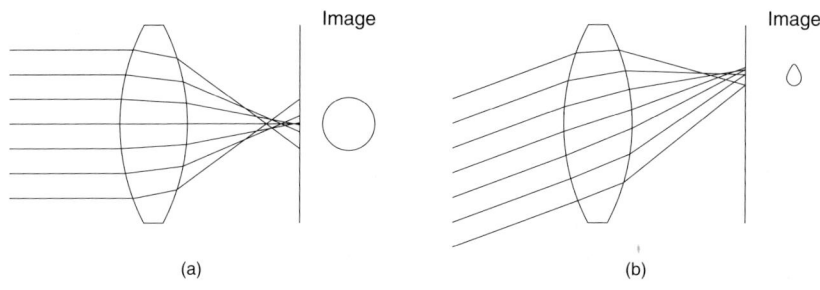

Fig. 2.31 (a) Spherical aberration: light rays far from the optical axis do not intersect at the same point as light rays close to the optical axis. (b) Coma: light rays that pass through the lens at an angle to the optical axis do not intersect at the same point.

A similar aberration is shown in Figure 2.31(b). Here, rays that pass through the lens at an angle to the optical axis do not intersect at the same point. In this case, the image is not a circle, but a comet-shaped figure. Therefore, this aberration is called coma. Note that, in contrast to the spherical aberration, the coma is asymmetric. This may cause feature extraction algorithms like edge extraction to return wrong positions. The coma can be reduced by setting the aperture stop to a larger f-number.

Figure 2.32 shows a different kind of aberration. Here, light rays in the tangential plane, defined by the object point and the optical axis, and the sagittal plane, perpendicular to the tangential plane, do not intersect at points, but in lines perpendicular to the respective plane. Because of this property, this aberration is called astigmatism. The two lines are called the tangential image I_T and a sagittal image I_S. Between them, a circle of least confusion occurs. Astigmatism can be reduced by setting the aperture stop to a larger f-number or by carefully designing the lens surfaces.

A closely related aberration is shown in Figure 2.33(a). Above, we have seen that the tangential and sagittal images do not necessarily lie at the same image distance. In fact, the surfaces in which the tangential and sagittal images lie (the tangential and sagittal focal surface) may not even be planar. This aberration is called curvature of field. It leads to the fact that it is impossible to bring the entire image into focus. Figure 2.33(b) shows an image taken with a lens that exhibits significant curvature of field. The center of the image is in focus, while the borders of the image are severely defocused. Like the other aberrations, curvature of field can be reduced by setting the aperture stop to a larger f-number or by carefully designing the lens surfaces.

In addition to the above aberrations, which, apart from coma, mainly cause problems with focusing the image, lens aberrations may cause the image to be distorted,

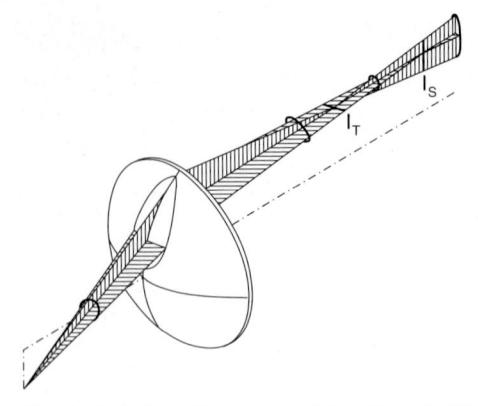

Fig. 2.32 Astigmatism: tangential and sagittal light rays do not intersect at the same point, creating a tangential image I_T and a sagittal image I_S that consists of two orthogonal lines.

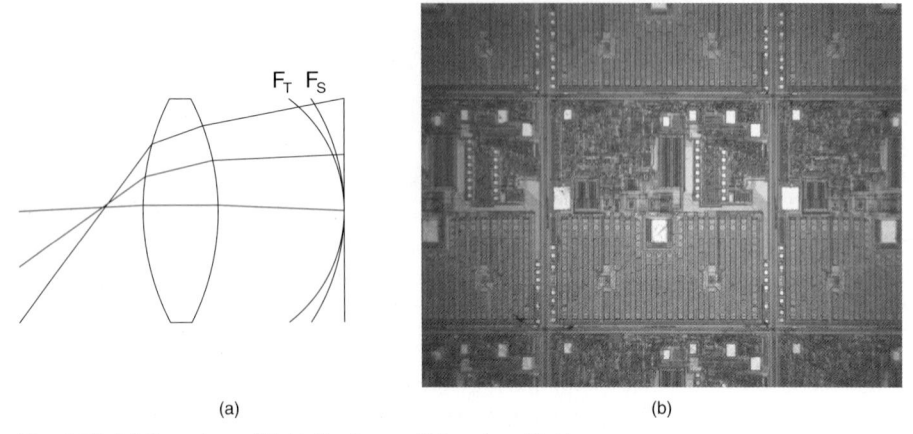

Fig. 2.33 (a) Curvature of field: the tangential and sagittal images lie in two curved focal planes F_T and F_S. (b) Image taken with a lens that exhibits significant curvature of field. The center of the image is in focus, while the borders of the image are severely defocused.

leading to the fact that straight lines not passing through the optical axis will no longer be imaged as straight lines. This is shown in Figure 2.34. Because of the characteristic shapes of the images of rectangles, two kinds of distortion can be identified: pincushion and barrel distortion. Note that distortion will not affect lines through the optical axis. Furthermore, circles with the optical axis as center will produce larger or smaller circles. Therefore, these distortions are called radial distortions. In addition, if the elements of the lens are not centered properly, decentering distortions may also occur. Note that there is an ISO standard that defines how distortions are to be measured and reported by lens manufacturers [13].

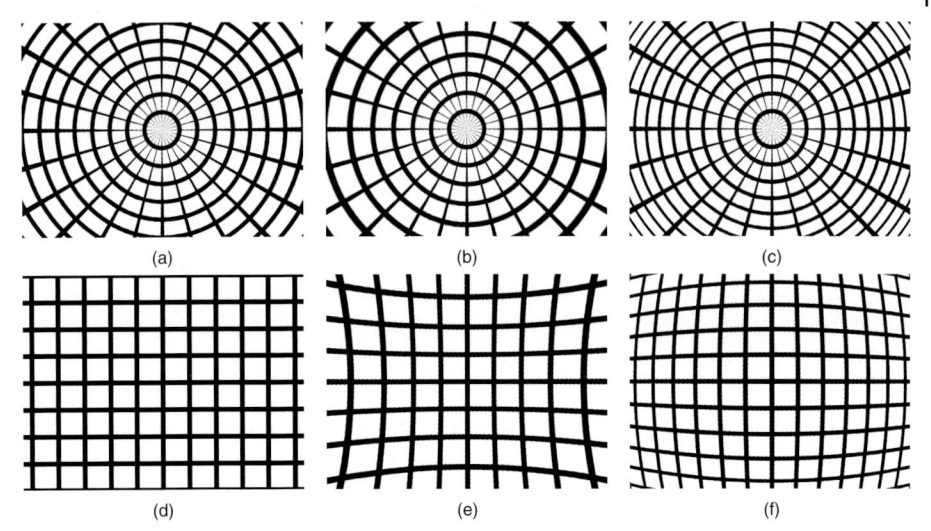

Fig. 2.34 (a), (d) Images without distortion. (b), (e) Images with pincushion distortion. (c), (f) Images with barrel distortion.

All of the above aberrations already occur for monochromatic light. If the object is illuminated with light that contains multiple wavelengths, e.g., white light, chromatic aberrations may occur, which lead to the fact that light rays having different wavelengths do not intersect at a single point, as shown in Figure 2.35. With color cameras, chromatic aberrations cause colored fringes at the objects' edges. With black-and-white cameras, chromatic aberrations lead to blurring. Furthermore, because they are not symmetric, they can cause position errors in feature extraction algorithms like edge extraction. Chromatic aberrations can be reduced by setting the aperture stop to a larger f-number. Furthermore, it is possible to design lenses in such a way that two different wavelengths will be focused at the same point. Such lenses are called achromatic lenses. It is even possible to design lenses in which three specific wavelengths will be focused at the same point. Such lenses are called apochromatic lenses. Nevertheless, the remaining wavelengths still do not intersect at the same point, leading to residual chromatic aberrations. Note that there is an ISO standard that defines how chromatic aberrations are to be measured and reported by lens manufacturers [14].

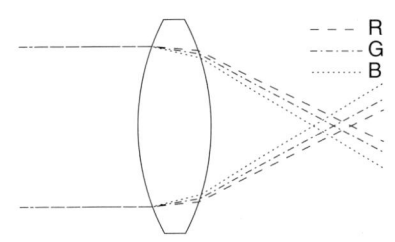

Fig. 2.35 Chromatic aberration: light rays with different wavelengths do not intersect at a single point.

It is generally impossible to design a lens system that is free from all of the above aberrations, as well as higher-order aberrations that we have not discussed [9]. Therefore, the lens designer must always make a trade-off between the different aberrations. While the residual aberrations may be small, they will often be noticeable in highly accurate feature extraction algorithms. This happens because some of the aberrations create asymmetric gray value profiles, which will influence all subpixel algorithms. Figure 2.36 shows an example of an ideal step edge profile that would be produced by a perfect diffraction-limited system as well as edge profiles that were measured with a real lens for an edge through the optical axis and an off-axis edge lying on a circle located at a distance of 5 mm from the optical axis (i.e., the edge profile for the off-axis edge is perpendicular to the circle). Note that the ideal profile and the on-axis profile are symmetric, and consequently cause no problems. The off-axis profile, however, is asymmetric, which causes subpixel-accurate edge detection algorithms to return incorrect positions, as discussed in Section 3.7.4.

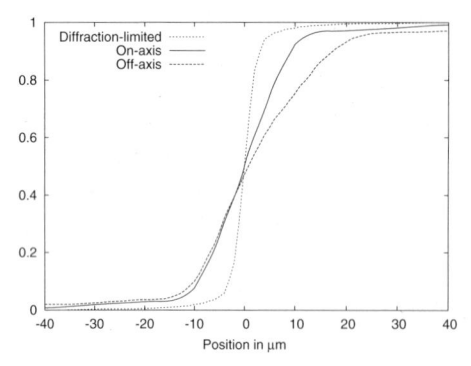

Fig. 2.36 Edge profile of a perfect diffraction-limited system as well as for a real lens for an edge through the optical axis and an off-axis edge in the direction perpendicular to a circle located at a distance of 5 mm from the optical axis. Note that the off-axis edge profile is not symmetric. (Adapted from [10].)

We conclude this section by discussing an effect that is not considered an aberration per se, but nevertheless influences the image quality. As can be seen from Figure 2.33(b), for some lenses there can be a significant drop in the image brightness toward the border of the image. This effect is called vignetting. It can have several causes. First of all, the lens may have been designed in such a way that for a small f-number the iris diaphragm no longer is the aperture stop for light rays that form a large angle with the optical axis. Instead, as shown in Figure 2.37, one of the lenses or the lens barrel becomes the aperture stop for a part of the light rays. Consequently, the effective entrance pupil for the off-axis rays will be smaller than for rays that lie on the optical axis. Thus, the amount of light that reaches the sensor is smaller for off-axis points. This kind of vignetting can be removed by increasing the f-number so that the iris diaphragm becomes the aperture stop for all light rays.

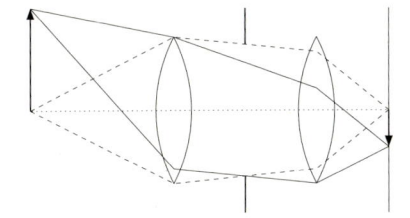

Fig. 2.37 Vignetting occurs because the iris diaphragm no longer is the aperture stop for a part of the off-axis light rays.

The second cause for vignetting is the natural light fall-off for off-axis rays. By looking at Figure 2.22, we can see that light rays that leave the exit pupil form an image-side field angle ω' with the optical axis. This means that for off-axis points with $\omega' \neq 0$ the exit pupil will no longer be a circle, but an ellipse that is smaller than the circle by a factor of $\cos \omega'$. Consequently, off-axis image points receive less light than axial image points. Furthermore, the length of the light ray to the image plane will be longer than for axial rays by a factor of $1/\cos \omega'$. By the inverse square law of illumination [9], the energy received at the image plane is inversely proportional to the square of the length of the ray from the light source to the image point. Hence, the light received by off-axis points will be reduced by another factor of $(\cos \omega')^2$. Finally, the light rays fall on the image plane at an angle of ω'. This further reduces the amount of light by a factor of $\cos \omega'$. In total, we can see that the amount of light for off-axis points is reduced by a factor of $(\cos \omega')^4$. A more thorough analysis [12] shows that, for large aperture stops (f-numbers larger than about 5), the light fall-off is slightly less than $(\cos \omega')^4$. This kind of vignetting cannot be reduced by increasing the f-number.

2.3 Cameras

The camera's purpose is to create an image from the light focused in the image plane by the lens. The most important component of the camera is a digital sensor. In this section, we will discuss the two main sensor technologies: CCD (charge-coupled device) and CMOS (complementary metal–oxide–semiconductor). They differ primarily in their readout architecture, i.e., in the manner in which the image is read out from the chip [15]. Furthermore, we will discuss how to characterize the performance of a camera. Finally, since we do not discuss this in this chapter, it should be mentioned that the camera contains the necessary electronics to create a video signal that can be transmitted to the computer as well as electronics that permit the image acquisition to be controlled externally by a trigger.

2.3.1
CCD Sensors

To describe the architecture of CCD sensors, we start by examining the simplest case: a linear sensor, as shown in Figure 2.38. The CCD sensor consists of a line of light-sensitive photodetectors. Typically they are photogates or photodiodes [16]. We will not describe the physics involved in photodetection. For our purposes, it suffices to think of a photodetector as a device that converts photons into electrons and corresponding positively charged holes and collects them to form a charge. Each type of photodetector has a maximum charge that can be stored, depending, among other factors, on its size. The charge is accumulated during the exposure of the photodetector to light. To read out the charges, they are moved through transfer gates to serial readout registers (one for each photodetector). The serial readout registers are also light-sensitive, and thus must be covered with a metal shield to prevent the image from receiving additional exposure during the time it takes to perform the readout. The readout is performed by shifting the charges to the charge conversion unit, which converts the charge into a voltage, and by amplifying the resulting voltage [16]. The transfer gates and serial readout registers are charge-coupled devices. Each CCD consists of several gates (up to four) that are used to transport the charge in a given direction. Details of the charge transportation process can be found, for example, in [16]. After the charge has been converted into a voltage and has been amplified, it can be converted into an analog or digital video signal. In the latter case, the voltage would be converted into a digital number through an analog-to-digital converter (ADC).

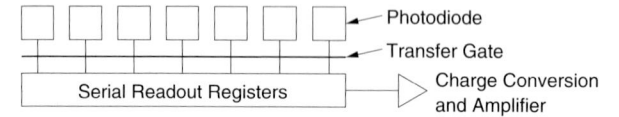

Fig. 2.38 A linear CCD sensor. Light is converted into charge in the photodetectors, which typically are photodiodes for line sensors, transferred to the serial readout registers, and read out sequentially through the charge converter and amplifier.

A line sensor as such would create an image that is one pixel high, which would not be very useful in practice. Therefore, typically many lines are assembled into a 2D image. Of course, for this image to contain any useful content, the sensor must move with respect to the object that is to be imaged. In one scenario, the line sensor is mounted above the moving object, e.g., a conveyor belt. In a second scenario, the object remains stationary while the camera is moved across the object, e.g., to image a printed circuit board. This is the principle that is also used in flatbed scanners. In this respect, a flatbed scanner is a line sensor with an integrated illumination.

To acquire images with a line sensor, the line sensor itself must be parallel to the plane in which the objects lie and must be perpendicular to the movement direction to ensure that rectangular pixels are obtained. Furthermore, the frequency with which the lines are acquired must match the speed of the relative movement of the camera and the object, and the resolution of the sensor to ensure that square pixels result. It must be assumed that the speed is constant to ensure that all the pixels have the same size. If this is not the case, an encoder must be used, which essentially is a device that triggers the acquisition of each image line. It is driven, for example, by the stepper of the motor that causes the relative movement. Since a perfect alignment of the sensor to the object and movement direction is relatively difficult to achieve, in some applications the camera must be calibrated with the methods described in Section 3.9 to ensure high measurement accuracy.

The line readout rates of line sensors vary between 14 and 140 kHz. This obviously limits the exposure time of each line. Consequently, line scan applications require very bright illumination. Furthermore, the diaphragm of the lens must typically be set to a relatively small f-number, which can severely limit the depth of field. Therefore, line scan applications often pose significant challenges for obtaining a suitable setup.

We now turn our attention to area sensors. The logical extension of the principle of the line sensor is the full frame sensor, shown in Figure 2.39. Here, the light is converted into charge in the photodetectors and is shifted row by row into the serial readout registers, from where it is converted into a video signal in the same manner as in the line sensor.

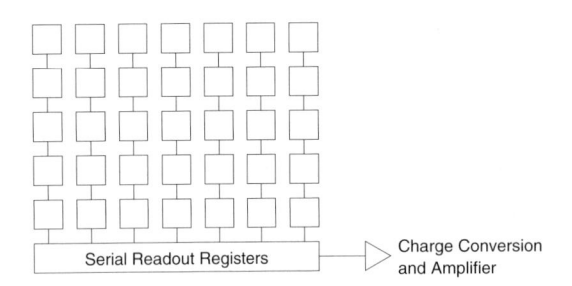

Fig. 2.39 A full frame CCD sensor. Light is converted into charge in the photodetectors, transferred to the serial readout registers row by row, and read out sequentially through the charge converter and amplifier.

During the readout, the photodetectors are still exposed to light, and thus continue to accumulate charge. Because the upper pixels are shifted through all the lower pixels, they accumulate information from the entire scene, and consequently appear smeared. To avoid the smear, a mechanical shutter or a strobe light must be used. This is also the biggest disadvantage of the full frame sensor. Its biggest advantage is that the fill factor (the ratio of the light-sensitive area of a pixel to its total area) can reach 100%, maximizing the sensitivity of the pixels to light and minimizing aliasing.

To reduce the smearing problem in the full frame sensor, the frame transfer sensor, shown in Figure 2.40, uses an additional sensor that is covered by a metal light shield, and can thus be used as a storage area. In this sensor type, the image is created in

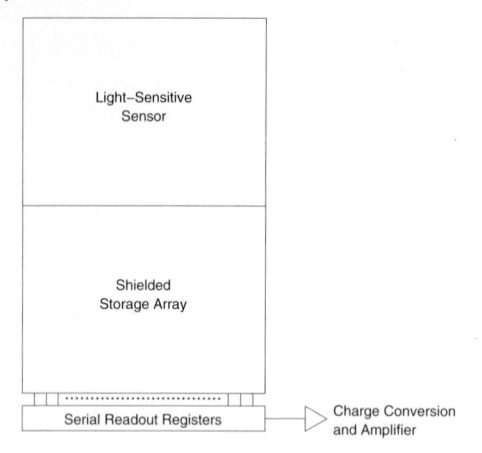

Fig. 2.40 A frame transfer CCD sensor. Light is converted into charge in the light-sensitive sensor, quickly transferred to the shielded storage array, and read out row by row from there.

the light-sensitive sensor, and then transferred into the shielded storage array, from which it can be read out at leisure. Since the transfer between the two sensors is quick (usually less than 500 µs [16]), smear is significantly reduced.

The biggest advantage of the frame transfer sensor is that it can have fill factors of 100%. Furthermore, no mechanical shutter or strobe light needs to be used. Nevertheless, a residual smear may remain because the image is still exposed during the short time required to transfer the charges to the second sensor. The biggest disadvantage of the frame transfer sensor is its high cost, since it essentially consists of two sensors.

Because of the above characteristics (high sensitivity and smearing), full frame and frame transfer sensors are most often used in scientific applications for which the exposure time is long compared to the readout time, e.g., in astronomical applications.

The final type of CCD sensor, shown in Figure 2.41, is the interline transfer sensor. In addition to the photodetectors, which in most cases are photodiodes, there are vertical transfer registers that are covered by an opaque metal shield. After the image has been exposed, the accumulated charges are shifted through transfer gates (not shown in Figure 2.41) to the vertical transfer registers. This can typically be done in less than 1 µs [16]. To create the video signal, the charges are then shifted through the vertical transfer registers into the serial readout registers, and read out from there.

Because of the quick transfer from the photodiodes into the shielded vertical transfer registers, there is no smear in the image, and consequently no mechanical shutter or strobe light needs to be used. The biggest disadvantage of the interline transfer sensor is that the transfer registers take up space on the sensor, causing fill factors that may be as low as 20%. Consequently, aliasing effects may increase. To increase the fill factor, microlenses are typically located in front of the sensor to focus the light onto

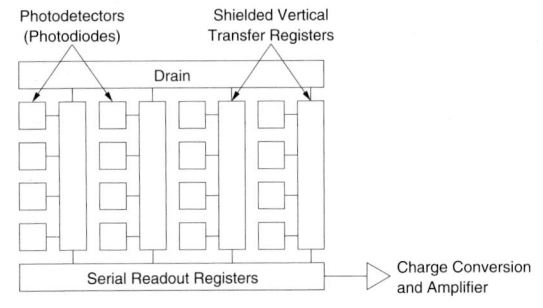

Fig. 2.41 An interline transfer CCD sensor. Light is converted into charge in the light-sensitive sensor, quickly transferred to the shielded vertical transfer registers, transferred row by row to the serial readout registers, and read out from there.

the light-sensitive photodiodes, as shown in Figure 2.42. Nevertheless, fill factors of 100% typically cannot be achieved.

One problem with CCD sensors is an effect called blooming: when the charge capacity of a photodetector is exhausted, the charge spills over into the adjacent photodetectors. Thus, bright areas in the image are significantly enlarged. To prevent this problem, anti-bloom drains can be built into the sensor [16]. The drains form a electrostatic potential barrier that causes extraneous charge from the photodetectors to flow into the drain. The drains can be located either next to the pixels on the surface of the sensor (lateral overflow drains), e.g., on the opposite side of the vertical transport registers, or can be buried in the substrate of the device (vertical overflow drains). Figure 2.41 displays a vertical overflow drain, which must be imagined to lie underneath the transfer registers.

An interesting side-effect of building anti-bloom drains into the sensor is that they can be used to create an electronic shutter for the camera. By setting the drain potential to zero, the photodetectors discharge. Afterwards, the potential can be set to a high value during the exposure time to accumulate the charge until it is read out. The anti-bloom drains also facilitate the construction of sensors that can immediately acquire images after receiving a trigger signal. Here, the entire sensor is reset immediately after receiving the trigger signal. Then, the image is exposed and read out as usual. This operation mode is called asynchronous reset.

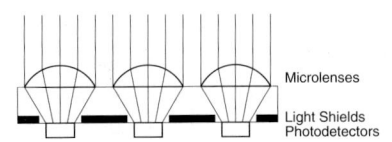

Fig. 2.42 Microlenses are typically used in interline transfer sensors to increase the fill factor by focusing the light onto the light-sensitive photodiodes.

We conclude this section by describing a readout mode that is often implemented in analog CCD cameras because the image is transmitted with one of the analog video standards described in Section 2.4.1. The analog video standards require an image to be transmitted as two fields, one containing the odd lines of the image and one containing the even lines. This readout mode is called interlaced scan. Because of the readout architecture of CCD sensors, this means that the image will have to be exposed twice. After the first exposure, the odd rows are shifted into the transfer registers; while after the second exposure, the even rows are shifted and read out. If the object moves between the two exposures, its image will appear serrated, as shown in Figure 2.43(a). The mode of reading out the rows of the CCD sensor sequentially is called progressive scan. From Figure 2.43(b), it is clear that this mode is essential for capturing correct images of moving objects.

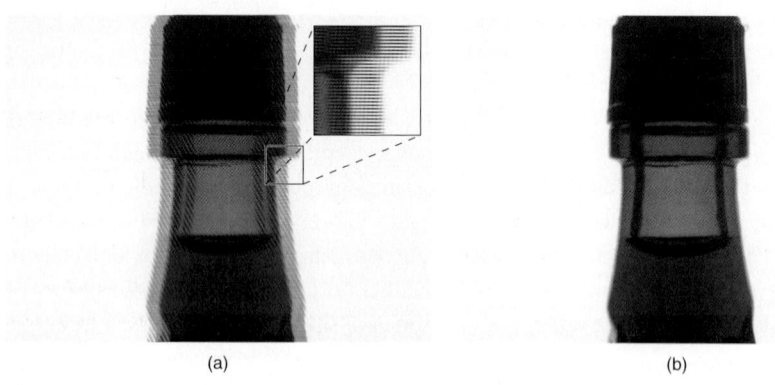

Fig. 2.43 Comparison of using (a) an interlaced camera and (b) a progressive scan camera for acquiring an image of a moving object. Note that progressive scan is essential for capturing correct images of moving objects.

2.3.2
CMOS Sensors

CMOS sensors, shown in Figure 2.44, typically use photodiodes for photodetection [15, 17]. In contrast to CCD sensors, the charge of the photodiodes is not transported sequentially to a readout register. Instead, each row of the CMOS sensor can be selected directly for readout through the row and column select circuits. In this respect, a CMOS sensor acts like a random access memory. Furthermore, as shown in Figure 2.44, each pixel has its own amplifier. Hence, this type of sensor is also called an active pixel sensor (APS). CMOS sensors typically produce a digital video signal. Therefore, the pixels of each image row are converted in parallel to digital numbers through a set of ADCs.

Since the amplifier and row and column select circuits typically use a significant amount of the area of each pixel, CMOS sensors, like interline transfer CCD sensors,

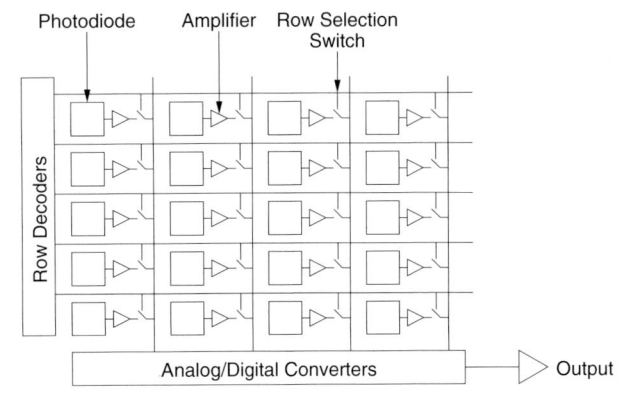

Fig. 2.44 A CMOS sensor. Light is converted into charge in the photodiodes. Each row of the CMOS sensor can be selected directly for readout through the row and column select circuits.

have low fill factors and therefore normally use microlenses to increase the fill factor and to reduce aliasing (see Figure 2.42).

The random access behavior of CMOS sensors facilitates an easy readout of rectangular areas of interest (AOI) from the image. This gives them a significant advantage over CCD sensors in some applications since it enables much higher frame rates for small AOIs. While AOIs can also be implemented for CCDs, their readout architecture requires that all rows above and below the AOI are transferred and then discarded. Since discarding rows is faster than reading them out, this results in a speed increase. However, typically no speed increase can be obtained by making the AOI smaller horizontally since the charges must be transferred through the charge conversion unit. Another big advantage is the parallel analog-to-digital conversion that is possible in CMOS sensors. This can give CMOS sensors a speed advantage even if AOIs are not used. It is even possible to integrate the ADCs into each pixel to further increase the readout speed [15]. Such sensors are also called digital pixel sensors (DPS).

Although Figure 2.44 shows an area sensor, the same principle can also be applied to construct a line sensor. Here, the main advantage of using a CMOS sensor is also the readout speed.

Since each row in a CMOS sensor can be read out individually, the simplest strategy to acquire an image is to expose each line individually and to read it out. Of course, exposure and readout can be overlapped for consecutive image lines. This is called a rolling shutter. Obviously, this readout strategy creates a significant time lag between the acquisition of the first and last image lines. As shown in Figure 2.45(a), this causes sizeable distortion when moving objects are acquired. For these kinds of applications, sensors with a global shutter must be used [18]. This requires a separate storage area for each pixel, and thus further lowers the fill factor. As shown in Figure 2.45(b), the global shutter results in a correct image for moving objects.

Fig. 2.45 Comparison of using (a) a rolling shutter and (b) a global shutter for acquiring an image of a moving object. Note that the rolling shutter significantly distorts the object.

Because of their architecture, it is easy to support asynchronous reset for triggered acquisition in CMOS cameras.

The integrating CMOS sensor we have discussed so far, like the CCD sensor, provides a linear response, which is essential for accurate edge detection (see Section 3.7.4). However, in some applications, e.g., online welding inspection, it is desirable to acquire images from scenes that have intensity differences of six orders of magnitude or more. To fit these huge intensity differences into a gray value image, a nonlinear gray value response must be used. For this reason, CMOS sensors with a logarithmic response [19, 20] and a combined linear–logarithmic response [21, 22, 23] have been developed. Most of them work by feeding the photocurrent created by the photodetector into a resistor with a logarithmic current–voltage characteristic, and thus are non-integrating sensors, which means that they must use a rolling shutter.

2.3.3
Color Cameras

CCD and CMOS sensors are sensitive to light with wavelengths ranging from near ultraviolet (200 nm) through the visible range (380–780 nm) into the near infrared (1100 nm). A sensor responds to the incoming light with its spectral response function. The gray value produced by the sensor is obtained by multiplying the spectral distribution of the incoming light by the spectral response of the sensor and then integrating over the range of wavelengths for which the sensor is sensitive. Figure 2.46 displays the spectral responses of typical CCD and CMOS sensors and the human visual system (HVS) under photopic conditions (daylight viewing). Note that the spectral response of each sensor is much wider than that of the HVS. This can be used advantageously in some applications by illuminating objects with infrared strobe lights and using an infrared pass filter that suppresses the visible spectrum and only lets infrared light pass to the sensor. Since the HVS does not perceive the infrared

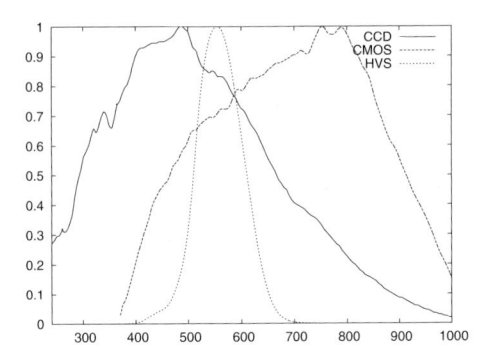

Fig. 2.46 Spectral responses of typical CCD and CMOS sensors and the human visual system (HVS) under photopic conditions (daylight viewing) as a function of the wavelength in nanometers. The responses have been normalized so that the maximum response is unity.

light, it may be possible to use the strobe without a screen. On the other hand, despite the fact that the sensors are also sensitive to ultraviolet light, often no special filter is required because lenses are usually made of glass, which blocks ultraviolet light.

Since CCD and CMOS sensors respond to all frequencies in the visible spectrum, they are unable to produce color images. To construct a color camera, a color filter array (CFA) can be placed in front of the sensor, which allows only light of a specific range of wavelengths to pass to each photodetector. Since this kind of camera only uses one chip to obtain the color information, it is called a single-chip camera. Figure 2.47 displays the most commonly used Bayer CFA [24]. Here, the CFA consists of three kinds of filters, one for each primary color that the HVS can perceive (red, green, and blue). Note that green is sampled at twice the frequency of the other colors because the HVS is most sensitive to colors in the green range of the visible spectrum. Also note that the colors are subsampled by factors of 2 (green) and 4 (red and blue). Since this can lead to severe aliasing problems, often an optical anti-aliasing filter is placed before the sensor [16]. Furthermore, to obtain each color with the full

Fig. 2.47 In a single-chip color camera, a color filter array is placed in front of the sensor, which allows only light of a specific range of wavelengths to pass to each photodetector. Here, the Bayer CFA is shown.

resolution, the missing samples must be reconstructed through a process called demosaicking. The simplest method to do this is to use methods such as bilinear or bicubic interpolation. This can, however, cause significant color artifacts, such as colored fringes. Therefore, new methods for demosaicking are still being actively investigated (see, e.g., [25, 26]).

A second method to construct a color camera is shown in Figure 2.48. Here, the light beam coming from the lens is split into three beams by a beam splitter or prism and sent to three sensors that have different color filters placed in front of them [16]. Hence, these cameras are called three-chip cameras. This construction obviously prevents the aliasing problems of the single-chip cameras. However, since three sensors must be used and carefully aligned, three-chip cameras are much more expensive than single-chip cameras.

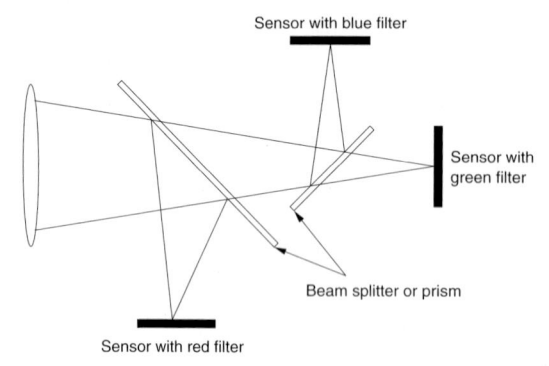

Fig. 2.48 In a three-chip color camera, the light beam coming from the lens is split into three beams by a beam splitter or prism and sent to three sensors that have different color filters.

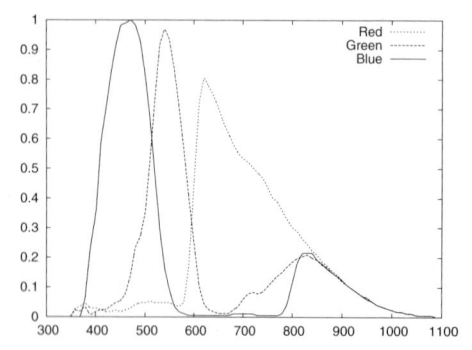

Fig. 2.49 Spectral responses of a typical color CCD sensor as a function of the wavelength in nanometers. Note the sensitivity in the near infrared range. The responses have been normalized so that the maximum response is unity.

Figure 2.49 shows the spectral response of a typical color CCD sensor. Note that the sensor is still sensitive to near infrared light. Since this may cause unexpected colors in the image, an infrared cut filter should be used.

2.3.4
Sensor Sizes

CCD and CMOS are manufactured in various sizes, which typically are specified in inches. The most common sensor sizes and their typical widths, heights, and diagonals are shown in Table 2.2. Note that the size classification is a remnant of the days when video camera tubes were used in television. The size defined the outer diameter of the video tube. The usable area of the image plane of the tubes was approximately two-thirds of their diameter. Consequently, the diagonals of the sensors in Table 2.2 roughly are two-thirds of the respective sensor sizes. An easier rule to remember is that the width of the sensors is approximately half the sensor size.

Tab. 2.2 Typical sensor sizes and the corresponding pixel pitches for an image size of 640 × 480 pixels.

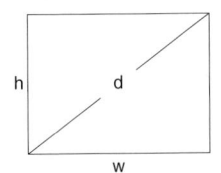

Size	Width	Height	Diag.	Pitch
1 inch	12.8 mm	9.6 mm	16 mm	20 μm
2/3 inch	8.8 mm	6.6 mm	11 mm	13.8 μm
1/2 inch	6.4 mm	4.8 mm	8 mm	10 μm
1/3 inch	4.8 mm	3.6 mm	6 mm	7.5 μm
1/4 inch	3.2 mm	2.4 mm	4 mm	5 μm

It is imperative to use a lens that has been designed for a sensor size that is at least as large as the sensor that is actually used. If this is not done, the outer parts of the sensor will receive no light. For example, a lens designed for a $\frac{1}{2}$ inch chip cannot be used with a $\frac{2}{3}$ inch chip.

Table 2.2 also displays typical pixel pitches for an image size of 640 × 480 pixels. If the sensor has a higher resolution, the pixel pitches decrease by the corresponding factor, e.g., by a factor of 2 for an image size of 1280 × 960.

CCD and CMOS area sensors are manufactured in different resolutions, from 640 × 480 to 4008 × 2672 and more. The resolutions are often specified by the underlying video norm for analog cameras (e.g., RS-170: 640 × 480; CCIR: 768 × 576) or by the display resolutions of video cards (VGA: 640 × 480; XGA: 1024 × 768; SXGA: 1280 × 1024; UXGA: 1600 × 1200; QXGA: 2048 × 1536, etc.). The resolution of line sensors ranges from 512 pixels to 12 888 pixels and probably more in the future. As a general rule, the maximum frame and line rates decrease as the sensors get larger.

2.3.5
Camera Performance

When selecting a camera for a particular application, it is essential that we are able to compare the performance characteristics of different cameras. For machine vision applications, the most interesting questions are related to noise of the camera. Therefore, in this section we take a closer look at the noise that the camera produces. Our presentation follows the EMVA standard 1288 [27, 28]. (There is also an ISO standard for noise measurement [29]. However, it is tailored to consumer cameras, which generally have nonlinear gray value responses. Therefore, we will not consider this standard here.)

We begin by looking at how a gray value in an image is produced. During the exposure time, a number n_p of photons fall on the pixel area (including the light-insensitive area) and create a number n_e of electrons and holes (for CCD sensors) or discharge the reset charge, i.e., annihilate n_e electron–hole pairs (for CMOS sensors). The ratio of electrons created or annihilated per photon is called the total quantum efficiency η. The electron–hole pairs form a charge that is converted into a voltage, which is amplified and digitized by an ADC, resulting in the gray value y that is related to the number of electrons by the overall system gain K.

Note that this process assumes that the camera produces digital data directly. For an analog camera, we would consider the system consisting of camera and frame grabber.

Different sources of noise modify the gray value that is obtained. First of all, the photons arrive at the sensor, not at regular time intervals, but according to a stochastic distribution that can be modeled by the Poisson distribution. This random fluctuation of the number of photons is called photon noise. The photon noise is characterized by the variance σ_p^2 of the number of photons. For the Poisson distribution, σ_p^2 is identical to the mean number μ_p of photons arriving at the pixel. Note that this means that light has an inherent signal-to-noise ratio $\text{SNR}_p = \mu_p/\sigma_p = \sqrt{\mu_p}$. Consequently, using more light typically results in a better image quality.

During the exposure, each photon creates an electron with a certain probability η that depends on the optical fill factor of the sensor (including the increase in fill factor caused by the microlenses), the material of the sensor (its quantum efficiency), and the wavelength of the light. The product of the optical fill factor and the quantum efficiency of the sensor material is called the total quantum efficiency. Hence, the number of electrons $n_e = \eta n_p$ is also Poisson distributed with $\mu_e = \eta \mu_p$ and $\sigma_e^2 = \mu_e = \eta \mu_p = \eta \sigma_p^2$.

During the readout of the charge from the pixel, different effects in the electronics create random fluctuations of the resulting voltages [16]. Reset noise occurs because a pixel's charge may not be completely reset during the readout. It can be eliminated through correlated double sampling, i.e., subtracting the voltage before the readout from the voltage after the readout. Dark current noise occurs because thermal excitation also causes electron–hole pairs to be created. This is only a problem for very long exposure times, and can usually be ignored in machine vision applications. If it is an issue, the camera must be cooled. Furthermore, the amplifier creates amplifier noise [16]. All these noise sources are typically combined into a single noise value called the noise floor or dark noise, because it is present even if the sensor receives no light. Dark noise can be modeled by a Gaussian distribution with mean μ_d and variance σ_d^2. Finally, the voltage is converted to a digital number (DN) by the ADC. This introduces quantization noise, which we will include in the dark noise for the purposes of this section. For a more detailed analysis, see [28].

The process of converting the charge into a DN can be modeled by a conversion factor K called the overall system gain. Its inverse $1/K$ can be interpreted as the number of electrons required to produce unit change in the gray value. With this, the mean gray value is given by $\mu_y = K(\eta \mu_p + \mu_d)$. If the noise sources are assumed to be stochastically independent, we also have $\sigma_y^2 = K^2(\eta \sigma_p^2 + \sigma_d^2) = K^2(\eta \mu_p + \sigma_d^2)$.

In [27, 28], the above noise sources are called temporal noise, because they can be averaged out over time (for temporal averaging, see Section 3.2.3).

The noise sources we have examined so far are assumed to be identical for each pixel. Manufacturing tolerances, however, cause two other effects that produce gray value changes. In contrast to temporal noise, these gray value fluctuations cannot be averaged out over time. Hence, they really are systematic errors. Nevertheless, since they look like noise, they are called spatial noise in [27, 28] or pattern noise. One effect is that the dark signal may not be identical for each pixel. This is called offset noise in [27, 28], fixed pattern noise, or dark signal non-uniformity (DSNU). The second effect is that the responsivity of the pixels to light may be different for each pixel. This is called gain noise in [27, 28] or photoresponse non-uniformity (PRNU). CMOS sensors typically have a significantly larger spatial noise because each pixel has its own amplifier, which may have a different gain and offset. Spatial noise can be modeled by a light-independent noise with a variance of σ_o^2 and a light-dependent noise with a variance of $\sigma_g^2 = S_g^2 \eta^2 \mu_p^2$. Here, S_g^2 is the variance coefficient of the gain noise. The offset and gain noise are amplified by the overall system gain K.

The total noise of the sensor is given by the combination of all of the above noise sources:

$$\sigma_y^2 = K^2(\eta \mu_p + \sigma_d^2 + S_g^2 \eta^2 \mu_p^2 + \sigma_o^2) \tag{2.17}$$

With this, the signal-to-noise ratio at a particular illumination μ_p is given by [27, 28]

$$\text{SNR} = \frac{\eta \mu_p}{\sqrt{\eta \mu_p + \sigma_d^2 + S_g^2 \eta^2 \mu_p^2 + \sigma_o^2}} \tag{2.18}$$

Note that the overall system gain K occurs in the numerator and denominator and consequently cancels out. For some CMOS sensors, the SNR can be dominated by the gain noise [28] and Eq. (2.18) becomes $\text{SNR} \approx 1/S_g$. The SNR is often specified in decibels (dB): $\text{SNR}_{dB} = 20 \log \text{SNR}$, where $\log x$ is the base-10 logarithm. Alternatively, it can be specified as the number of significant bits: $\text{SNR}_{bit} = \lg \text{SNR}$, where $\lg x = \log_2 x$ is the base-2 logarithm. Note that the SNR increases as the illumination μ_p increases. It is maximized just before the pixels reach their saturation capacity $\mu_{p.sat}$.

Another interesting performance metric can be derived if we ignore the spatial noise and acquire images with very little light ($\eta \mu_p \ll \sigma_d^2$). Here, $\text{SNR} = \eta \mu_p / \sigma_d$. The illumination $\mu_{\text{p.min}}$ for which this SNR is 1 is usually taken as the smallest detectable amount of light and is called the absolute sensitivity threshold:

$$\mu_{\text{p.min}} = \sigma_d/\eta \tag{2.19}$$

From it, we can derive the dynamic range of the camera:

$$\text{DYN} = \frac{\mu_{\text{p.sat}}}{\mu_{\text{p.min}}} \tag{2.20}$$

Like the SNR, the dynamic range is often specified in decibels or bits. Note that this is really the input dynamic range (called DYN_{in} in [27]). For sensors with a linear response, it is identical to the output dynamic range. For nonlinear sensors, a different definition must be used for the output dynamic range.

Detailed instructions about how the signal-to-noise ratio and dynamic range (and the parameters required to determine them) can be measured are specified in [27, 28].

It should also be noted that the above derivation assumes monochromatic light. For white light or other light with mixed wavelengths, the derivation needs to take into account that the quantum efficiency is a function of the wavelength: $\eta = \eta(\lambda)$. Details can be found in [28].

We conclude this section by looking at the performance data of typical CCD and CMOS sensors. The CCD sensor has a pixel size of 6.45 µm × 6.45 µm. Its noise characteristics can be described by $\eta = 0.54$, $\sigma_d = 9$, $\mu_{\text{e.sat}} = 19\,000$ (i.e., $\mu_{\text{p.sat}} = \mu_{\text{e.sat}}/\eta = 35\,185$), $\sigma_o = 1.6$, and $S_g = 0.007$. The inverse of the overall system gain is $1/K = 4.7$. The CMOS sensor has a pixel size of 9.2 µm × 9.2 µm. Its noise characteristics can be described by $\eta = 0.32$, $\sigma_d = 121$, $\mu_{\text{e.sat}} = 53\,000$ (i.e., $\mu_{\text{p.sat}} = 165\,625$), $\sigma_o = 44.8$, and $S_g = 0.011$. The inverse of the overall system gain is $1/K = 56.5$. The units of the standard deviations and of $\mu_{\text{e.sat}}$ are electrons; $\mu_{\text{p.sat}}$ is given in photons; $1/K$ is specified in electrons per gray value; and all other values are dimensionless. Figure 2.50 displays the SNR of both cameras as a function of the number of photons, with and without the spatial noise included in Eq. (2.18). As mandated by [27], the number of photons is specified in bits, while the SNR is shown both in bits and in decibels. Note that spatial noise can significantly lower the SNR, especially for CMOS sensors. Also note that the CCD sensor is significantly more sensitive to light (its curve lies to the left of the curve of the CMOS sensor). Furthermore, according to Eq. (2.19), the absolute sensitivity threshold for the CCD sensor is 17, while that of the CMOS sensor is 378. Finally, according to Eq. (2.20), the dynamic range of the CCD sensor is 11.0 bits, while that of the CMOS sensor is 8.8 bits.

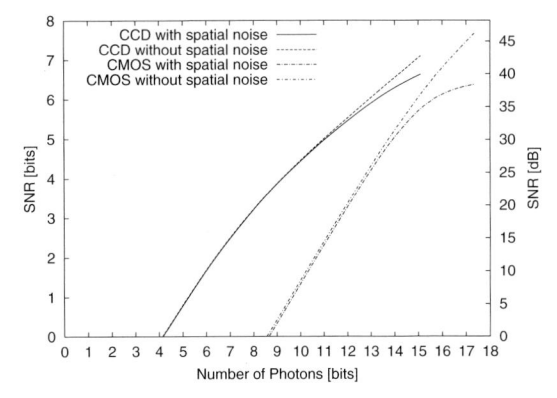

Fig. 2.50 Comparison of a CCD and a CMOS sensor. The graph shows the SNR of both sensors with and without spatial noise included in Eq. (2.18). The graphs are plotted until the sensors reach their saturation capacity. Note that spatial noise can significantly lower the SNR, especially for CMOS sensors.

2.4 Camera–Computer Interfaces

As described in the previous section, the camera acquires an image and produces either an analog or a digital video signal. In this section, we will take a closer look at how the image is transmitted to the computer and how it is reconstructed as a matrix of gray or color values. We start by examining how the image can be transmitted via an analog signal. This requires a special interface card to be present on the computer, which is conventionally called a frame grabber. We then describe the different means by which the image can be transmitted in digital form – digital video signals that require a frame grabber, IEEE 1394, USB 2.0, and Gigabit Ethernet. We also briefly discuss the different bus systems through which the image must be transported inside the computer. Finally, we discuss the different acquisition modes with which images are typically acquired.

2.4.1 Analog Video Signals

Analog video standards have been defined since the early 1940s. Because of the long experience with these standards, the components required to build an analog camera are relatively inexpensive. Therefore, many cameras still use analog video signals for transmitting images. While there are several analog video standards for television [30], only four of them are important for machine vision. Table 2.3 displays their characteristics. EIA-170 and CCIR are black-and-white video standards, while NTSC and PAL are color video standards. The primary difference between them is that EIA-170 and NTSC have a frame rate of 30 Hz with 525 lines per image, while CCIR and PAL have a frame rate of 25 Hz with 625 lines per image. In all four standards, it

2 Image Acquisition

Tab. 2.3 Analog video standards.

Standard	Type	Frame rate (frame s^{-1})	Lines	Line period (µs)	Line rate (line s^{-1})	Pixel period (ns)	Image size (pixels)
EIA-170	B/W	30.00	525	63.49	15 750	82.2	640 × 480
CCIR	B/W	25.00	625	64.00	15 625	67.7	768 × 576
NTSC	Color	29.97	525	63.56	15 734	82.3	640 × 480
PAL	Color	25.00	625	64.00	15 625	67.7	768 × 576

takes roughly the same time to transmit each line. Of the 525 and 625 lines, nominally 40 and 50 lines, respectively, are used for synchronization, e.g., to indicate the start of a new frame. Furthermore, 10.9 µs (EIA-170 and NTSC) and 12 µs (CCIR and PAL) of each line are also used for synchronization. This usually results in an image size of 640 × 480 (EIA-170 and NTSC) and 768 × 576 (CCIR and PAL). From these characteristics, it can be seen that a pixel must be roughly sampled every 82 ns or 68 ns, respectively.

Figure 2.51 displays an overview of the EIA-170 video signal. The CCIR signal is very similar. We will point out the differences in the text.

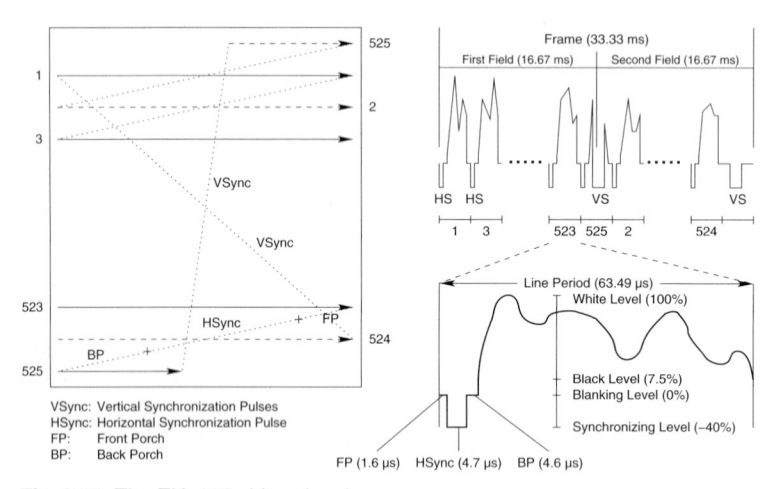

Fig. 2.51 The EIA-170 video signal.

As already mentioned at the end of Section 2.3.1, an image (a frame) is transmitted interlaced as two fields. The first field consists of the even-numbered lines and the second field of the odd-numbered lines for EIA-170 (for CCIR, the order is odd lines, then even lines). Each line consists of a horizontal blanking interval that contains the horizontal synchronization information and the active line period that contains the actual image signal of the line. The horizontal blanking interval consists of a front porch, where the signal is set to the blanking level, a horizontal synchronization pulse, where the signal is set to the synchronizing level (−40% for EIA-170, −43% for

CCIR), and a back porch, where again the signal is set to the blanking level. For CCIR, the length of the front and back porches is 1.5 μs and 5.8 μs, respectively. The purpose of the front porch was to allow voltage levels to stabilize in older televisions, preventing interference between lines. The purpose of the horizontal synchronization pulse is to indicate the start of the valid signal of each line. The purpose of the back porch was to allow the slow electronics in early televisions time to respond to the synchronization pulse and to prepare for the active line period. During the active line period, the signal varies between the white and black levels. For CCIR, the black level is 0%, while for EIA-170 it is 7.5%.

Each field starts with a series of vertical synchronization pulses. In Figure 2.51, for simplicity, they are drawn as single pulses. In reality, they consist of a series of three different kinds of pulses, each of which spans multiple lines during the vertical blanking interval. The vertical blanking interval lasts for 20 lines in EIA-170 and for 25 lines in CCIR.

The vertical blanking interval was originally needed to allow the beam to return from bottom to top in cathode-ray-tube televisions because of the inductive inertia of the magnetic coils that deflect the electron beam vertically. The magnetic field, and hence the position of the spot on the screen, cannot change instantly. Likewise, the horizontal blanking interval allowed the beam to return from right to left.

Color can be transmitted in three different manners. First of all, the color information (called chrominance) can be added to the standard video signal, which carries the luminance (brightness) information, using quadrature amplitude modulation. Chrominance is encoded using two signals that are 90° out of phase, known as I (in-phase) and Q (quadrature) signals. To enable the receiver to demodulate the chrominance signals, the back porch contains a reference signal, known as the color burst. This encoding is called composite video. It has the advantage that a color signal can be decoded by a black-and-white receiver simply by ignoring the chrominance signal. Furthermore, the signal can be transmitted over a cable with a single wire. A disadvantage of this encoding is that the luminance signal must be low-pass filtered to prevent crosstalk between high-frequency luminance information and the color information. In S-Video ("separate video"), also called Y/C, the two signals are transmitted separately in two wires. Therefore, low-pass filtering is unnecessary, and the image quality is higher. Finally, color information can also be transmitted directly as an RGB signal in three wires. This results in an even better image quality. In RGB video, the synchronization signals are transmitted either in the green channel or through a separate wire.

While interlaced video can fool the human eye into perceiving a flicker-free image (which is the reason why it is used in television), we have already seen at the end of Section 2.3.1 that it causes severe artifacts for moving objects since the image must be exposed twice. For machine vision, images should be acquired in progressive scan mode. The above video standards can be extended easily to handle this transmission mode. Furthermore, the video standards can also be extended for image sizes larger than those specified in the standards. Note, however, that in both cases the signal is no longer a standard video signal that can be displayed on a monitor.

To reconstruct the image in the computer from the video signal, a frame grabber card is required. A frame grabber consists of components that separate the synchronization signals from the video signal in order to create a pixel clock signal that is used to control an analog-to-digital converter (ADC) that samples the video signal. Furthermore, analog frame grabbers typically also contain an amplifier for the video signal. It is often possible to multiplex video signals from different cameras, i.e., to switch the acquisition between different cameras, if the frame grabber has multiple connectors, but only a single ADC. If the frame grabber has as many ADCs as connectors, it is also possible to acquire images from several cameras simultaneously. This is interesting, for example, for stereo reconstruction or for applications where an object must be inspected from multiple sides at the same time. After sampling the video signal, the frame grabber transfers the image to the computer's main memory through direct memory access (DMA), i.e., without using up valuable processing cycles of the CPU.

During the image acquisition, the frame grabber must reconstruct the pixel clock with which the camera has stored its pixels in the video signal, since this is not explicitly encoded in the video signal. This is typically performed by a phase-locked loop (PLL) circuitry in the frame grabber.

Reconstructing the pixel clock can create two problems. First of all, the frequencies of the pixel clock in the camera and the frame grabber may not match exactly. If this is the case, the pixels will be sampled at different rates. Consequently, the aspect ratio of the pixels will change. For example, square pixels on the camera may no longer be square in the image. Although this is not desirable, it can be corrected by camera calibration (see Section 3.9).

The second problem is that the PLL may not be able to reconstruct the start of the active line period with perfect accuracy. As shown in Figure 2.52, this causes the sampling of each line to be offset by a time Δt. This effect is called line jitter or pixel jitter. Depending on the behavior of the PLL, the offset may be random or systematic. From Figure 2.3, we can see that $\Delta t = \pm 7$ ns will already cause an error of approximately ± 0.1 pixels. In areas of constant or slowly changing gray values, this usually causes no problems. However, for pixels that contain edges, the gray value may change by as much as 10% of the amplitude of the edge. If the line jitter is truly random, i.e., averages out for each line across multiple images of the same scene, the precision with which the edges can be extracted will decrease, while the accuracy does not change, as explained in Section 3.7.4. If the line jitter is systematic, i.e., does not average out for each line across multiple images of the same scene, the accuracy of the edge positions will decrease, while the precision does not change.

Line jitter can be detected easily by acquiring multiple images of the same object, computing the temporal average of the images (see Section 3.2.3), and subtracting the temporal average image from each image. Figures 2.53(a) and (c) display two images from a sequence of 20 images that show horizontal and vertical edges. Figures 2.53(b)

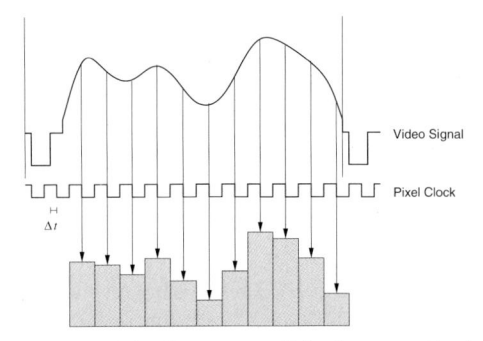

Fig. 2.52 Line jitter occurs if the frame grabber's pixel clock is not perfectly aligned with the active line period. The sampling of each line is offset by a random or systematic time Δt.

and (d) show the result of subtracting the temporal average image of the 20 images from the images in Figures 2.53(a) and (c). Note that, as predicted above, there are substantial gray value differences at the vertical edges, while there are no gray value differences in the homogeneous areas and at the horizontal edges. This clearly shows that the lines are offset by line jitter. At first glance, the effect of line jitter is increased noise at the vertical edges. However, if you look closely you will see that the gray value differences look very much like a sine wave in the vertical direction. Although this is a systematic error in this particular image, it averages out over multiple images (the sine wave will have a different phase). Hence, for this frame grabber the precision of the edges will decrease significantly.

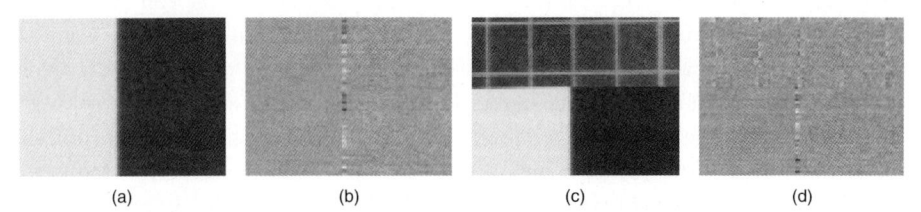

Fig. 2.53 Images of (a) a vertical edge and (c) vertical and horizontal edges. (b) (d) Gray value fluctuations of (a) and (c) caused by line jitter. The gray value fluctuations have been scaled by a factor of 5 for better visibility. Note that the fluctuations only appear for the vertical edges, indicating that in reality each line is offset by line jitter.

To prevent line jitter and non-square pixels, the pixel clock of the camera can be fed into the frame grabber (if the camera outputs its pixel clock and the frame grabber can use this signal). However, in these cases it is typically better to transmit the video signal digitally.

2.4.2
Digital Video Signals: Camera Link

In contrast to analog video signals, in which the synchronization information is embedded into the signal, digital video signals make this information explicit, as shown in Figure 2.54. The frame valid signal, which replaces the vertical synchronization signal, is asserted for the duration of a frame. Similarly, the line valid signal, which replaces the horizontal synchronization signal, is asserted for the duration of a line. The pixel clock is transmitted explicitly, preventing all the problems inherent in analog video that were described at the end of the previous section. For digital cameras, the aspect ratio of the pixels of the image is identical to the aspect ratio of the pixels on the camera. Furthermore, there is no line jitter. To create the digital video signal, the camera performs an analog-to-digital conversion of the voltage of the sensor and transmits the resulting digital number in a parallel or serial manner to the frame grabber.

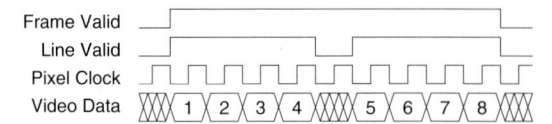

Fig. 2.54 A typical digital video signal. In contrast to analog video, the synchronization information is made explicit through the frame valid, line valid, and pixel clock signals.

In the past (i.e., until the beginning of the 21st century), the machine vision industry lacked a standard even for the physical connector between the camera and the frame grabber, not to mention the lack of a standard digital video format. Camera manufacturers used a plethora of different connectors, making cable production for frame grabber manufacturers very burdensome and the cables extremely expensive. In October 2000, the connector problem was addressed through the introduction of the Camera Link specification, which defines a 26-pin MDR connector as the standard connector. The current version of the Camera Link specification is 1.1 [31].

The Camera Link specification defines not only the connector, but also the physical means with which digital video is transmitted. The basic technology is low-voltage differential signaling (LVDS). LVDS transmits two different voltages, which are compared at the receiver. This difference in voltage between the two wires is used to encode the information. Hence, two wires are required to transmit a single signal. The big advantage of LVDS is that very high transmission speeds can be achieved.

Camera Link is based on Channel Link, a solution that was developed for transmitting video signals to flat panel displays and then extended for general-purpose data transmission. Camera Link consists of a driver and receiver pair. The driver is a chip on the camera that accepts 28 single-ended data signals and a single-ended clock signal and serializes the data 7 : 1, i.e., the 28 data signals are transmitted serially over four wire pairs with LVDS. The clock is transmitted over a fifth wire pair. The receiver is a similar chip on the frame grabber that accepts the four data signals and the clock signal, and reconstructs the original 28 data signals from them. Camera Link specifies that four of the 28 data signals are used for so-called enable signals (frame valid, line valid, data valid, as well as a signal reserved for future use). Hence, 24 data bits can be transmitted with a single chip. The chips can run at up to 85 MHz, resulting in a maximum data rate of 255 MB s^{-1} (megabytes per second). Camera Link also specifies that four additional LVDS pairs are reserved for general-purpose camera control signals, e.g., triggering. How these signals are used is up to the camera manufacturer. Finally, two LVDS pairs are used for serial communication to and from the camera, e.g., to configure the camera. Unfortunately, the standard leaves the protocol for configuring the camera up to the manufacturer. All of the above signals are transmitted over a single MDR connector.

The above configuration is called the base configuration. Camera Link also defines two additional configurations for even higher data rates. They require a second MDR connector. If a second driver–receiver chip pair is added, 24 additional data bits and four enable signals can be transmitted, resulting in a maximum data rate of 510 MB s^{-1}. Finally, if a third driver–receiver chip pair is added, 16 additional data bits and four enable signals can be transmitted, resulting in a maximum data rate of 680 MB s^{-1}.

Camera Link has become the standard for high-speed digital image acquisition (line scan cameras and high-speed or high-resolution area scan cameras). At the same time, one of its drawbacks is that it is so powerful: it requires a frame grabber to reconstruct the image in the computer, which may increase the cost of an application. In some applications, transmitting the digital video signal through standard means already present in most computers may lead to a less expensive alternative.

2.4.3
Digital Video Signals: IEEE 1394

IEEE 1394, also known as FireWire, is a standard for a high-speed serial bus system. The original standard, IEEE 1394, was released in 1995 [32]. It defines data rates of 93.304, 196.608, and 393.216 Mb s^{-1} (megabits per second), i.e., 12.288, 24.576, and 49.152 MB s^{-1} (megabytes per second). Annex J of the standard stipulates that these data rates should be referred to as 100, 200, and 400 Mb s^{-1}. A six-pin connector is used. The data is transmitted over a cable with two twisted-pair wires used for signals, and one wire each for power and ground. Hence, low-power devices can be used without having to use an external power source. A latched version of the connector is also defined, which is important in industrial applications to prevent accidental unplugging of the cable. IEEE 1394a [33] adds various clarifications to the standard and defines a four-pin connector that does not include power. The latest version of the standard, IEEE 1394b [34], defines data rates of 800 and 1600 Mb s^{-1} as well as the architecture for 3200 Mb s^{-1} (these are again really multiples of 93.304 Mb s^{-1}). A nine-pin connector as well as a fiber optic connector and cable are defined.

A standard that describes a protocol to transmit digital video for consumer applications was released as early as 1998 [35]. This led to the widespread adoption of IEEE 1394 in devices such as camcorders. Since this standard did not address the requirements for machine vision, a specification for the use of IEEE 1394 for industrial cameras was developed by the 1394 Trade Association Instrumentation and Industrial Control Working Group, Digital Camera Sub Working Group [36]. This standard is commonly referred to as IIDC.

IIDC defines various standard video formats, which define the resolution, frame rate, and pixel data that can be transmitted. The standard resolutions range from 160×120 up to 1600×1200. The frame rates range from 3.75 frames per second up to 240 frames per second. The pixel data includes monochrome (8 and 16 bits per pixel), RGB (8 and 16 bits per channel, i.e., 24 and 48 bits per pixel), YUV in various chrominance compression ratios (4 : 1 : 1, 12 bits per pixel; 4 : 2 : 2, 16 bits per pixel; 4 : 4 : 4, 24 bits per pixel), as well as the raw Bayer images (8 and 16 bits per pixel). Not all combinations of resolution, frame rate, and pixel data are supported. In addition to the standard formats, arbitrary resolutions, including rectangular areas of interest, can be used through a special video format called Format 7.

Furthermore, IIDC defines standard means to control the settings of the camera, e.g., exposure (called shutter in IIDC), iris, gain, trigger, trigger delay, and even advanced control of pan-tilt cameras, to name just a few.

IEEE 1394 defines two modes of data transfer: asynchronous and isochronous. Asynchronous transfer uses data acknowledge packets to ensure that a data packet is received correctly. If necessary, the data packet is sent again. Consequently, asynchronous mode cannot guarantee that a certain bandwidth is available for the signal, and hence it is not used for transmitting digital video signals.

Isochronous data transfer, on the other hand, guarantees a desired bandwidth, but does not ensure that the data is received correctly. In isochronous mode, each device requests a certain bandwidth. One of the devices on the bus, typically the computer to which the IEEE 1394 devices are connected, acts as the cycle master. The cycle master sends a cycle start request every 125 µs. During each cycle, it is guaranteed that each device that has requested a certain isochronous bandwidth can transmit a single packet. Approximately 80% of the bandwidth can be used for isochronous data; the rest is used for asynchronous and control data. The maximum isochronous packet payload per cycle is 1024 bytes for 100 Mb s^{-1}, and is proportional to the bus speed (i.e., it is 4096 bytes for 400 Mb s^{-1}). An example of two isochronous devices on the bus is shown in Figure 2.55.

IIDC uses asynchronous mode for transmitting control data to and from the camera. The digital video data is sent in isochronous mode by the camera. Since cameras are forced to send their data within the limits of the maximum isochronous packet payload, the maximum data rate that a camera can use is 31.25 MB s^{-1} for a bus

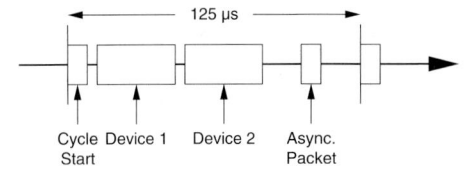

Fig. 2.55 One IEEE 1394 data transfer cycle. The cycle starts with the cycle master sending a cycle start request. Each isochronous device is allowed to send a packet of a certain size. The remainder of the cycle is used for asynchronous packets.

speed of 400 Mb s^{-1}. Hence, the maximum frame rate supported by a 400 Mb s^{-1} bus is 106 frames per second for a 640 × 480 image. If images are acquired from multiple cameras at the same time, the maximum frame rate drops accordingly.

2.4.4 Digital Video Signals: USB 2.0

The Universal Serial Bus (USB) was originally intended to replace the various serial and parallel ports that were available at the time the USB 1.0 specification was released in 1996. Hence, it was designed to support relatively low transfer speeds of 1.5 Mb s^{-1} and 12 Mb s^{-1} (megabits per second), i.e., 0.1875 MB s^{-1} and 1.5 MB s^{-1} (megabytes per second). While this allowed accessing peripheral devices such as keyboards, mice, and mass storage devices at reasonable speeds, accessing scanners was relatively slow, and acquiring images from video devices such as webcams offered only small image sizes and low frame rates. The most recent version of the specification, USB 2.0 [37], supports transfer speeds of up to 480 Mb s^{-1} (60 MB s^{-1}), making it very attractive for machine vision.

USB uses four-pin connectors that use one pair of wires for signal transmission and one wire each for power and ground. Hence, low-power devices can be used without having to use an external power source. There is no standard for a connector that can be latched or otherwise fixed to prevent accidental unplugging of the cable.

While there is a USB video class specification [38], its intention is to standardize access to webcams, digital camcorders, analog video converters, analog and digital television tuners, and still-image cameras that support video streaming. It does not cover the requirements for machine vision. Consequently, manufacturers of USB machine vision cameras typically use proprietary transfer protocols and their own device drivers to transfer images.

USB is a polled bus. Each bus has a host controller, typically the USB device in the computer to which the devices are attached, that initiates all data transfers. When a device is attached to the bus, it requests a certain bandwidth from the host controller. The host controller periodically polls all attached devices. The devices can respond with data that they want to send or can indicate that they have no data to transfer. The USB architecture defines four types of data transfers. Control transfers are used to configure devices when they are first attached to the bus. Bulk data transfers typically consist of larger amounts of data, such as data used by printers or scanners. Bulk data transfers use the bandwidth that is left over by the other three transfer types. Interrupt data transfers are limited-latency transfers that typically consist of event notifications, such as characters entered on a keyboard or mouse coordinates. For these three transfer types, data delivery is lossless. Finally, isochronous data transfers can use a pre-negotiated amount of USB bandwidth with a pre-negotiated delivery latency. To achieve the bandwidth requirements, packets can be delivered corrupted or can be lost. No error correction through retries is performed. Video data is typically transferred using isochronous or bulk data transfers.

USB 2.0 divides time into chunks of 125 μs, called microframes. Up to 20% of each microframe is reserved for control transfers. On the other hand, up to 80% of a microframe can be allocated for periodic (isochronous and interrupt) transfers. The remaining bandwidth is used for bulk transfers.

For isochronous data transfers, each end point can request up to three 1024-byte packets of bandwidth per microframe. (An end point is defined by the USB specification as "a uniquely addressable portion of a USB device that is the source or sink of information in a communication flow between the host and device.") Hence, the maximum data rate for an isochronous end point is $192\,\text{Mb}\,\text{s}^{-1}$ ($24\,\text{MB}\,\text{s}^{-1}$). Consequently, the maximum frame rate a 640×480 camera could support over a single end point with USB 2.0 is 78 frames per second. To work around this limitation, the camera could define multiple end points. Alternatively, if a guaranteed data rate is not important, the camera could try to send more data or the entire data using bulk transfers. Since bulk transfers are not limited to a certain number of packets per microframe, this can result in a higher data rate if there are no other devices that have requested bandwidth on the bus.

2.4.5
Digital Video Signals: Gigabit Ethernet

Ethernet was invented in the 1970s [39] as a physical layer for local area networks. The original experimental Ethernet provided a data rate of $2.94\,\text{Mb}\,\text{s}^{-1}$ (megabits per second). Ethernet was first standardized in 1985 with a data rate of $10\,\text{Mb}\,\text{s}^{-1}$. The current version of the Ethernet standard, IEEE 802.3 [40], defines data rates of $10\,\text{Mb}\,\text{s}^{-1}$, $100\,\text{Mb}\,\text{s}^{-1}$ ("Fast Ethernet"), $1\,\text{Gb}\,\text{s}^{-1}$ ("Gigabit Ethernet"), and $10\,\text{Gb}\,\text{s}^{-1}$. Currently, Gigabit Ethernet is in widespread use. It is used as the physical layer of virtually all local area networks. Its wide availability and high data rate make it very attractive for machine vision. Another attractive feature of the Ethernet is that it uses widely available, and therefore cheap, cables and connectors. The only drawback is that the cable does not provide power, which requires extra cabling on the camera.

To communicate on the Ethernet, each device, called host, that wants to send a data packet first checks whether the medium is idle. If not, the host waits until the medium becomes idle plus a time interval called the interframe gap (e.g., 96 ns for Gigabit Ethernet). If the medium is idle, the host simply sends its packet and checks whether a collision has occurred, i.e., whether another host has sent a packet at the same time. If this is the case, each host calculates a random back-off period and tries to transmit its data again. The back-off periods increase exponentially until a maximum number of transmission attempts has been reached. This access architecture is called carrier sense multiple access with collision detection (CSMA/CD). One of the consequences of this access architecture is that a certain bandwidth cannot be guaranteed if there are multiple hosts on the Ethernet that are transmitting data at the same time.

The Ethernet comprises the physical and data link layers of the TCP/IP model (or Internet reference model) for communications and computer network protocol design. As such, it is not used to directly transfer application data. On top of these two layers, there is a network layer, which provides the functional and procedural means of transferring variable-length data sequences from a source to a destination via one or more networks. This is typically implemented using the Internet Protocol (IP) [41]. One additional layer, the transport layer, provides transparent transfer of data between applications, e.g., by segmenting and merging packets as required by the lower layers. The Transmission Control Protocol (TCP) [42] and User Datagram Protocol (UDP) [43] are probably the best-known protocols since they form the basis of almost all Internet networking software. TCP provides connection-oriented, reliable transport, i.e., the data will arrive completely and in the same order it was sent. UDP, on the other hand, provides connectionless, unreliable transport, i.e., datagrams may be lost or duplicated, or may arrive in a different order in which they were sent. UDP requires less overhead than TCP. The final layer in the TCP/IP model is the application layer. This is the layer from which an application actually sends and receives its data. Probably the best-known application-layer protocol is the Hypertext Transfer Protocol (HTTP) [44], which forms the basis of the World Wide Web (WWW).

An application-layer protocol for machine vision cameras was standardized in 2006 under the name GigE Vision [45]. Although its name refers to Gigabit Ethernet, the standard explicitly states that it can be applied to lower or higher Ethernet speeds.

While IEEE 1394 and USB are plug-and-play buses, i.e., devices announce their presence on the bus and have a standardized manner to describe themselves, things are not quite as simple for Ethernet. The first hurdle that a camera must take when it is connected to the Ethernet is that it must obtain an IP address. GigE Vision cameras must do this through the Dynamic Host Configuration Protocol (DHCP) [46] or through the dynamic configuration of Link-Local Addresses (LLA) [47]. Optionally, a fixed IP address may be stored in the camera. DHCP requires that the camera's Ethernet MAC address is entered into the DHCP server. LLA requires the use of a special address range for the subnet, and therefore possibly requires a reconfiguration of the entire network.

A machine vision application can enquire which cameras are connected to the Ethernet through a process called device enumeration. To do so, it must send a special UDP broadcast message to the Ethernet and collect the responses by the cameras.

To control the camera, GigE Vision specifies an application-layer protocol called GVCP (GigE Vision Control Protocol). It is based on UDP. Since UDP is unreliable, explicit reliability and error recovery mechanisms are defined in GVCP. With this mechanism, GVCP establishes a control channel over the connectionless UDP protocol. The control channel is used to control the camera by writing and reading of registers and memory locations in the camera. One noteworthy feature is that each camera must describe its capabilities via an XML file according to the GenICam standard [48].

GVCP can be used to create a message channel to the camera. The message channel can be used to transmit events from the camera to the application, e.g., a timestamp when the acquisition of an image has finished.

GVCP can also be used to create from 1 to 512 stream channels to the camera. The stream channels are used to transfer the actual image data. For this purpose, GigE Vision defines a special application-layer protocol called GVSP (GigE Vision Streaming Protocol). It is based on UDP and by default does not use any reliability and error recovery mechanisms. This is done to maximize the data rate that can be transmitted. However, applications can optionally request the camera to resend lost packets. Consequently, the camera must have a memory buffer if it wants to support this feature. GVSP also defines the image data that can be transmitted. The pixel data includes monochrome (8 (unsigned and signed), 10, 12, and 16 bits per pixel), raw Bayer (8, 10, and 12 bits per pixel), RGB (8, 10, and 12 bits per channel), or YUV values in different chrominance compression ratios (4 : 1 : 1, 12 bits per pixel; 4 : 2 : 2, 16 bits per pixel; 4 : 4 : 4, 24 bits per pixel). It is also possible to transmit RGB values in three different planes using three stream channels.

Because of the CSMA/CD architecture, it is relatively difficult to estimate the actual data rate that can be achieved with Ethernet. For networks with no load apart from the data sent by the camera (e.g., if the camera is attached to the computer via a crossover cable), a UDP data rate between 600 and 840 Mb s^{-1} (75–105 MB s^{-1}) can be achieved for Gigabit Ethernet. From this, we need to subtract the GVSP protocol overhead. Hence, the maximum frame rate for a 640×480 camera would be around 320 frames per second. For 10-Gigabit Ethernet, the frame rate would increase accordingly. (We say "would" because there are no 10-Gigabit Ethernet cameras on the market at the moment.) It should also be noted that the GigE Vision standard specifies that a camera may have up to four network interfaces. If the computer or network to which the camera is attached also has four network interfaces, the data rate can be increased by a factor of up to 4.

2.4.6 Image Acquisition Modes

We conclude this section by looking at the different timing modes with which images can be acquired. We start with probably the simplest modes of all, shown in Figure 2.56. Here, the camera is free-running, i.e., it delivers its images with a fixed frame rate that is not influenced by any external events. Cameras that use analog video signals are often configured to use this mode by default. To acquire an image the application issues a command to start the acquisition to the device driver. Since the camera is free-running, the driver must wait for the start of the frame, e.g., by waiting for the vertical synchronization signal for analog video. The driver can then start to reconstruct the image i that is transmitted by the camera. For interline transfer and frame transfer CCD cameras and CMOS cameras with a global shutter, the camera has exposed the image during the previous frame cycle. After the image has been created in memory by the driver, the application can process the image. After the processing of image i has finished, the acquisition cycle starts again. Since the acquisition and processing must wait for the image to be transferred, we call this acquisition mode synchronous acquisition.

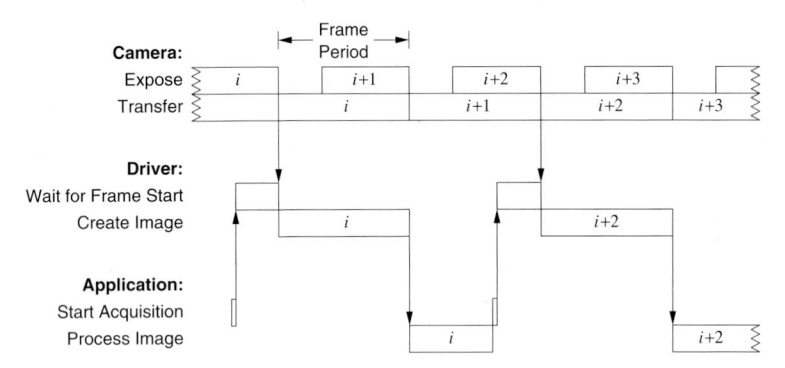

Fig. 2.56 Synchronous image acquisition from a free-running camera.

As we can see from Figure 2.56, synchronous acquisition has the big disadvantage that the application spends most of its time waiting for the image acquisition to complete. In the best case, when processing the image takes less than the frame period, only every second image can be processed.

Note that synchronous acquisition could also be used with asynchronous reset cameras. This would allow the driver and application to omit the waiting for the frame start. Nevertheless, this would not gain much of an advantage since the application would still have to wait for the frame to be transferred and the image to be created in memory.

All device drivers deliver the data to memory through direct memory access (DMA), i.e., without involving any processing by the CPU. We can make use of this capability by processing the previous image during the acquisition of the current image, as shown

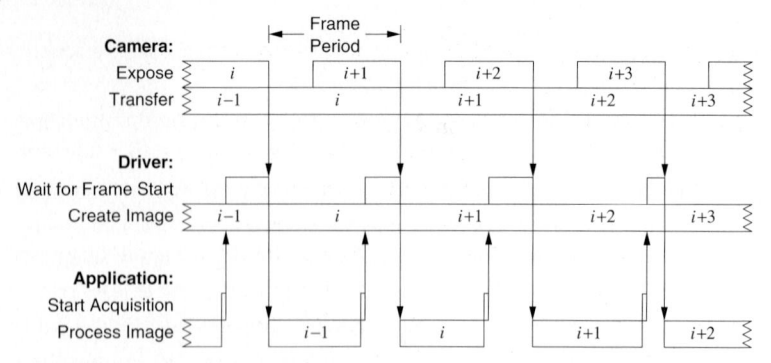

Fig. 2.57 Asynchronous image acquisition from a free-running camera.

in Figure 2.57 for a free-running camera. Here, image i is processed while image $i+1$ is acquired and transferred to memory using DMA. This mode has the big advantage that every frame can be processed if the processing time is less than the frame period.

Asynchronous grabbing can also be used with asynchronous reset cameras. However, this would be of little use as such since asynchronous grabbing enables the application to process every frame anyway (at least if the processing is fast enough). Asynchronous reset cameras and asynchronous grabbing are typically used if the image acquisition must be synchronized with some external event. Here, a trigger device, e.g., a proximity sensor or a photoelectric sensor, creates a trigger signal if the object of which the image should be acquired is in the correct location. This acquisition mode is shown in Figure 2.58. The process starts when the application instructs the device driver to start the acquisition. The driver instructs the camera or frame grabber to wait for the trigger. At the same time, the image of the previous acquisition command is returned to the application. If necessary, the driver waits for the image to be created completely in memory. After receiving the trigger, the camera resets, exposes the image, and transfers it to the device driver, which reconstructs the image in mem-

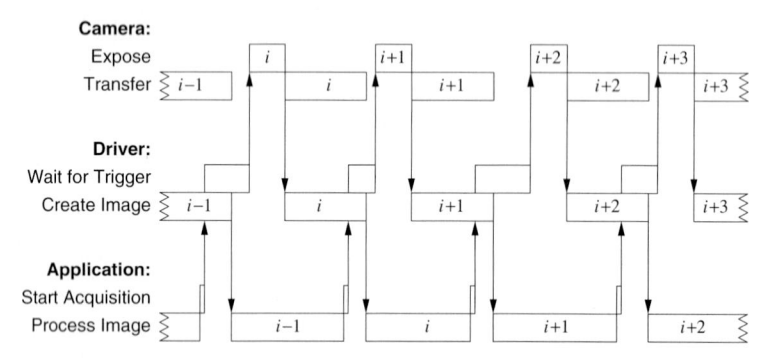

Fig. 2.58 Triggered asynchronous image acquisition from an asynchronous reset camera.

ory through DMA. Like in standard asynchronous acquisition mode, the application can process image i while image $i+1$ is being acquired.

It should be noted that all of the above modes also apply to the acquisition from line scan cameras since, as discussed in Section 2.3.1, line scan cameras typically are configured to return a 2D image. Nevertheless, here it is often essential that the acquisition does not miss a single line. Since the command to start the acquisition takes a finite amount of time, it is essential to pre-queue the acquisition commands with the driver. This mode is also sometimes necessary for area scan cameras to ensure that no trigger signal or frame is missed. This acquisition mode is often called queued acquisition. An alternative is to configure the camera and/or driver into a continuous acquisition mode. Here, the application no longer needs to issue the commands to start the acquisition. Instead, the camera and/or driver automatically prepare for the next acquisition, e.g., through a trigger signal, as soon as the current image has been transferred.

3
Machine Vision Algorithms

In the previous chapter, we have examined the different hardware components that are involved in delivering an image to the computer. Each of the components plays an essential role in the machine vision process. For example, illumination is often crucial to bring out the objects we are interested in. Triggered frame grabbers and cameras are essential if the image is to be taken at the right time with the right exposure. Lenses are important for acquiring a sharp and aberration-free image. Nevertheless, none of these components can "see," i.e., extract the information we are interested in from the image. This is analogous to human vision. Without our eyes we cannot see. Yet, even with eyes we could not see anything without our brain. The eye is merely a sensor that delivers data to the brain for interpretation. To extend this analogy a little further, even if we are myopic we can still see – only worse. Hence, we can see that the processing of the images delivered to the computer by the sensors is truly the core of machine vision. Consequently, in this chapter, we will discuss the most important machine vision algorithms.

3.1
Fundamental Data Structures

Before we can delve into the study of the machine vision algorithms, we need to examine the fundamental data structures that are involved in machine vision applications. Therefore, in this section we will take a look at the data structures for images, regions, and subpixel-precise contours.

3.1.1
Images

An image is the basic data structure in machine vision, since this is the data that an image acquisition device typically delivers to the computer's memory. As we saw in Section 2.3, a pixel can be regarded as a sample of the energy that falls on the sensor element during the exposure, integrated over the spectral distribution of the light and the spectral response of the sensor. Depending on the camera type, typically the spectral response of the sensor will comprise the entire visible spectrum and optionally a

---

Machine Vision Algorithms and Applications. Carsten Steger, Markus Ulrich, and Christian Wiedemann
Copyright © 2008 WILEY-VCH Verlag GmbH & Co. KGaA, Weinheim
ISBN: 978-3-527-40734-7

part of the near infrared spectrum. In this case, the camera will return one sample of the energy per pixel, i.e., a single-channel gray value image. RGB cameras, on the other hand, will return three samples per pixel, i.e., a three-channel image. These are the two basic types of sensors that are encountered in machine vision applications. However, in other application areas, e.g., remote sensing, images with a very large number samples per pixel, which sample the spectrum very finely, are possible. For example, the HYDICE sensor collects 210 spectral samples per pixel [49]. Therefore, to handle all possible applications, an image can be considered as a set of an arbitrary number of channels.

Intuitively, an image channel can simply be regarded as a two-dimensional (2D) array of numbers. This is also the data structure that is used to represent images in a programming language. Hence, the gray value at the pixel (r,c) can be interpreted as an entry of a matrix: $g = f_{r,c}$. In a more formalized manner, we can regard an image channel f of width w and height h as a function from a rectangular subset $R = \{0,\ldots,h-1\} \times \{0,\ldots,w-1\}$ of the discrete 2D plane $\mathbb{Z}^2$ (i.e., $R \subset \mathbb{Z}^2$) to a real number: thus $f : R \mapsto \mathbb{R}$, with the gray value g at the pixel position (r,c) defined by $g = f(r,c)$. Likewise, a multichannel image can be regarded as a function $f : R \mapsto \mathbb{R}^n$, where n is the number of channels.

In the above discussion, we have assumed that the gray values are given by real numbers. In almost all cases, the image acquisition device will not only discretize the image spatially, but also discretize the gray values to a fixed number of gray levels. In most cases, the gray values will be discretized to 8 bits (1 byte), i.e., the set of possible gray values will be $\mathbb{G}_8 = \{0,\ldots,255\}$. In some cases, a higher bit depth will be used, e.g., 10, 12, or even 16 bits. Consequently, to be perfectly accurate, a single-channel image should be regarded as a function $f : R \mapsto \mathbb{G}_b$, where $\mathbb{G}_b = \{0,\ldots,2^b-1\}$ is the set of discrete gray values with b bits. However, in many cases this distinction is unimportant, so we will regard an image as a function to the set of real numbers.

Up to now, we have regarded an image as a function that is sampled spatially, because this is the manner in which we receive the image from an image acquisition device. For theoretical considerations, it is sometimes convenient to regard the image as a function in an infinite continuous domain, i.e., $f : \mathbb{R}^2 \mapsto \mathbb{R}^n$. We will use this convention occasionally in this chapter. It will be obvious from the context which of the two conventions is used.

3.1.2
Regions

One of the tasks in machine vision is to identify regions in the image that have certain properties, e.g., by performing a threshold operation (see Section 3.4). Therefore, at the least we need a representation for an arbitrary subset of the pixels in an image. Furthermore, for morphological operations, we will see in Section 3.6.1 that it will be essential that regions can also extend beyond the image borders to avoid artifacts. Therefore, we define a region as an arbitrary subset of the discrete plane: $R \subset \mathbb{Z}^2$.

The choice of the letter R is intentionally identical to the R that is used in the previous section to denote the rectangle of the image. In many cases, it is extremely useful to restrict the processing to a certain part of the image that is specified by a region of interest (ROI). In this context, we can regard an image as a function from the region of interest to a set of numbers: thus $f : R \mapsto \mathbb{R}^n$. The region of interest is sometimes also called the domain of the image because it is the domain of the image function f. We can even unify the two views: we can associate a rectangular ROI with every image that uses the full number of pixels. Therefore, from now on, we will silently assume that every image has an associated region of interest, which will be denoted by R.

In Section 3.4.2, we will also see that often we will need to represent multiple objects in an image. Conceptually, this can simply be achieved by considering sets of regions.

From an abstract point of view, it is therefore simple to talk about regions in the image. It is not immediately clear, however, how best to represent regions. Mathematically, we can describe regions as sets, as in the above definition. An equivalent definition is to use the characteristic function of the region:

$$\chi_R(r,c) = \begin{cases} 1 & (r,c) \in R \\ 0 & (r,c) \notin R \end{cases} \tag{3.1}$$

This definition immediately suggests the use of binary images to represent regions. A binary image has a gray value of 0 for points that are not included in the region and 1 (or any other number different from 0) for points that are included in the region. As an extension to this, we could represent multiple objects in the image as label images, i.e., as images in which the gray value encodes the region to which the point belongs. Typically, a label of 0 would be used to represent points that are not included in any region, while numbers >0 would be used to represent the different regions.

The representation of regions as binary images has one obvious drawback: it needs to store (sometimes very many) points that are not included in the region. Furthermore, the representation is not particularly efficient: we need to store at least one bit for every point in the image. Often, the representation actually uses one byte per point because it is much easier to access bytes than bits. This representation is also not particularly efficient for runtime purposes: to determine which points are included in the region, we need to perform a test for every point in the binary image. In addition, it is a little awkward to store regions that extend to negative coordinates as binary images, which also leads to cumbersome algorithms. Finally, the representation of multiple regions as label images leads to the fact that overlapping regions cannot be represented, which will cause problems if morphological operations are performed on the regions. Therefore, a representation that only stores the points included in a region in an efficient manner would be very useful.

Table 3.1 shows a small example region. We first note that, either horizontally or vertically, there are extended runs in which adjacent pixels belong to the region. This is typically the case for most regions. We can use this property and only store the necessary data for each run. Since images are typically stored line by line in memory, it is better to use horizontal runs. Therefore, the minimum amount of data for each run is the row coordinate of the run and the start and end columns of the run. This method of storing a region is called a run-length representation or a run-length encoding. With this representation, the example region can be stored with just four runs, as shown in Table 3.1. Consequently, the region can also be regarded as the union of all of its runs:

$$R = \bigcup_{i=1}^{n} \mathbf{r}_i \tag{3.2}$$

Here, $\mathbf{r}_i$ denotes a single run, which can also be regarded as a region. Note that the runs are stored sorted in lexicographic order according to their row and start column coordinates. This means that there is an order of the runs $\mathbf{r}_i = (r_i, cs_i, ce_i)$ in R defined by: $\mathbf{r}_i \prec \mathbf{r}_j \Leftrightarrow r_i < r_j \vee r_i = r_j \wedge cs_i < cs_j$. This order is crucial for the execution speed of algorithms that use run-length encoded regions.

Tab. 3.1 Run-length representation of a region.

Run	Row	Start column	End column
1	1	1	4
2	2	2	2
3	2	4	5
4	3	2	5

In the above example, the binary image could be stored with 35 bytes if one byte per pixel is used or with 5 bytes if one bit per pixel is used. If the coordinates of the region are stored as 2-byte integers, the region can be represented with 24 bytes in the run-length representation. This is already a saving, albeit a small one, compared to binary images stored with one byte per pixel, but no saving if the binary image is stored as compactly as possible with one bit per pixel. To get an impression of how much this representation really saves, we can note that we are roughly storing the boundary of the region in the run-length representation. On average, the number of points on the boundary of the region will be proportional to the square root of the area of the region. Therefore, we can typically expect a very significant saving from the run-length representation compared to binary images, which must at least store every pixel in the surrounding rectangle of the region. For example, a full rectangular ROI of a $w \times h$ image can be stored with h runs instead of $w \times h$ pixels in a binary image (i.e., wh or $\lceil w/8 \rceil h$ bytes, depending on whether one byte or one bit per pixel is used). Similarly, a circle with diameter d can be stored with d runs as opposed to at least $d \times d$ pixels. We can see that the run-length representation often leads to an enormous reduction in memory consumption. Furthermore, since this representation only stores the points actually contained in the region, we do not need to perform a test to see whether a point lies in the region or not. These two features can save a significant amount of execution time. Also, with this representation it is straightforward to have regions with negative coordinates. Finally, to represent multiple regions, lists or arrays of run-length encoded regions can be used. Since in this case each region is treated separately, overlapping regions do not pose any problems.

3.1.3
Subpixel-Precise Contours

The data structures we have considered so far have been pixel-precise. Often, it is important to extract subpixel-precise data from an image because the application requires an accuracy that is higher than the pixel resolution of the image. The subpixel data can, for example, be extracted with subpixel thresholding (see Section 3.4.3) or subpixel edge extraction (see Section 3.7.3). The results of these operations can be described with subpixel-precise contours. Figure 3.1 displays several example contours. As we can see, the contours can basically be represented as a polygon, i.e., an ordered set of control points (r_i, c_i), where the ordering defines which control points are connected to each other. Since the extraction typically is based on the pixel grid, the distance between the control points of the contour is approximately one pixel on average. In the computer, the contours are simply represented as arrays of floating-point row and column coordinates. From Figure 3.1 we can also see that there is a rich topology associated with the contours. For example, contours can be closed (contour 1) or open (contours 2–5). Closed contours are usually represented by having the first contour point identical to the last contour point or by a special attribute that is stored with the contour. Furthermore, we can see that several contours can meet at a junction point, e.g., contours 3, 4, and 5. It is sometimes useful to explicitly store this topological information with the contours.

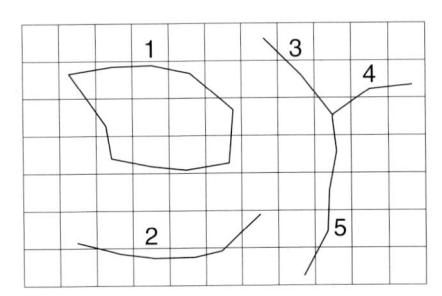

Fig. 3.1 Different subpixel-precise contours. Contour 1 is a closed contour, while contours 2–5 are open contours. Contours 3, 4, and 5 meet at a junction point.

3.2
Image Enhancement

In Chapter 2, we have seen that we have various means at our disposal to obtain a good image quality. The illumination, lenses, cameras, and image acquisition devices all play a crucial role here. However, although we try very hard to select the best possible hardware setup, sometimes the image quality is not sufficient. Therefore, in this section we will take a look at several common techniques for image enhancement.

3.2.1
Gray Value Transformations

Despite our best efforts in controlling the illumination, in some cases it is necessary to modify the gray values of the image. One of the reasons for this may be a weak contrast. With controlled illumination, this problem usually only occurs locally. Therefore, we may only need to increase the contrast locally. Another possible reason for adjusting the gray values may be that the contrast or brightness of the image has changed from the settings that were in effect when we set up our application. For example, illuminations typically age and produce a weaker contrast after some time.

A gray value transformation can be regarded as a point operation. This means that the transformed gray value $t_{r,c}$ only depends on the gray value $g_{r,c}$ in the input image at the same position: $t_{r,c} = f(g_{r,c})$. Here, $f(g)$ is a function that defines the gray value transformation to apply. Note that the domain and range of $f(g)$ typically are $\mathbb{G}_b$, i.e., they are discrete. Therefore, to increase the transformation speed, gray value transformations can be implemented as a look-up table (LUT) by storing the output gray value for each possible input gray value in a table. If we denote the LUT as f_g, we have $t_{r,c} = f_g[g_{r,c}]$, where the [ ] operator denotes the table look-up.

The most important gray value transformation is a linear gray value scaling: $f(g) = ag + b$. If $g \in \mathbb{G}_b$, we need to ensure that the output value is also in $\mathbb{G}_b$. Hence, we must clip and round the output gray value as follows:

$$f(g) = \min\left(\max(\lfloor ag + b + 0.5 \rfloor, 0), 2^b - 1\right)$$

For $|a| > 1$ the contrast is increased, while for $|a| < 1$ the contrast is decreased. If $a < 0$ the gray values are inverted. For $b > 0$ the brightness is increased, while for $b < 0$ the brightness is decreased.

Figure 3.2(a) shows a small part of an image of a printed circuit board. The entire image was acquired such that the full range of gray values is used. Three components are visible in the image. As we can see, the contrast of the components is not as good as it could be. Figures 3.2(b)–(e) show the effect of applying a linear gray value transformation with different values for a and b. As we can see from Figure 3.2(e), the component can be seen more clearly for $a = 2$.

The parameters of the linear gray value transformation must be selected appropriately for each application and must be adapted to changed illumination conditions.

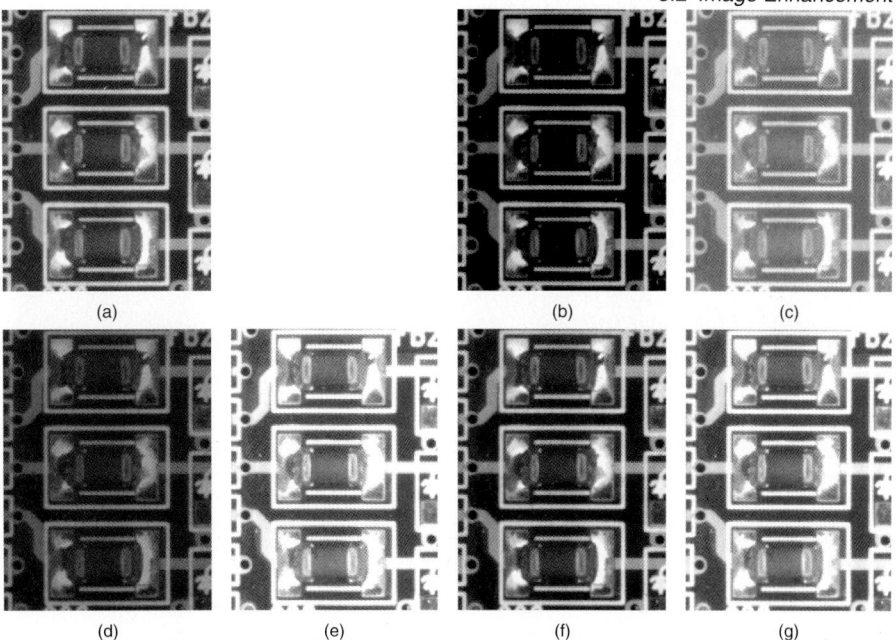

Fig. 3.2 Examples of linear gray value transformations. (a) Original image. (b) Decreased brightness ($b = -50$). (c) Increased brightness ($b = 50$). (d) Decreased contrast ($a = 0.5$). (e) Increased contrast ($a = 2$). (f) Gray value normalization. (g) Robust gray value normalization ($p_l = 0$, $p_u = 0.8$).

Since this can be quite cumbersome, ideally we would like to have a method that selects a and b automatically based on the conditions in the image. One obvious method to do this is to select the parameters such that the maximum range of the gray value space $\mathbb{G}_b$ is used. This can be done as follows: let $g_{\min}$ and $g_{\max}$ be the minimum and maximum gray value in the ROI under consideration. Then, the maximum range of gray values will be used if $a = (2^b - 1)/(g_{\max} - g_{\min})$ and $b = -ag_{\min}$. This transformation can be thought of as a normalization of the gray values. Figure 3.2(f) shows the effect of the gray value normalization of the image in Figure 3.2(a). As we can see, the contrast is not much better than in the original image. This happens because there are specular reflections on the solder, which have the maximum gray value, and because there are very dark parts in the image with a gray value of almost 0. Hence, there is not much room to improve the contrast.

The problem with the gray value normalization is that a single pixel with a very bright or dark gray value can prevent us from using the desired gray value range. To get a better understanding of this point, we can take a look at the gray value histogram of the image. The gray value histogram is defined as the frequency with which a particular gray value occurs. Let n be the number of points in the ROI under consid-

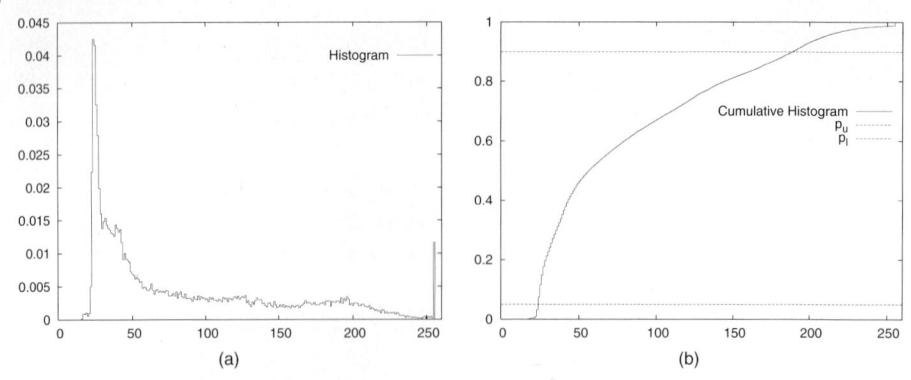

Fig. 3.3 (a) Histogram of the image in Figure 3.2(a). (b) Corresponding cumulative histogram with probability thresholds p_u and p_l superimposed.

eration and n_i be the number of pixels that have the gray value i. Then, the gray value histogram is a discrete function with domain $\mathbb{G}_b$ that has the values

$$h_i = \frac{n_i}{n} \tag{3.3}$$

In probabilistic terms, the gray value histogram can be regarded as the probability density of the occurrence of gray value i. We can also compute the cumulative histogram of the image as follows:

$$c_i = \sum_{j=0}^{i} h_j \tag{3.4}$$

This corresponds to the probability distribution of the gray values. Figure 3.3 shows the histogram and cumulative histogram of the image in Figure 3.2(a). We can see that the specular reflections on the solder create a peak in the histogram at gray value 255. Furthermore, we can see that the smallest gray value in the image is 16. This explains why the gray value normalization did not increase the contrast significantly. We can also see that the dark part of the gray value range contains the most information about the components, while the bright part contains the information corresponding to the specular reflections as well as the printed rectangles on the board. Therefore, to get a more robust gray value normalization, we can simply ignore a part of the histogram that includes a fraction p_l of the darkest gray values and a fraction $1 - p_u$ of the brightest gray values. This can easily be done based on the cumulative histogram by selecting the smallest gray value for which $c_i \geq p_l$ and the largest gray value for which $c_i \leq p_u$. Conceptually, this corresponds to intersecting the cumulative histogram with the lines $p = p_l$ and $p = p_u$. Figure 3.3(b) shows two example probability thresholds superimposed on the cumulative histogram. For the example image in Figure 3.2(a), it is best to ignore only the bright gray values that correspond to the reflections and print on the board to get a robust gray value normalization. Figure 3.2(g) shows the result that is obtained with $p_l = 0$ and $p_u = 0.8$. As we can see, the contrast of the components is significantly improved.

The robust gray value normalization is an extremely powerful method that is used, for example, as a feature extraction method for optical character recognition (OCR; see Section 3.12), where it can be used to make the OCR features invariant to illumination changes. However, it requires transforming the gray values in the image, which is computationally expensive. If we want to make an algorithm robust to illumination changes, it is often possible to adapt the parameters to the changes in the illumination, e.g., as described in Section 3.4.1 for the segmentation of images.

3.2.2
Radiometric Calibration

Many image processing algorithms rely on the fact that there is a linear correspondence between the energy that the sensor collects and the gray value in the image, namely $G = aE + b$, where E is the energy that falls on the sensor and G is the gray value in the image. Ideally, $b = 0$, which means that twice as much energy on the sensor leads to twice the gray value in the image. However, $b = 0$ is not necessary for measurement accuracy. The only requirement is that the correspondence is linear. If the correspondence is nonlinear, the accuracy of the results returned by these algorithms typically will degrade. Examples of this are the subpixel-precise threshold (see Section 3.4.3), the gray value features (see Section 3.5.2), and, most notably, subpixel-precise edge extraction (see Section 3.7, in particular Section 3.7.4). Unfortunately, sometimes the gray value correspondence is nonlinear, i.e., either the camera or the frame grabber produces a nonlinear response to the energy. If this is the case, and we want to perform accurate measurements, we must determine the nonlinear response and invert it. If we apply the inverse response to the images, the resulting images will have a linear response. The process of determining the inverse response function is known as radiometric calibration.

In laboratory settings, traditionally calibrated targets (typically gray ramps such as the target shown in Figure 3.4) are used to perform the radiometric calibration.

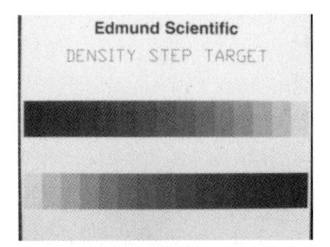

Fig. 3.4 Example of a calibrated density target that is traditionally used for radiometric calibration in laboratory settings.

Consequently, the corresponding algorithms are called chart-based. The procedure is to measure the gray values in the different patches and to compare them to the known reflectance of the patches [50]. This yields a small number of measurements (e.g., 15 independent measurements in the target in Figure 3.4), through which a function is fitted, e.g., a gamma response function that includes gain and offset, given by

$$f(g) = (a + bg)^\gamma \tag{3.5}$$

There are several problems with this approach. First of all, it requires a very even illumination throughout the entire field of view in order to be able to determine the gray values of the patches correctly. While this may be achievable in laboratory settings, it is much harder to achieve in a production environment, where the calibration often must be performed. Furthermore, effects like vignetting may lead to an apparent light drop-off toward the border, which also prevents the extraction of the correct gray values. This problem is always present, independent of the environment. Another problem is that the format of the calibration targets is not standardized, and hence it is difficult to implement a general algorithm for finding the patches on the targets and to determine their correspondence to the true reflectances. In addition, the reflectances on the density targets are often tailored for photographic applications, which means that the reflectances are specified as a linear progression in density, which is related logarithmically to the reflectance. For example, the target in Figure 3.4 has a linear density progression, i.e., a logarithmic gray value progression. This means that the samples for the curve fitting are not evenly distributed, which can cause the fitted response to be less accurate in the parts of the curve that contain the samples with the larger spacing. Finally, the range of functions that can be modeled for the camera response is limited to the single function that is fitted through the data.

Because of the above problems, a radiometric calibration algorithm that does not require any calibration target is highly desirable. These algorithms are called chart-less radiometric calibration. They are based on taking several images of the same scene with different exposures. The exposure can be varied by changing the aperture stop of the lens or by varying the exposure time of the camera. Since the aperture stop can be set less accurately than the exposure time, and since the exposure time of most industrial cameras can be controlled very accurately in software, varying the exposure time is the preferred method of acquiring images with different exposures. The advantages of this approach are that no calibration targets are required and that they do not require an even illumination. Furthermore, the range of possible gray values can be covered with multiple images instead of a single image, as required by the algorithms that use calibration targets. The only requirement on the image content is that there should be no gaps in the histograms of the different images within the gray value range that each image covers. Furthermore, with a little extra effort, even overexposed (i.e., saturated) images can be handled.

To derive an algorithm for chart-less calibration, let us examine what two images with different exposures tell us about the response function. We know that the gray value G in the image is a nonlinear function r of the energy E that falls onto the sensor during the exposure e [51]:

$$G = r(eE) \tag{3.6}$$

Note that e is proportional to the exposure time and proportional to the area of the entrance pupil of the lens, i.e., proportional to $(1/F)^2$, where F is the f-number of the lens. As described above, in industrial applications, we typically leave the aperture stop constant and vary the exposure time. Therefore, we can think of e as the exposure time.

The goal of the radiometric calibration is to determine the inverse response $q = r^{-1}$. The inverse response can be applied to an image via an LUT to achieve a linear response.

Now, let us assume that we have acquired two images with different exposures e_1 and e_2. Hence, we know that $G_1 = r(e_1 E)$ and $G_2 = r(e_2 E)$. By applying the inverse response q to both equations, we obtain $q(G_1) = e_1 E$ and $q(G_2) = e_2 E$. We can now divide the two equations to eliminate the unknown energy E, and obtain

$$\frac{q(G_1)}{q(G_2)} = \frac{e_1}{e_2} = e_{1,2} \tag{3.7}$$

As we can see, q depends only on the gray values in the images and on the ratio $e_{1,2}$ of the exposures, and not on the exposures e_1 and e_2 themselves. Equation (3.7) is the defining equation for all chart-less radiometric calibration algorithms.

One way to determine q based on Eq. (3.7) is to discretize q in a look-up table. Thus, $q_i = q(G_i)$. To derive a linear algorithm to determine q, we can take logarithms on both sides of Eq. (3.7) to obtain $\log(q_1/q_2) = \log e_{1,2}$, i.e., $\log(q_1) - \log(q_2) = \log e_{1,2}$ [51]. If we set $Q_i = \log(q_i)$ and $E_{1,2} = \log e_{1,2}$, each pixel in the image pair yields one linear equation for the inverse response function Q:

$$Q_1 - Q_2 = E_{1,2} \tag{3.8}$$

Hence, we obtain a linear equation system $AQ = E$, where Q is a vector of the LUT for the logarithmic inverse response function, while A is a matrix with 256 columns for byte images. Matrices A and E have as many rows as pixels in the image, e.g., 307 200 for a 640 × 480 image. Therefore, this equation system is much too large to be solved in an acceptable time. To derive an algorithm that solves the equation system in an acceptable time, we can note that each row of the equation system has the following form:

$$(0 \ \dots \ 0 \ 1 \ 0 \ \dots \ 0 \ -1 \ 0 \ \dots \ 0) Q = E_{1,2} \tag{3.9}$$

The indices of the 1 and -1 entries in the above equation are determined by the gray values in the first and second image. Note that each pair of gray values that occurs multiple times leads to several identical rows in A. Also note that $AQ = E$ is an overdetermined equation system, which can be solved through the normal equations $A^\top A Q = A^\top E$. This means that each row that occurs k times in A will have the weight k in the normal equations. The same behavior is obtained by multiplying the row (3.9) that corresponds to the gray value pair by $\sqrt{k}$ and to include that row only once in A. This typically reduces the number of rows in A from several hundred thousand to a few thousand, and thus makes the solution of the equation system feasible.

The simplest method to determine k is to compute the 2D histogram of the image pair. The 2D histogram determines how often gray value i occurs in the first image, while gray value j occurs in the second image at the same position. Hence, for byte images the 2D histogram is a 256×256 image in which the column coordinate indicates the gray value in the first image, while the row coordinate indicates the gray value in the second image. It is obvious that the 2D histogram contains the required values of k. We will see examples for the 2D histograms below.

Note that the discussion so far has assumed that the calibration is performed from a single image pair. It is, however, very simple to include multiple images in the calibration since additional images provide the same type of equations as in (3.9), and can thus simply be added to A. This makes it much easier to cover the entire range of gray values. Thus, we can start with a fully exposed image and successively reduce the exposure time until we reach an image in which the smallest possible gray values are assumed. We could even start with a slightly overexposed image to ensure that the highest gray values are assumed. However, in this case we have to take care that the overexposed (saturated) pixels are excluded from A because they violate the defining equation (3.7). This is a very tricky problem to solve in general since some cameras exhibit a bizarre saturation behavior. Suffice it to say that for many cameras it is sufficient to exclude pixels with the maximum gray value from A.

Despite the fact that A has many more rows than columns, the solution Q is not uniquely determined because we cannot determine the absolute value of the energy E that falls onto the sensor. Hence, the rank of A is at most 255 for byte images. To solve this problem, we could arbitrarily require $q(255) = 255$, i.e., scale the inverse response function such that the maximum gray value range is used. Since the equations are solved in a logarithmic space, it is slightly more convenient to require $q(255) = 1$ and to scale the inverse response to the full gray value range later. With this, we obtain one additional equation of the form

$$(0 \quad \ldots \quad 0 \quad k) Q = 0 \tag{3.10}$$

To enforce the constraint $q(255) = 1$, the constant k must be chosen such that Eq. (3.10) has the same weight as the sum of all other equations (3.9), i.e., $k = \sqrt{wh}$, where w and h are the width and height of the image.

Even with this normalization, we still face some practical problems. One problem is that, if the images contain very little noise, the equations (3.9) can become decoupled, and hence do not provide a unique solution for Q. Another problem is that, if the possible range of gray values is not completely covered by the images, there are no equations for the range of gray values that are not covered. Hence, the equation system will become singular. Both problems can be solved by introducing smoothness constraints for Q, which couple the equations and enable an extrapolation of Q into the range of gray values that is not covered by the images. The smoothness constraints require that the second derivative of Q should be small. Hence, for byte images they lead to 254 equations of the form

$$(0 \ \ldots \ 0 \ s \ -2s \ s \ 0 \ \ldots \ 0) Q = 0 \tag{3.11}$$

The parameter s determines the amount of smoothness that is required. Like for Eq. (3.10), s must be chosen such that Eq. (3.11) has the same weight as the sum of all the other equations, i.e., $s = c\sqrt{wh}$, where c is a small number. Empirically, $c = 4$ works well for a wide range of cameras.

The approach of tabulating the inverse response q has two slight drawbacks. First of all, if the camera has a resolution of more than 8 bits, the equation system and 2D histograms will become very large. Second, the smoothness constraints will lead to straight lines in the logarithmic representation of q, i.e., exponential curves in the normal representation of q in the range of gray values that is not covered by the images. Therefore, sometimes it may be preferable to model the inverse response as a polynomial, e.g., as in [52]. This model also leads to linear equations for the coefficients of the polynomial. Since polynomials are not very robust in the extrapolation into areas in which no constraints exist, we also have to add smoothness constraints in this case by requiring that the second derivative of the polynomial is small. Because this is done in the original representation of q, the smoothness constraints will extrapolate straight lines into the gray value range that is not covered.

Let us now consider two cameras: one with a linear response, and one with a strong gamma response, i.e., with a small γ in Eq. (3.5), and hence with a large γ in the inverse response q. Figure 3.5 displays the 2D histograms of two images taken with each camera with an exposure ratio of 0.5. Note that in both cases the values in the 2D histogram correspond to a line. The only difference is the slope of the line. A different slope, however, could also be caused by a different exposure ratio. Hence, we can see that it is quite important to know the exposure ratios precisely if we want to perform the radiometric calibration.

To conclude this section, we give two examples of the radiometric calibration. The first camera is a linear camera. Here, five images were acquired with exposure times of 32, 16, 8, 4, and 2 ms, as shown in Figure 3.6(a). The calibrated inverse response curve is shown in Figure 3.6(b). Note that the response is linear, but the camera has set a slight offset in the amplifier, which prevents very small gray values from being assumed. The second camera is a camera with a gamma response. In this case, six images were taken with exposure times of 30, 20, 10, 5, 2.5, and 1.25 ms, as shown in Figure 3.6(c). The calibrated inverse response curve is shown in Figure 3.6(d). Note the strong gamma response of the camera. The 2D histograms in Figure 3.5 were computed from the second and third brightest images in both sequences.

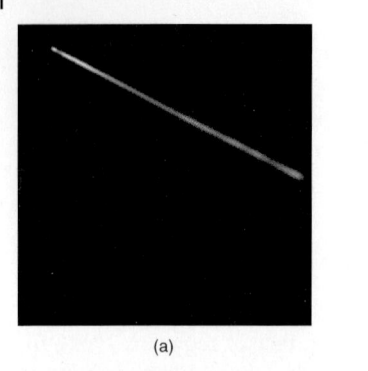

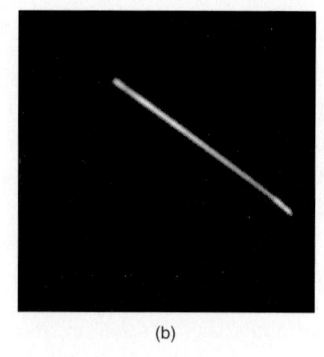

Fig. 3.5 (a) A 2D histogram of two images taken with an exposure ratio of 0.5 with a linear camera. (b) A 2D histogram of two images taken with an exposure ratio of 0.5 with a camera with a strong gamma response curve. For better visualization, the 2D histograms are displayed with a square root LUT. Note that in both cases the values in the 2D histogram correspond to a line. Hence, linear responses cannot be distinguished from gamma responses without knowing the exact exposure ratio.

3.2.3
Image Smoothing

Every image contains some degree of noise. For the purposes of this chapter, noise can be regarded as random changes in the gray values, which occur for various reasons, e.g., because of the randomness of the photon flux. In most cases, the noise in the image will need to be suppressed by using image smoothing operators.

In a more formalized manner, noise can be regarded as a stationary stochastic process [53]. This means that the true gray value $g_{r,c}$ is disturbed by a noise term $n_{r,c}$ to get the observed gray value: $\hat{g}_{r,c} = g_{r,c} + n_{r,c}$. We can regard the noise $n_{r,c}$ as a random variable with mean 0 and variance σ^2 for every pixel. We can assume a mean of 0 for the noise because any mean different from 0 would constitute a systematic bias of the observed gray values, which we could not detect anyway. Stationary means that the noise does not depend on the position in the image, i.e., is identically distributed for each pixel. In particular, σ^2 is assumed constant throughout the image. The last assumption is a convenient abstraction that does not necessarily hold because the variance of the noise sometimes depends on the gray values in the image. However, we will assume that the noise is always stationary.

Figure 3.7 shows an image of an edge from a real application. The noise is clearly visible in the bright patch in Figure 3.7(a) and in the horizontal gray value profile in Figure 3.7(b). Figures 3.7(c) and (d) show the actual noise in the image. How the noise has been calculated is explained below. It can be seen that there is slightly more noise in the dark patch of the image.

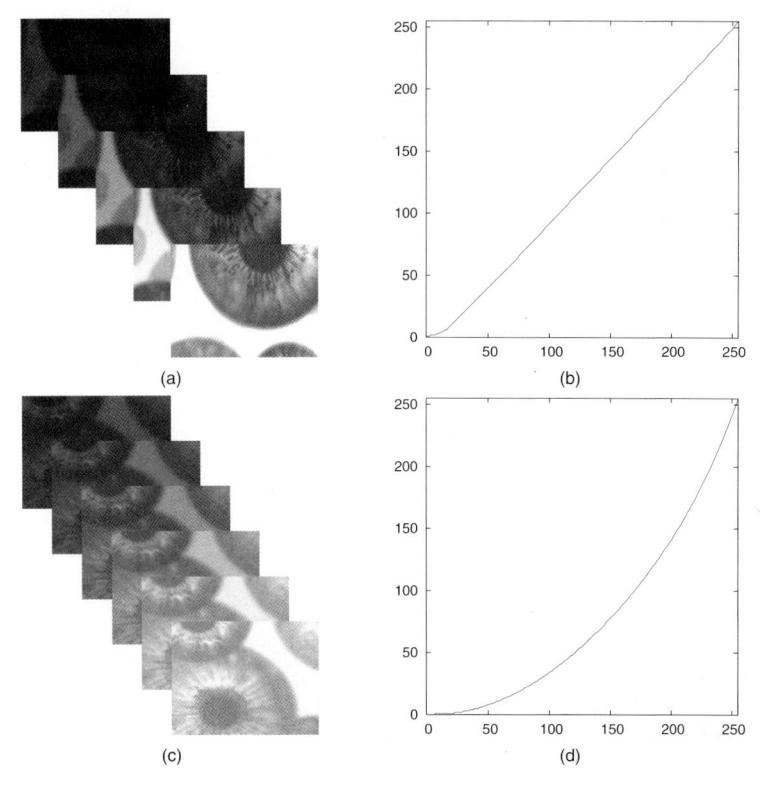

Fig. 3.6 (a) Five images taken with a linear camera with exposure times of 32, 16, 8, 4, and 2 ms. (b) Calibrated inverse response curve. Note that the response is linear, but the camera has set a slight offset in the amplifier, which prevents very small gray values from being assumed. (c) Six images taken with a camera with a gamma response with exposure times of 30, 20, 10, 5, 2.5, and 1.25 ms. (d) Calibrated inverse response curve. Note the strong gamma response of the camera.

With the above discussion in mind, noise suppression can be regarded as a stochastic estimation problem, i.e., given the observed noisy gray values $\hat{g}_{r,c}$, we want to estimate the true gray values $g_{r,c}$. An obvious method to reduce the noise is to acquire multiple images of the same scene and to simply average these images. Since the images are taken at different times, we will refer to this method as temporal averaging or the temporal mean. If we acquire n images, the temporal average is given by

$$g_{r,c} = \frac{1}{n} \sum_{i=1}^{n} \hat{g}_{r,c;i} \tag{3.12}$$

where $\hat{g}_{r,c;i}$ denotes the noisy gray value at position (r, c) in image i. This approach is frequently used in X-ray inspection systems, which inherently produce quite noisy images. From probability theory [53], we know that the variance of the noise is reduced by a factor of n by this estimation: $\sigma_m^2 = \sigma^2/n$. Consequently, the standard deviation of the noise is reduced by a factor of $\sqrt{n}$. Figure 3.8 shows the result of

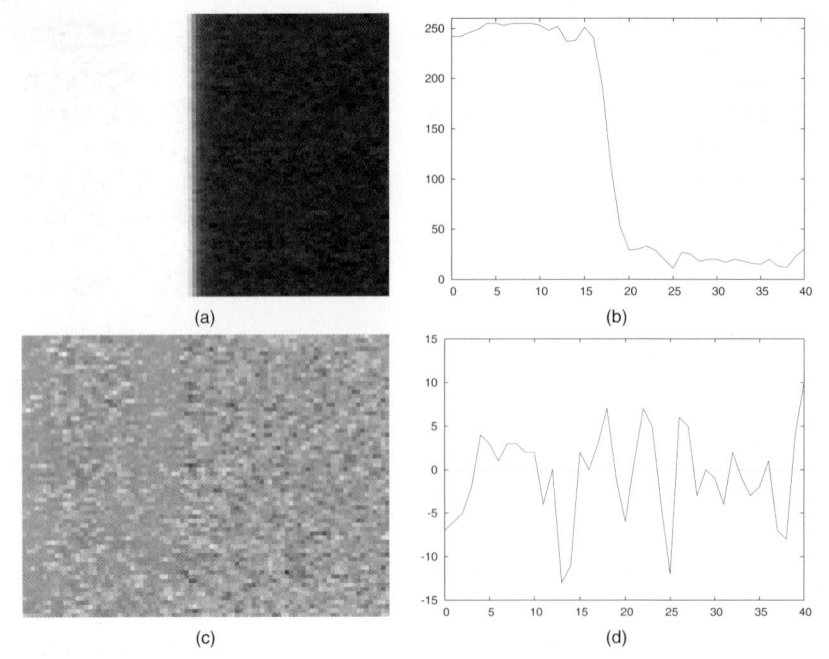

Fig. 3.7 (a) An image of an edge. (b) Horizontal gray value profile through the center of the image. (c) The noise in (a) scaled by a factor of 5. (d) Horizontal gray value profile of the noise.

acquiring 20 images of an edge and computing the temporal average. Compared to Figure 3.7(a), which shows one of the 20 images, the noise has been reduced by a factor of $\sqrt{20} \approx 4.5$, as can be seen from Figure 3.8(b). Since this temporally averaged image is a very good estimate for the true gray values, we can subtract it from any of the images that were used in the averaging to obtain the noise in that image. This is how the image in Figure 3.7(c) was computed.

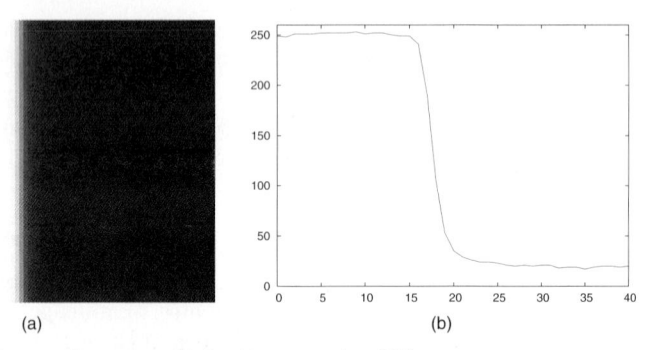

Fig. 3.8 (a) An image of an edge obtained by averaging 20 images of the edge. (b) Horizontal gray value profile through the center of the image.

One of the drawbacks of the temporal averaging is that we have to acquire multiple images to reduce the noise. This is not very attractive if the speed of the application is important. Therefore, other means for reducing the noise are required in most cases. Ideally, we would like to use only one image to estimate the true gray value. If we turn to the theory of stochastic processes again, we see that the temporal averaging can be replaced with a spatial averaging if the stochastic process, i.e., the image, is ergodic [53]. This is precisely the definition of ergodicity, and we will assume for the moment that it holds for our images. Then, the spatial average or spatial mean can be computed over a window (also called mask) of $(2n+1) \times (2m+1)$ pixels as follows:

$$g_{r,c} = \frac{1}{(2n+1)(2m+1)} \sum_{i=-n}^{n} \sum_{j=-m}^{m} \hat{g}_{r-i,c-j} \qquad (3.13)$$

This spatial averaging operation is also called a mean filter. Like for temporal averaging, the noise variance is reduced by a factor that corresponds to the number of measurements that are used to calculate the average, i.e., by $(2n+1)(2m+1)$. Figure 3.9 shows the result of smoothing the image of Figure 3.7 with a 5×5 mean filter. The standard deviation of the noise is reduced by a factor of 5, which is approximately the same as the temporal averaging in Figure 3.8. However, we can see that the edge is no longer as sharp as for temporal averaging. This happens, of course, because the images are not ergodic in general, only in areas of constant intensity. Therefore, in contrast to the temporal mean, the spatial mean filter blurs edges.

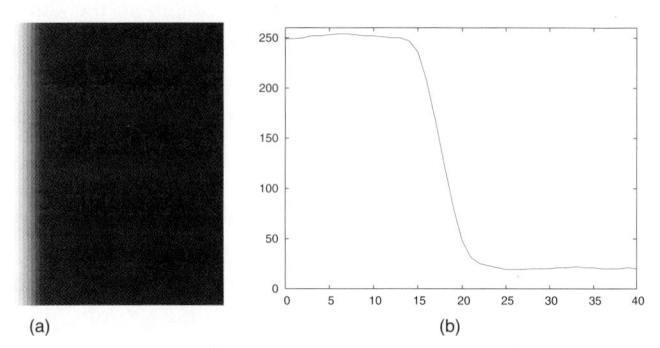

Fig. 3.9 (a) An image of an edge obtained by smoothing the image of Figure 3.7(a) with a 5×5 mean filter. (b) Horizontal gray value profile through the center of the image.

In Eq. (3.13), we have ignored the fact that the image has finite extent. Therefore, if the mask is close to the image border, it will partially stick out of the image, and consequently will access undefined gray values. To solve this problem, several approaches are possible. A very simple approach is to calculate the filter only for pixels for which the mask lies completely within the image. This means that the output image is smaller than the input image, which is not very helpful if multiple filtering operations are applied in sequence. We could also define that the gray values outside the image are 0. For the mean filter, this would mean that the result of the filter would become progressively darker as the pixels get closer to the image border. This is also not desirable. Another approach would be to use the closest gray value on the image border for pixels outside the image. This approach would still create unwanted edges at the image border. Therefore, typically the gray values are mirrored at the image border. This creates the least amount of artifacts in the result.

As was mentioned above, noise reduction from a single image is preferable for speed reasons. Therefore, let us take a look at the number of operations involved in the calculation of the mean filter. If the mean filter is implemented based on Eq. (3.13), the number of operations will be $(2n+1)(2m+1)$ for each pixel in the image, i.e., the calculation will have the complexity $O(whmn)$, where w and h are the width and height of the image, respectively. For $w = 640$, $h = 480$, and $m = n = 5$ (i.e., an 11×11 filter), the algorithm will perform 37 171 200 additions and 307 200 divisions. This is quite a substantial number of operations, so we should try to reduce the operation count as much as possible. One way to do this is to use the associative law of the addition of real numbers as follows:

$$g_{r,c} = \frac{1}{(2n+1)(2m+1)} \sum_{i=-n}^{n} \left(\sum_{j=-m}^{m} \hat{g}_{r-i,c-j} \right) \qquad (3.14)$$

This may seem like a trivial observation, but if we look closer we can see that the term in parentheses only needs to be computed once and can be stored, e.g., in a temporary image. Effectively, this means that we are first computing the sums in the column direction of the input image, save them in a temporary image, and then compute the sums in the row direction of the temporary image. Hence, the double sum in Eq. (3.13) of complexity $O(nm)$ is replaced by two sums of total complexity $O(n+m)$. Consequently, the complexity drops from $O(whmn)$ to $O(wh(m+n))$. With the above numbers, now only 6 758 400 additions are required. The above transformation is so important that it has its own name. Whenever a filter calculation allows a decomposition into separate row and column sums, the filter is called separable. It is obviously of great advantage if a filter is separable, and it is often the best speed improvement that can be achieved. In this case, however, it is not the best we can do. Let us take a look at the column sum, i.e., the part in parentheses in Eq. (3.14), and let the result of the column sum be denoted by $t_{r,c}$. Then, we have

$$t_{r,c} = \sum_{j=-m}^{m} \hat{g}_{r,c-j} = t_{r,c-1} + \hat{g}_{r,c+m} - \hat{g}_{r,c-m-1} \qquad (3.15)$$

i.e., the sum at position (r, c) can be computed based on the already computed sum at position $(r, c-1)$ with just two additions. The same holds, of course, also for the row sums. The result of this is that we need to compute the complete sum only once for the first column or row, and can then update it very efficiently. With this, the total complexity is $O(wh)$. Note that the mask size does not influence the runtime in this implementation. Again, since this kind of transformation is so important, it has a special name. Whenever a filter can be implemented with this kind of updating scheme based on previously computed values, it is called a recursive filter. For the above example, the mean filter requires just 1 238 880 additions for the entire image. This is more than a factor of 30 faster for this example than the naive implementation based on Eq. (3.13). Of course, the advantage becomes even bigger for larger mask sizes.

In the above discussion, we have called the process of spatial averaging a mean filter without defining what is meant by the word filter. We can define a filter as an operation that takes a function as input and produces a function as output. Since images can be regarded as functions (see Section 3.1.1), for our purposes a filter transforms an image into another image.

The mean filter is an instance of a linear filter. Linear filters are characterized by the following property: applying a filter to a linear combination of two input images yields the same result as applying the filter to the two images and then computing the linear combination. If we denote the linear filter by h, and the two images by f and g, we have:

$$h\{af(p) + bg(p)\} = ah\{f(p)\} + bh\{g(p)\}$$

where $p = (r, c)$ denotes a point in the image and the $\{\ \}$ operator denotes the application of the filter. Linear filters can be computed by a convolution. For a one-dimensional (1D) function on a continuous domain, the convolution is given by

$$f * h = (f * h)(x) = \int_{-\infty}^{\infty} f(t) h(x - t) \, dt \qquad (3.16)$$

Here, f is the image function and the filter h is specified by another function called the convolution kernel of the filter mask. Similarly, for 2D functions we have

$$f * h = (f * h)(r, c) = \int_{-\infty}^{\infty} \int_{-\infty}^{\infty} f(u, v) h(r - u, c - v) \, du \, dv \qquad (3.17)$$

For functions with discrete domains, the integrals are replaced by sums:

$$f * h = \sum_{i=-\infty}^{\infty} \sum_{j=-\infty}^{\infty} f_{i,j} h_{r-i, c-j} \qquad (3.18)$$

The integrals and sums are formally taken over an infinite domain. Of course, to be able to compute the convolution in a finite amount of time, the filter $h_{r,c}$ must be 0 for sufficiently large r and c. For example, the mean filter is given by

$$h_{r,c} = \begin{cases} \dfrac{1}{(2n+1)(2m+1)} & |r| \leq n \wedge |c| \leq m \\ 0 & \text{otherwise} \end{cases} \qquad (3.19)$$

The notion of separability can be extended for arbitrary linear filters. If $h(r,c)$ can be decomposed as $h(r,c) = s(r)t(c)$ (or as $h_{r,c} = s_r t_c$), then h is called separable. As for the mean filter, we can factor out s in this case to get a more efficient implementation:

$$\begin{aligned} f * h &= \sum_{i=-\infty}^{\infty} \sum_{j=-\infty}^{\infty} f_{i,j} h_{r-i,c-j} = \sum_{i=-\infty}^{\infty} \sum_{j=-\infty}^{\infty} f_{i,j} s_{r-i} t_{c-j} \\ &= \sum_{i=-\infty}^{\infty} s_{r-i} \left(\sum_{j=-\infty}^{\infty} f_{i,j} t_{c-j} \right) \end{aligned} \quad (3.20)$$

Obviously, separable filters have the same speed advantage as the separable implementation of the mean filter. Therefore, separable filters are preferred over non-separable filters. There is also a definition for recursive linear filters, which we cannot cover in detail. The interested reader is referred to [54]. Recursive linear filters have the same speed advantage as the recursive implementation of the mean filter, i.e., the runtime does not depend on the filter size. Unfortunately, many interesting filters cannot be implemented as recursive filters. Usually they can only be approximated by a recursive filter.

Although the mean filter produces good results, it is not the optimum smoothing filter. To see this, we can note that noise primarily manifests itself as high-frequency fluctuations of the gray values in the image. Ideally, we would like a smoothing filter to remove these high-frequency fluctuations. To see how well the mean filter performs this task, we can examine how the mean filter responds to certain frequencies in the image. The theory of how to do this is provided by the Fourier transform (see Section 3.2.4). Figure 3.10(a) shows the frequency response of a 3×3 mean filter. In this plot, the row and column coordinates of 0 correspond to a frequency of 0, which represents the medium gray value in the image, while the row and column coordinates of ± 0.5 represent the highest possible frequencies in the image. For example, the frequencies with column coordinate 0 and row coordinate ± 0.5 correspond to a grid with alternating one-pixel-wide vertical bright and dark lines. From Figure 3.10(a), we can see that the 3×3 mean filter removes certain frequencies completely. These are the points for which the response has a value of 0. They occur for relatively high frequencies. However, we can also see that the highest frequencies are not removed completely. To illustrate this, Figure 3.10(b) shows an image with one-pixel-wide lines spaced three pixels apart. From Figure 3.10(c), we can see that this frequency is completely removed by the 3×3 mean filter: the output image has a constant gray value. If we change the spacing of the lines to two pixels, as in Figure 3.10(d), we can see from Figure 3.10(e) that this higher frequency is not removed completely. This is an undesirable behavior since it means that noise is not removed completely by the mean filter. Note also that the polarity of the lines has been reversed by the mean filter, which is also undesirable. This is caused by the negative parts of the frequency response. Furthermore, from Figure 3.10(a) we can see that the frequency response

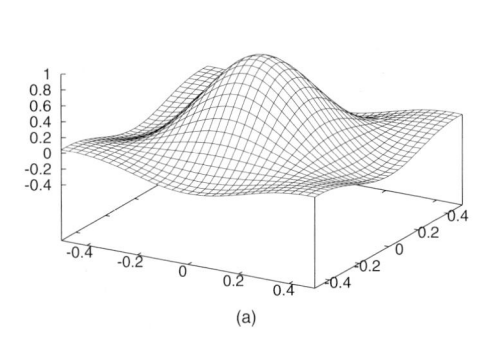

Fig. 3.10 (a) Frequency response of the 3 × 3 mean filter. (b) Image with one-pixel-wide lines spaced three pixels apart. (c) Result of applying the 3 × 3 mean filter to the image in (b). Note that all the lines have been smoothed out. (d) Image with one-pixel-wide lines spaced two pixels apart. (e) Result of applying the 3 × 3 mean filter to the image in (d). Note that the lines have not been completely smoothed out, although they have a higher frequency than the lines in (b). Note also that the polarity of the lines has been reversed.

of the mean filter is not rotationally symmetric, i.e., it is anisotropic. This means that diagonal structures are smoothed differently than horizontal or vertical structures.

Because the mean filter has the above drawbacks, the question of which smoothing filter is optimal arises. One way to approach this problem is to define certain natural criteria that the smoothing filter should fulfill, and then to search for the filters that fulfill the desired criteria. The first natural criterion is that the filter should be linear. This is natural because we can imagine an image being composed of multiple objects in an additive manner. Hence, the filter output should be a linear combination of the input. Furthermore, the filter should be position-invariant, i.e., it should produce the same results no matter where an object is in the image. This is automatically fulfilled for linear filters. Also, we would like the filter to be rotation-invariant, i.e., isotropic, so that it produces the same result independent of the orientation of the objects in the image. As we saw above, the mean filter does not fulfill this criterion. We would also like to control the amount of smoothing (noise reduction) that is being performed. Therefore, the filter should have a parameter t that can be used to control the smoothing, where higher values of t indicate more smoothing. For the mean filter, this corresponds to the mask sizes m and n. Above, we saw that the mean filter does not suppress all high frequencies, i.e., noise, in the image. Therefore, a criterion that describes the noise suppression of the filter in the image should be added. One such criterion is that, the larger t gets, the more local maxima in the image should be eliminated. This is a quite intuitive criterion, as can be seen in Figure 3.7(a), where many local maxima because of noise can be detected. Note that, because of linearity, we only need to require maxima to be eliminated. This automatically implies that local minima are eliminated as well. Finally, sometimes we would like to execute the smoothing filter several times in succession. If we do this, we would also have a simple means to predict the result of the combined filtering. Therefore, first filtering with t and then with s should be identical to a single filter operation with $t + s$. It can be shown that, among all smoothing filters, the Gaussian filter is the only filter that fulfills all of the above criteria [55]. Other natural criteria for a smoothing filter have been proposed [56, 57, 58], which also single out the Gaussian filter as the optimal smoothing filter. In 1D, the Gaussian filter is given by

$$g_\sigma(x) = \frac{1}{\sqrt{2\pi}\,\sigma} e^{-x^2/(2\sigma^2)} \qquad (3.21)$$

This is the function that also defines the probability density of a normally distributed random variable. In 2D, the Gaussian filter is given by

$$g_\sigma(r,c) = \frac{1}{2\pi\sigma^2} e^{-(r^2+c^2)/(2\sigma^2)} = \frac{1}{\sqrt{2\pi}\,\sigma} e^{-r^2/(2\sigma^2)} \frac{1}{\sqrt{2\pi}\,\sigma} e^{-c^2/(2\sigma^2)}$$
$$= g_\sigma(r)g_\sigma(c) \qquad (3.22)$$

Hence, the Gaussian filter is separable. Therefore, it can be computed very efficiently. In fact, it is the only isotropic separable smoothing filter. Unfortunately, it cannot be implemented recursively. However, some recursive approximations have been proposed [59, 60]. Figure 3.11 shows plots of the 1D and 2D Gaussian filters with $\sigma = 1$. The frequency response of a Gaussian filter is also a Gaussian function, albeit with σ inverted. Therefore, Figure 3.11(b) also gives a qualitative impression of the frequency response of the Gaussian filter. It can be seen that the Gaussian filter suppresses high frequencies much better than the mean filter.

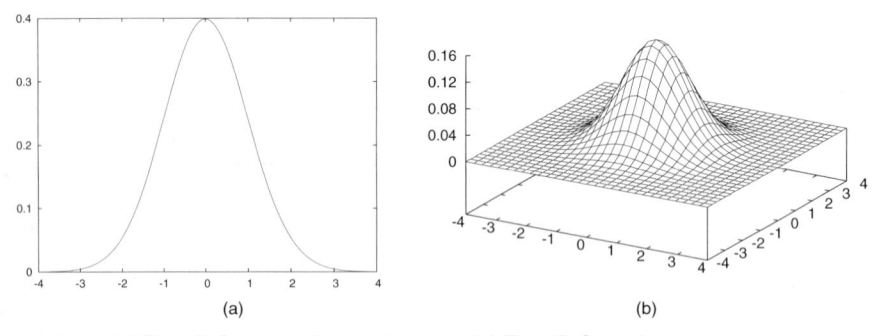

Fig. 3.11 (a) The 1D Gaussian filter with $\sigma = 1$. (b) The 2D Gaussian filter with $\sigma = 1$.

Like the mean filter, any linear filter will change the variance of the noise in the image. It can be shown that, for a linear filter $h(r,c)$ or $h_{r,c}$, the noise variance is multiplied by the following factor [53]:

$$\int_{-\infty}^{\infty}\int_{-\infty}^{\infty} h(r,c)^2\, dr\, dc \quad \text{or} \quad \sum_{i=-\infty}^{\infty}\sum_{j=-\infty}^{\infty} h_{r,c}^2 \qquad (3.23)$$

For a Gaussian filter, this factor is $1/(4\pi\sigma^2)$. If we compare this to a mean filter with a square mask with parameter n, we see that, to get the same noise reduction with the Gaussian filter, we need to set $\sigma = (2n+1)/(2\sqrt{\pi})$. For example, a 5×5 mean filter has the same noise reduction effect as a Gaussian filter with $\sigma \approx 1.41$.

Figure 3.12 compares the results of the Gaussian filter with those of the mean filter of an equivalent size. For small filter sizes ($\sigma = 1.41$ and 5×5), there is hardly any noticeable difference between the results. However, if larger filter sizes are used, it becomes clear that the mean filter turns the edge into a ramp, leading to a badly defined edge that is also visually quite hard to locate, whereas the Gaussian filter produces a much sharper edge. Hence, we can see that the Gaussian filter produces better results, and consequently it is usually the preferred smoothing filter if the quality of the results is the primary concern. If speed is the primary concern, then the mean filter is preferable.

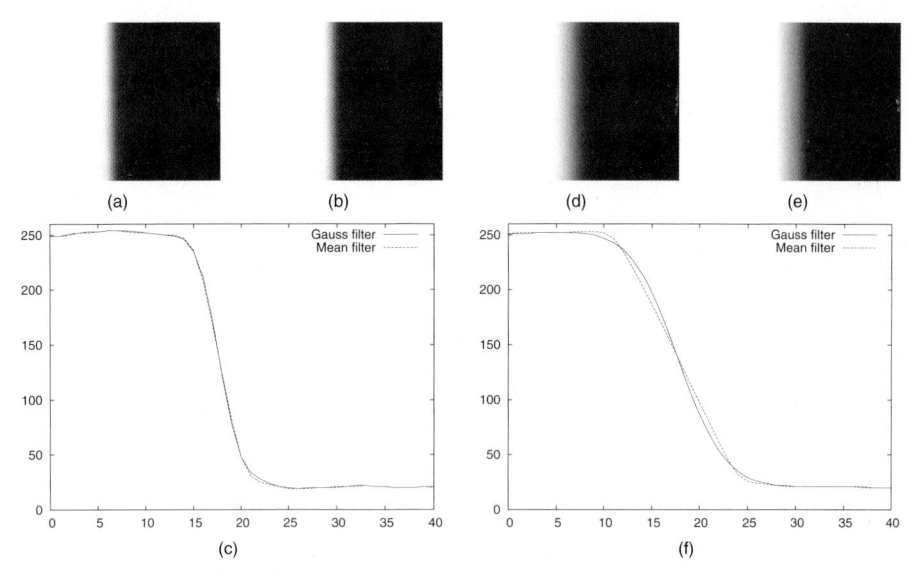

Fig. 3.12 Images of an edge obtained by smoothing the image of Figure 3.7(a). Results of (a) a Gaussian filter with $\sigma = 1.41$ and (b) a mean filter of size 5×5; and (c) the corresponding gray value profiles. Note that the two filters return very similar results in this example. Results of (d) a Gaussian filter with $\sigma = 3.67$ and (e) a 13×13 mean filter; and (f) the corresponding profiles. Note that the mean filter turns the edge into a ramp, leading to a badly defined edge, whereas the Gaussian filter produces a much sharper edge.

We close this section with a nonlinear filter that can also be used for noise suppression. The mean filter is a particular estimator for the mean value of a sample of random values. From probability theory, we know that other estimators are also possible, most notably the median of the samples. The median is defined as the value for which 50% of the values in the probability distribution of the samples are smaller than the median and 50% are larger. From a practical point of view, if the sample set contains n values

g_i, $i = 0, \ldots, n - 1$, we sort the values g_i in ascending order to get s_i, and then select the value $\text{median}(g_i) = s_{n/2}$. Hence, we can obtain a median filter by calculating the median instead of the mean inside a window around the current pixel. Let W denote the window, e.g., a $(2n + 1) \times (2m + 1)$ rectangle as for the mean filter. Then the median filter is given by

$$g_{r,c} = \underset{(i,j) \in W}{\text{median}}\, \hat{g}_{r-i, c-j} \qquad (3.24)$$

The median filter is not separable. However, with sophisticated algorithms it is possible to obtain a runtime complexity that is comparable to a separable linear filter: $O(whn)$ [61, 62]. The properties of the median filter are quite difficult to analyze. We can note, however, that it performs no averaging of the input gray values, but simply selects one of them. This can lead to surprising results. For example, the result of applying a 3×3 median filter to the image in Figure 3.10(b) would be a completely black image. Hence, the median filter would remove the bright lines because they cover less than 50% of the window. This property can sometimes be used to remove objects completely from the image. On the other hand, applying a 3×3 median filter to the image in Figure 3.10(d) would swap the bright and dark lines. This result is as undesirable as the result of the mean filter on the same image. On the edge image of Figure 3.7(a), the median filter produces quite good results, as can be seen from Figure 3.13. In particular, it should be noted that the median filter preserves the sharpness

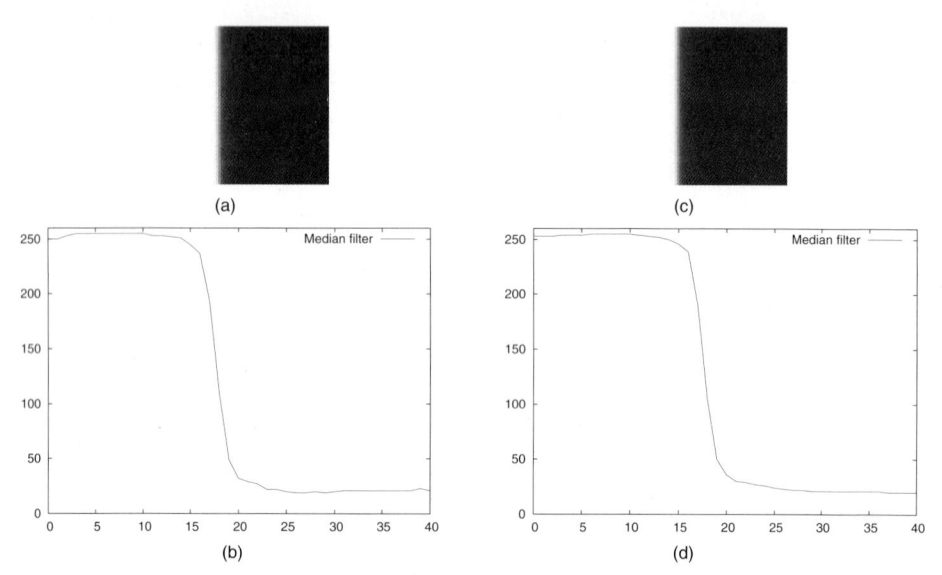

Fig. 3.13 Images of an edge obtained by smoothing the image of Figure 3.7(a). (a) Result with a median filter of size 5×5 (a); and (b) corresponding gray value profile. (c) Result of a 13×13 median filter; and (d) corresponding profile. Note that the median filter preserves the sharpness of the edge to a great extent.

of the edge even for large filter sizes. However, it cannot be predicted if and how much the position of the edge is changed by the median filter, which is possible for the linear filters. Furthermore, we cannot estimate how much noise is removed by the median filter, in contrast to the linear filters. Therefore, for high-accuracy measurements the Gaussian filter should be used.

Finally, it should be mentioned that the median is a special case of the more general class of rank filters. Instead of selecting the median $s_{n/2}$ of the sorted gray values, the rank filter would select the sorted gray value at a particular rank r, i.e., s_r. We will see other cases of rank operators in Section 3.6.2.

3.2.4
Fourier Transform

In the previous section, we considered the frequency responses of the mean and Gaussian filters. In this section, we will take a look at the theory that is used to derive the frequency response: the Fourier transform [63, 64]. The Fourier transform of a 1D function $h(x)$ is given by

$$H(f) = \int_{-\infty}^{\infty} h(x) e^{2\pi i f x} \, dx \tag{3.25}$$

It transforms the function $h(x)$ from the spatial domain into the frequency domain, i.e., $h(x)$, a function of the position x, is transformed into $H(f)$, a function of the frequency f. Note that $H(f)$ is in general a complex number. Because of Eq. (3.25) and the identity $e^{ix} = \cos x + i \sin x$, we can think of $h(x)$ as being composed of sine and cosine waves of different frequencies and different amplitudes. Then $H(f)$ describes precisely which frequency occurs with which amplitude and with which phase (overlaying sine and cosine terms of the same frequency simply leads to a phase-shifted sine wave). The inverse Fourier transform from the frequency domain to the spatial domain is given by

$$h(x) = \int_{-\infty}^{\infty} H(f) e^{-2\pi i f x} \, df \tag{3.26}$$

Because the Fourier transform is invertible, it is best to think of $h(x)$ and $H(f)$ as being two different representations of the same function.

In 2D, the Fourier transform and its inverse are given by

$$H(u, v) = \int_{-\infty}^{\infty} \int_{-\infty}^{\infty} h(r, c) e^{2\pi i (ur + vc)} \, dr \, dc$$

$$h(r, c) = \int_{-\infty}^{\infty} \int_{-\infty}^{\infty} H(u, v) e^{-2\pi i (ur + vc)} \, du \, dv \tag{3.27}$$

In image processing, $h(r,c)$ is an image, for which the position (r,c) is given in pixels. Consequently, the frequencies (u,v) are given in cycles per pixel.

Among the many interesting properties of the Fourier transform, probably the most interesting one is that a convolution in the spatial domain is transformed into a simple multiplication in the frequency domain: $(g*h)(r,c) \Leftrightarrow G(u,v)H(u,v)$, where the convolution is given by Eq. (3.17). Hence, a convolution can be performed by transforming the image and the filter into the frequency domain, multiplying the two results, and transforming the result back into the spatial domain.

Note that the convolution attenuates the frequency content $G(u,v)$ of the image $g(r,c)$ by the frequency response $H(u,v)$ of the filter. This justifies the analysis of the smoothing behavior of the mean and Gaussian filters that we have performed in Section 3.2.3. To make this analysis more precise, we can compute the Fourier transform of the mean filter in Eq. (3.19). It is given by

$$H(u,v) = \frac{1}{(2n+1)(2m+1)} \operatorname{sinc}((2n+1)u) \operatorname{sinc}((2m+1)v) \qquad (3.28)$$

where $\operatorname{sinc} x = (\sin \pi x)/(\pi x)$. See Figure 3.10 for a plot of the response of the 3×3 mean filter. Similarly, the Fourier transform of the Gaussian filter in Eq. (3.22) is given by

$$H(u,v) = e^{-2\pi^2\sigma^2(u^2+v^2)} \qquad (3.29)$$

Hence, the Fourier transform of the Gaussian filter is again a Gaussian function, albeit with σ inverted. Note that, in both cases, the frequency response becomes narrower if the filter size is increased. This is a relation that holds in general: $h(x/a) \Leftrightarrow |a|H(af)$.

Another interesting property of the Fourier transform is that it can be used to compute the correlation

$$g \star h = (g \star h)(r,c) = \int_{-\infty}^{\infty} \int_{-\infty}^{\infty} g(r+u, c+v) h(u,v) \, du \, dv \qquad (3.30)$$

Note that the correlation is very similar to the convolution in Eq. (3.17). The correlation is given in the frequency domain by $(g \star h)(r,c) \Leftrightarrow G(u,v)H(-u,-v)$. If $h(r,c)$ contains real numbers, which is the case for image processing, then $H(-u,-v) = \overline{H(u,v)}$, where the bar denotes complex conjugation. Hence, $(g \star h)(r,c) \Leftrightarrow G(u,v)\overline{H(u,v)}$.

Up to now, we have assumed that the images are continuous. Real images are, of course, discrete. This trivial observation has profound implications for the result of the Fourier transform. As noted above, the frequency variables u and v are given in cycles per pixel. If a discrete image $h(r,c)$ is transformed, the highest possible frequency for any sine or cosine wave is $1/2$, i.e., one cycle per two pixels. The frequency $1/2$ is called the Nyquist critical frequency. Sine or cosine waves with higher frequencies look like sine or cosine waves with correspondingly lower frequencies. For example, a discrete cosine wave with frequency 0.75 looks exactly like a cosine wave with frequency 0.25, as shown in Figure 3.14. Effectively, values of $H(u,v)$ outside the square $[-0.5, 0.5] \times [-0.5, 0.5]$ are mapped to this square by repeated mirroring at the borders of the square. This effect is known as aliasing. To avoid aliasing, we must ensure that frequencies higher than the Nyquist critical frequency are removed before the image is sampled. During the image acquisition, this can be achieved by optical low-pass filters in the camera. Aliasing, however, may also occur when an image is scaled down (see Section 3.3.3). Here, it is important to apply smoothing filters before the image is sampled at the lower resolution to ensure that frequencies above the Nyquist critical frequency are removed.

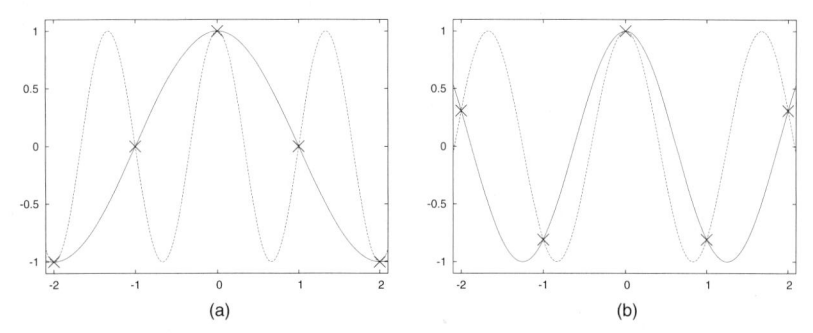

Fig. 3.14 Example of aliasing. (a) Two cosine waves, one with a frequency of 0.25 and one with a frequency of 0.75. (b) Two cosine waves, one with a frequency of 0.4 and one with a frequency of 0.6. Note that if both functions are sampled at integer positions, denoted by the crosses, the discrete samples will be identical.

Real images are not only discrete, they are also only defined within a rectangle of dimension $w \times h$, where w is the image width and h is the image height. This means that the Fourier transform is no longer continuous, but can be sampled at discrete frequencies $u_k = k/h$ and $v_l = l/w$. As discussed above, sampling the Fourier transform is only useful in the Nyquist interval $-1/2 \leq u_k, v_l < 1/2$. With this, the discrete Fourier transform (DFT) is given by

$$H_{k,l} = H(u_k, v_l) = \sum_{r=0}^{h-1} \sum_{c=0}^{w-1} h_{r,c} e^{2\pi i (u_k r + v_l c)}$$

$$= \sum_{r=0}^{h-1} \sum_{c=0}^{w-1} h_{r,c} e^{2\pi i (kr/h + lc/w)} \quad (3.31)$$

Analogously, the inverse discrete Fourier transform is given by

$$h_{r,c} = \frac{1}{wh} \sum_{k=0}^{h-1} \sum_{l=0}^{w-1} H_{k,l} e^{-2\pi i (kr/h + lc/w)} \quad (3.32)$$

As noted above, conceptually, the frequencies u_k and v_l should be sampled from the interval $(-1/2, 1/2]$, i.e., $k = -h/2 + 1, \ldots, h/2$ and $l = -w/2 + 1, \ldots, w/2$. Since we want to represent $H_{k,l}$ as an image, the negative coordinates are a little cumbersome. It is easy to see that Eqs. (3.31) and (3.32) are periodic with period h and w. Therefore, we can map the negative frequencies to their positive counterparts, i.e., $k = -h/2 + 1, \ldots, -1$ is mapped to $k = h/2 + 1, \ldots, h - 1$; and likewise for l.

We noted above that, for real images, $H(-u, -v) = \overline{H(u, v)}$. This property still holds for the discrete Fourier transform, with the appropriate change of coordinates as defined above, i.e., $H_{h-k, w-l} = \overline{H_{k,l}}$ (for $k, l > 0$). In practice, this means that we do not need to compute and store the complete Fourier transform $H_{k,l}$ because it contains redundant information. It is sufficient to compute and store one half of $H_{k,l}$, e.g., the left half. This saves a considerable amount of processing time and memory. This type of Fourier transform is called the real-valued Fourier transform.

To compute the Fourier transform from Eqs. (3.31) and (3.32), it might seem that $O((wh)^2)$ operations are required. This would prevent the Fourier transform from being useful in image processing applications. Fortunately, the Fourier transform can be computed in $O(wh \log(wh))$ operations for $w = 2^n$ and $h = 2^m$ [64] as well as for arbitrary w and h [65]. This fast computation algorithm for the Fourier transform is aptly called the fast Fourier transform (FFT). With self-tuning algorithms [65], the FFT can be computed in real time on standard processors.

As discussed above, the Fourier transform can be used to compute the convolution with any linear filter in the frequency domain. While this can be used to perform filtering with standard filter masks, e.g., the mean or Gaussian filter, typically there is a speed advantage only for relatively large filter masks. The real advantage of using the Fourier transform for filtering lies in the fact that filters can be customized to remove specific frequencies from the image, which occur, for example, for repetitive textures.

Figure 3.15(a) shows an image of a map. The map is drawn on a highly structured paper that exhibits significant texture. The texture makes the extraction of the data in the map difficult. Figure 3.15(b) displays the Fourier transform $H_{u,v}$. Note that the Fourier transform is cyclically shifted so that the zero frequency is in the center of the image to show the structure of the data more clearly. Hence, Figure 3.15(b) displays the frequencies u_k and v_l for $k = -h/2 + 1, \ldots, h/2$ and $l = -w/2 + 1, \ldots, w/2$. Because of the high dynamic range of the result, $H_{u,v}^{1/16}$ is displayed. It can be seen that $H_{u,v}$ contains several highly significant peaks that correspond to the characteristic frequencies of the texture. Furthermore, there are two significant orthogonal lines that correspond to the lines in the map. A filter $G_{u,v}$ that removes the characteristic frequencies of the texture is shown in Figure 3.15(c), while the result of the convolution $H_{u,v} G_{u,v}$ in the frequency domain is shown in Figure 3.15(d). The result of the inverse Fourier transform, i.e., the convolution in the spatial domain, is shown in Figure 3.15(e). Figures 3.15(f) and (g) show details of the input and result

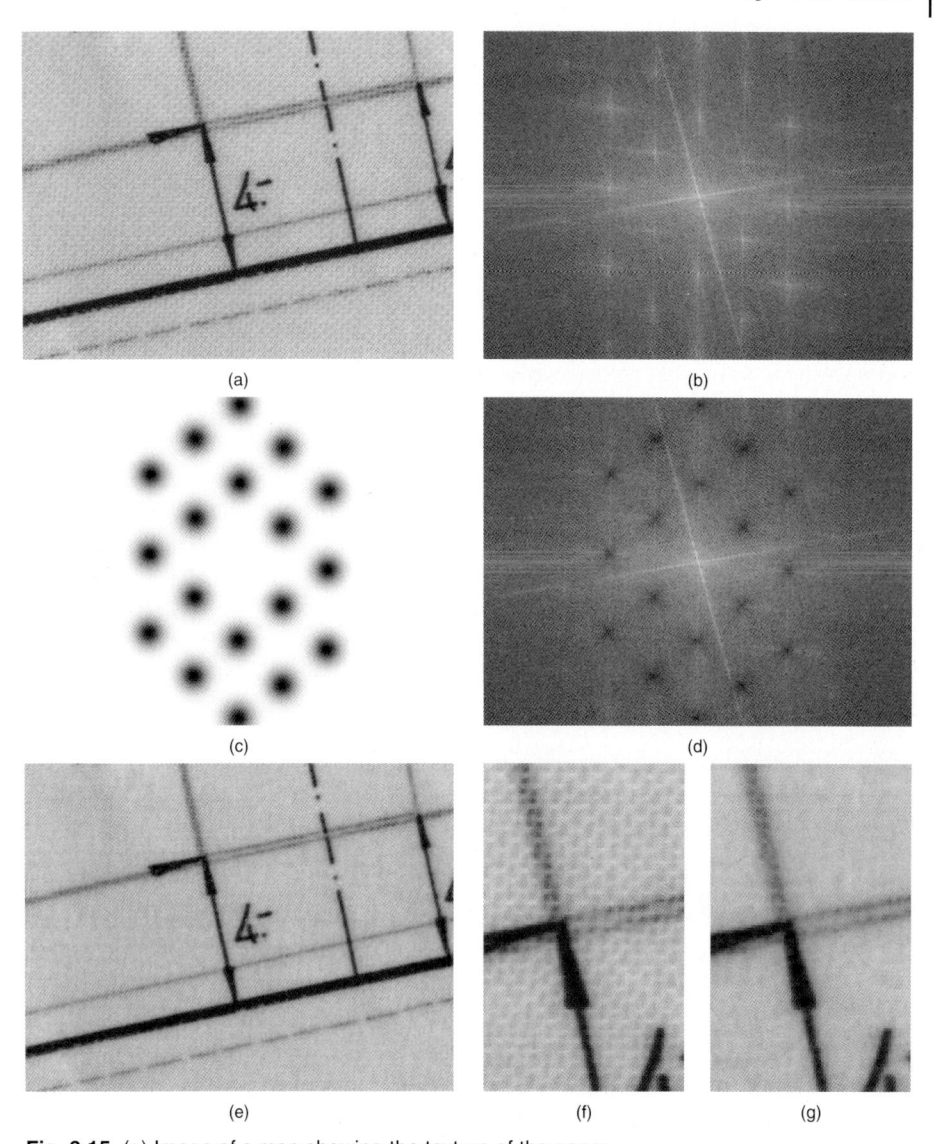

Fig. 3.15 (a) Image of a map showing the texture of the paper. (b) Fourier transform of (a). Because of the high dynamic range of the result, $H_{u,v}^{1/16}$ is displayed. Note the distinct peaks in $H_{u,v}$, which correspond to the texture of the paper. (c) Filter $G_{u,v}$ used to remove the frequencies that correspond to the texture. (d) Convolution $H_{u,v}G_{u,v}$. (e) Inverse Fourier transform of (d). (f), (g) Detail of (a) and (e). Note that the texture has been removed.

images, which show that the texture of the paper has been removed. Thus, it is now very easy to extract the map data from the image.

3.3
Geometric Transformations

In many applications, one cannot ensure that the objects to be inspected are always in the same position and orientation in the image. Therefore, the inspection algorithm must be able to cope with these position changes. Hence, one of the problems is to detect the position and orientation, also called the pose, of the objects to be examined. This will be the subject of later sections of this chapter. For the purposes of this section, we assume that we know the pose already. In this case, the simplest procedure to adapt the inspection to a particular pose is to align the ROIs appropriately. For example, if we know that an object is rotated by 45°, we could simply rotate the ROI by 45° before performing the inspection. In some cases, however, the image must be transformed (aligned) to a standard pose before the inspection can be performed. For example, the segmentation of the OCR is much easier if the text is either horizontal or vertical. Another example is the inspection of objects for defects based on a reference image. Here, we also need to align the image of the object to the pose in the reference image or vice versa. Therefore, in this section we will examine different geometric image transformations that are useful in practice.

3.3.1
Affine Transformations

If the position and rotation of the objects cannot be kept constant with the mechanical setup, we need to correct the rotation and translation of the object. Sometimes the distance of the object to the camera changes, leading to an apparent change in size of the object. These transformations are part of a very useful class of transformations called affine transformations, which are transformations that can be described by the following equation:

$$\begin{pmatrix} \tilde{r} \\ \tilde{c} \end{pmatrix} = \begin{pmatrix} a_{11} & a_{12} \\ a_{21} & a_{22} \end{pmatrix} \begin{pmatrix} r \\ c \end{pmatrix} + \begin{pmatrix} t_r \\ t_c \end{pmatrix} \tag{3.33}$$

Hence, an affine transformation consists of a linear part given by a 2 × 2 matrix and a translation. The above notation is a little cumbersome, however, since we always have to list the translation separately. To circumvent this, we can use a representation where we extend the coordinates with a third coordinate of 1, which enables us to write the transformation as a simple matrix multiplication:

$$\begin{pmatrix} \tilde{r} \\ \tilde{c} \\ 1 \end{pmatrix} = \begin{pmatrix} a_{11} & a_{12} & a_{13} \\ a_{21} & a_{22} & a_{23} \\ 0 & 0 & 1 \end{pmatrix} \begin{pmatrix} r \\ c \\ 1 \end{pmatrix} \tag{3.34}$$

Note that the translation is represented by the elements a_{13} and a_{23} of the matrix A. This representation with an added redundant third coordinate is called homogeneous coordinates. Similarly, the representation with two coordinates in Eq. (3.33) is called inhomogeneous coordinates. We will see the true power of the homogeneous representation below. Any affine transformation can be constructed from the following basic transformations, where the last row of the matrix has been omitted:

$$\begin{pmatrix} 1 & 0 & t_r \\ 0 & 1 & t_c \end{pmatrix} \quad \text{translation} \tag{3.35}$$

$$\begin{pmatrix} s_r & 0 & 0 \\ 0 & s_c & 0 \end{pmatrix} \quad \text{scaling in row and column direction} \tag{3.36}$$

$$\begin{pmatrix} \cos\alpha & -\sin\alpha & 0 \\ \sin\alpha & \cos\alpha & 0 \end{pmatrix} \quad \text{rotation by an angle of } \alpha \tag{3.37}$$

$$\begin{pmatrix} \cos\alpha & 0 & 0 \\ \sin\alpha & 1 & 0 \end{pmatrix} \quad \text{skew of row axis by an angle of } \theta \tag{3.38}$$

The first three basic transformations need no further explanation. The skew (or slant) is a rotation of only one axis, in this case the row axis. It is quite useful to rectify slanted characters in the OCR.

3.3.2
Projective Transformations

An affine transformation enables us to correct almost all relevant pose variations that an object may undergo. However, sometimes affine transformations are not general enough. If the object in question is able to rotate in 3D, it will undergo a general perspective transformation, which is quite hard to correct because of the occlusions that may occur. However, if the object is planar, we can model the transformation of the object by a 2D perspective transformation, which is a special 2D projective transformation [66, 67]. Projective transformations are given by

$$\begin{pmatrix} \tilde{r} \\ \tilde{c} \\ \tilde{w} \end{pmatrix} = \begin{pmatrix} h_{11} & h_{12} & h_{13} \\ h_{21} & h_{22} & h_{23} \\ h_{31} & h_{32} & h_{33} \end{pmatrix} \begin{pmatrix} r \\ c \\ w \end{pmatrix} \tag{3.39}$$

Note the similarity to the affine transformation in Eq. (3.34). The only changes that were made are that the transformation is now described by a full 3×3 matrix and that we have replaced the 1 in the third coordinate with a variable w. This representation is actually the true representation in homogeneous coordinates. It can also be used for affine transformations, which are special projective transformations. With this third coordinate, it is not obvious how we are able to obtain a transformed 2D coordinate, i.e., how to compute the corresponding inhomogeneous point. First of all, it must be noted that in homogeneous coordinates all points $p = (r, c, w)^\top$ are only defined up to a scale factor, i.e., the vectors p and λp ($\lambda \neq 0$) represent the same 2D point [66, 67]. Consequently, the projective transformation given by the matrix H is also only defined up to a scale factor, and hence has only eight independent parameters. To obtain an inhomogeneous 2D point from the homogeneous representation, we must divide the homogeneous vector by w. This requires $w \neq 0$. Such points are called finite points. Conversely, points with $w = 0$ are called points at infinity because they can be regarded as lying infinitely far away in a certain direction [66, 67].

Since a projective transformation has eight independent parameters, it can be uniquely determined from four corresponding points [66, 67]. This is how the projective transformations will usually be determined in machine vision applications. We will extract four points in an image, which typically represent a rectangle, and will rectify the image so that the four extracted points will be transformed to the four corners of the rectangle, i.e., to their corresponding points. Unfortunately, because of space limitations we cannot give the details of how the transformation is computed from the point correspondences. The interested reader is referred to [66, 67].

3.3.3
Image Transformations

After we have taken a look at how coordinates can be transformed with affine and projective transformations, we can consider how an image should be transformed. Our first idea might be to go through all the pixels in the input image, to transform their coordinates, and to set the gray value of the transformed point in the output image. Unfortunately, this simple strategy does not work. This can be seen by checking what happens if an image is scaled by a factor of 2: only one quarter of the pixels in the output image would be set. The correct way to transform an image is to loop through all the pixels in the output image and to calculate the position of the corresponding point in the input image. This is the simplest way to ensure that all relevant pixels in the output image are set. Fortunately, calculating the positions in the original image is simple: we only need to invert the matrix that describes the affine or projective transformation, which results again in an affine or projective transformation.

When the image coordinates are transformed from the output image to the input image, typically not all pixels in the output image transform back to coordinates that lie in the input image. This can be taken into account by computing a suitable ROI for the output image. Furthermore, we see that the resulting coordinates in the input image will typically not be integer coordinates. An example of this is given in Figure 3.16, where the input image is transformed by an affine transformation consisting of a translation, rotation, and scaling. Therefore, the gray values in the output image must be interpolated.

The interpolation can be done in several ways. Figure 3.17(a) displays a pixel in the output image that has been transformed back to the input image. Note that the transformed pixel center lies on a non-integer position between four adjacent pixel centers. The simplest and fastest interpolation method is to calculate the closest of the four adjacent pixel centers, which only involves rounding the floating-point coordinates of the transformed pixel center, and to use the gray value of the closest pixel in the input image as the gray value of the pixel in the output image, as shown in Figure 3.17(b).

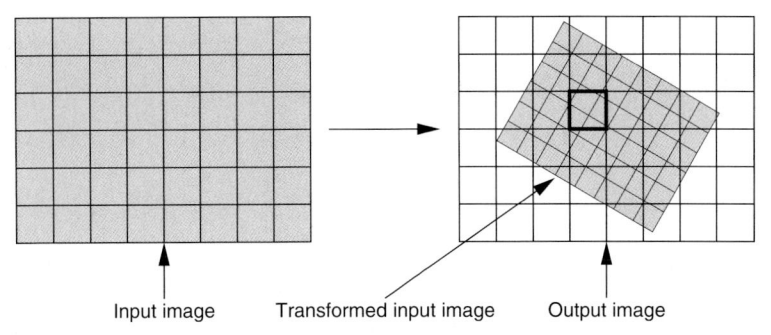

Fig. 3.16 An affine transformation of an image. Note that integer coordinates in the output image transform to non-integer coordinates in the original image, and hence must be interpolated.

This interpolation method is called nearest-neighbor interpolation. To see the effect of this interpolation, Figure 3.18(a) displays an image of a serial number of a bank note, where the characters are not horizontal. Figures 3.18(c) and (d) display the result of rotating the image such that the serial number is horizontal using this interpolation. Note that because the gray value is taken from the closest pixel center in the input image, the edges of the characters have a jagged appearance, which is undesirable.

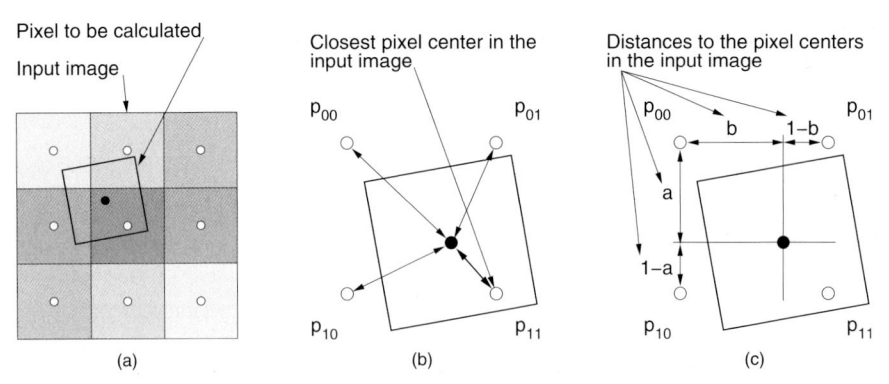

Fig. 3.17 (a) A pixel in the output image is transformed back to the input image. Note that the transformed pixel center lies at a non-integer position between four adjacent pixel centers. (b) Nearest-neighbor interpolation determines the closest pixel center in the input image and uses its gray value in the output image. (c) Bilinear interpolation determines the distances to the four adjacent pixel centers and weights their gray values using the distances.

The reason for the jagged appearance in the result of the nearest-neighbor interpolation is that essentially we are regarding the image as a piecewise constant function: every coordinate that falls within a rectangle of extent ± 0.5 in each direction is assigned the same gray value. This leads to discontinuities in the result, which cause the jagged edges. This behavior is especially noticeable if the image is scaled by a

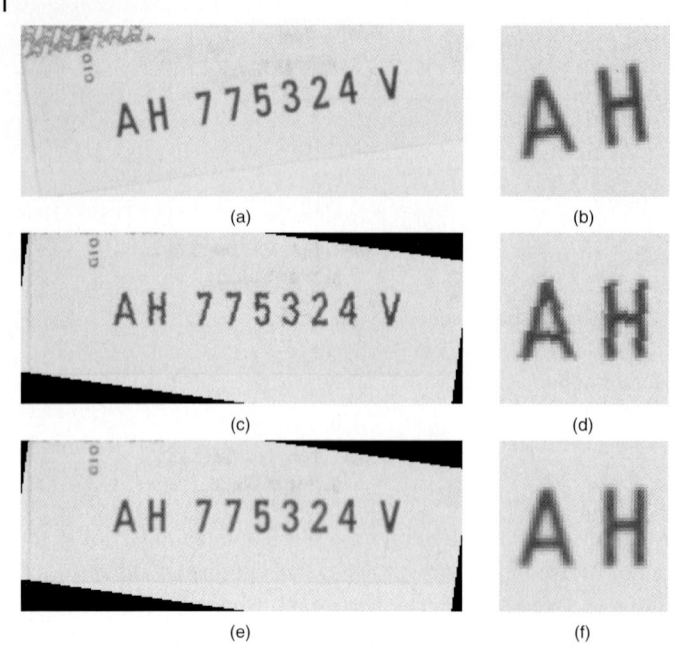

Fig. 3.18 (a) Image showing a serial number of a bank note. (b) Detail of (a). (c) Image rotated such that the serial number is horizontal using nearest-neighbor interpolation. (d) Detail of (c). Note the jagged edges of the characters. (e) Image rotated using bilinear interpolation. (f) Detail of (e). Note the smooth edges of the characters.

factor > 1. To get a better interpolation, we can use more information than the gray value of the closest pixel. From Figure 3.17(a), we can see that the transformed pixel center lies in a square of four adjacent pixel centers. Therefore, we can use the four corresponding gray values and weight them appropriately. One way to do this is to use bilinear interpolation, as shown in Figure 3.17(c). First, we compute the horizontal and vertical distances of the transformed coordinate to the adjacent pixel centers. Note that these are numbers between 0 and 1. Then, we weight the gray values according to their distances to get the bilinear interpolation:

$$\tilde{g} = b(ag_{11} + (1-a)g_{01}) + (1-b)(ag_{10} + (1-a)g_{00}) \tag{3.40}$$

Figures 3.18(e) and (f) display the result of rotating the image of Figure 3.18(a) using bilinear interpolation. Note that the edges of the characters now have a very smooth appearance. This much better result more than justifies the longer computation time (typically a factor of more than 5).

To conclude the discussion on interpolation, we discuss the effects of scaling an image down. In the bilinear interpolation scheme, we would interpolate from the closest four pixel centers. However, if the image is scaled down, adjacent pixel centers in the output image will not necessarily be close in the input image. Imagine a larger

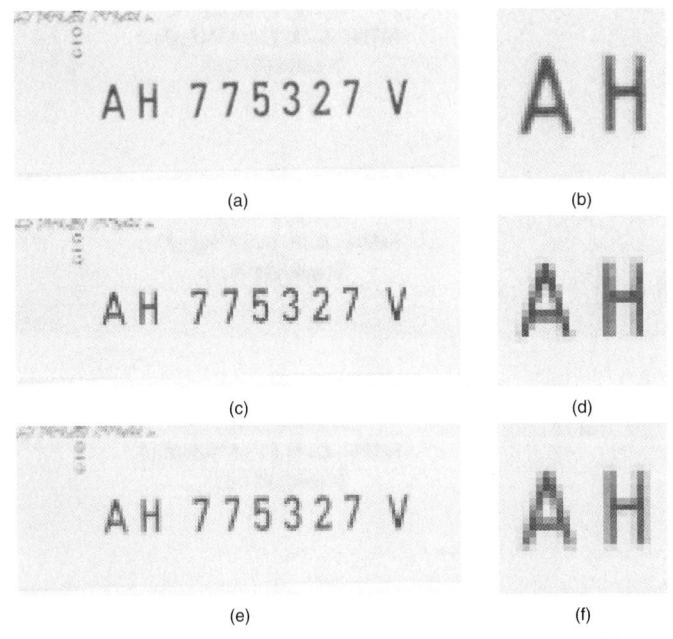

Fig. 3.19 (a) Image showing a serial number of a bank note. (b) Detail of (a). (c) The image of (a) scaled down by a factor of 3 using bilinear interpolation. (d) Detail of (c). Note the different stroke widths of the vertical strokes of the letter H. This is caused by aliasing. (e) Result of scaling the image down by integrating a smoothing filter (in this case a mean filter) into the image transformation. (f) Detail of (e).

version of the image of Figure 3.10(b) (one-pixel-wide vertical lines spaced three pixels apart) being scaled down by a factor of 4 using nearest-neighbor interpolation: we would get an image with one-pixel-wide lines that are four pixels apart. This is certainly not what we would expect. For bilinear interpolation, we would get similar unexpected results. If we scale down an image, we are essentially subsampling it. As a consequence, we may obtain aliasing effects (see Section 3.2.4). An example of aliasing can be seen in Figure 3.19. The image in Figure 3.19(a) is scaled down by a factor of 3 in Figure 3.19(c) using bilinear interpolation. Note that the stroke widths of the vertical strokes of the letter H, which are equally wide in Figure 3.19(a), now appear to be substantially different. This is undesirable. To improve the image transformation, the image must be smoothed before it is scaled down, e.g., using a mean or a Gaussian filter. Alternatively, the smoothing can be integrated into the gray value interpolation. Figure 3.19(e) shows the result of integrating a mean filter into the image transformation. Because of the smoothing, the strokes of the H now have the same width again.

In the above examples, we have seen the usefulness of the affine transformations for rectifying text. Sometimes, an affine transformation is not sufficient for this pur-

Fig. 3.20 (a), (b) Images of license plates. (c), (d) Result of a projective transformation that rectifies the perspective distortion of the license plates.

pose. Figures 3.20(a) and (b) show two images of license plates on cars. Because the position of the camera with respect to the car could not be controlled in this example, the images of the license plates show severe perspective distortions. Figures 3.20(c) and (d) show the result of applying projective transformations to the images that cut out the license plates and rectify them. Hence, the images in Figures 3.20(c) and (d) would result if we had looked at the license plates perpendicularly from in front of the car. Obviously, it is now much easier to segment and to read the characters on the license plates.

3.3.4
Polar Transformations

We conclude this section with another very useful geometric transformation: the polar transformation. This transformation is typically used to rectify parts of images that show objects that are circular or that are contained in circular rings in the image. An example is shown in Figure 3.21(a). Here, we can see the inner part of a CD that contains a ring with a bar code and some text. To read the bar code, one solution is to rectify the part of the image that contains the bar code. For this purpose, the polar transformation can be used, which converts the image into polar coordinates (d, ϕ), i.e., into the distance d to the center of the transformation and the angle ϕ of the vector to the center of the transformation. Let the center of the transformation be given by (m_r, m_c). Then, the polar coordinates of a point (r, c) are given by

$$d = \sqrt{(r - m_r)^2 + (c - m_c)^2}$$
$$\phi = \arctan\left(-\frac{r - m_r}{c - m_c}\right) \tag{3.41}$$

In the calculation of the arc tangent function, the correct quadrant must be used, based on the sign of the two terms in the fraction in the argument of arctan. Note that the transformation of a point into polar coordinates is quite expensive to compute because of the square root and the arc tangent. Fortunately, to transform an image, like for affine and projective transformations, the inverse of the polar transformation is used, which is given by

$$r = m_r - d \sin \phi$$
$$c = m_c + d \cos \phi \tag{3.42}$$

Here, the sines and cosines can be tabulated because they only occur for a finite number of discrete values, and hence only need to be computed once. Therefore, the polar transformation of an image can be computed efficiently. Note that by restricting the

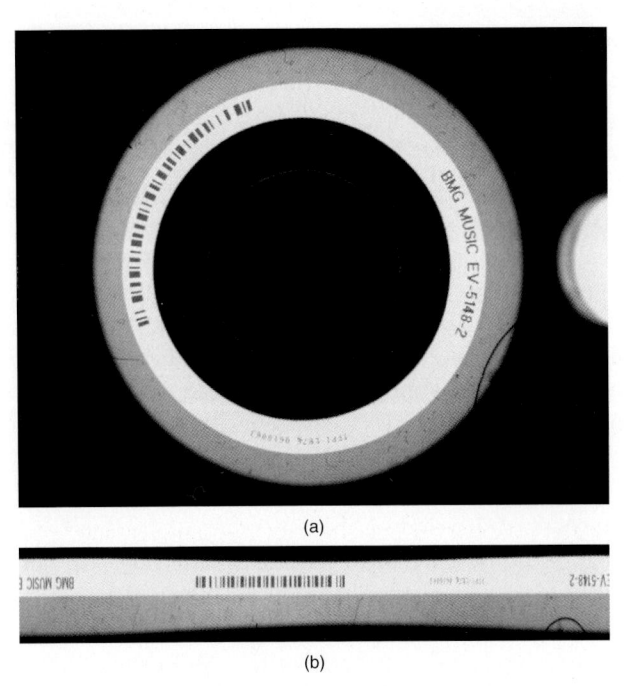

Fig. 3.21 (a) Image of the center of a CD showing a circular bar code. (b) Polar transformation of the ring that contains the bar code. Note that the bar code is now straight and horizontal.

ranges of d and ϕ, we can transform arbitrary circular sectors. Figure 3.21(b) shows the result of transforming a circular ring that contains the bar code in Figure 3.21(a). Note that because of the polar transformation the bar code is straight and horizontal, and consequently can be read easily.

3.4
Image Segmentation

In the preceding sections, we have looked at operations that transform an image into another image. These operations do not give us information about the objects in the image. For this purpose, we need to segment the image, i.e., extract regions from the image that correspond to the objects we are interested in. More formally, segmentation is an operation that takes an image as input and returns one or more regions or subpixel-precise contours as output.

3.4.1
Thresholding

The simplest segmentation algorithm is to threshold the image. The threshold operation is defined by

$$S = \{(r,c) \in R \mid g_{\min} \leq f_{r,c} \leq g_{\max}\} \qquad (3.43)$$

Hence, the threshold operation selects all points in the ROI R of the image that lie within a specified range of gray values into the output region S. Often, $g_{\min} = 0$ or $g_{\max} = 2^b - 1$ is used. If the illumination can be kept constant, the thresholds $g_{\min}$ and $g_{\max}$ are selected when the system is set up and are never modified. Since the threshold operation is based on the gray values themselves, it can be used whenever the object to be segmented and the background have significantly different gray values.

Figures 3.22(a) and (b) show two images of integrated circuits (ICs) on a printed circuit board (PCB) with a rectangular ROI overlaid in light gray. The result of thresholding the two images with $g_{\min} = 90$ and $g_{\max} = 255$ is shown in Figures 3.22(c) and (d). Since the illumination is kept constant the same threshold works for both images. Note also that there are some noisy pixels in the segmented regions. They can be removed, e.g., based on their area (see Section 3.5) or based on morphological operations (see Section 3.6).

The constant threshold only works well as long as the gray values of the object and the background do not change. Unfortunately, this occurs less frequently than one would wish, e.g., because of a changing illumination. Even if the illumination is kept constant, different gray value distributions on similar objects may prevent us from using a constant threshold. Figure 3.23 shows an example of this. In Figures 3.23(a) and (b) two different ICs on the same PCB are shown. Despite the identical illumination, the prints have a substantially different gray value distribution, which will not allow us to use the same threshold for both images. Nevertheless, the print and the

(a)   (b)

OKI JAPAN
M518222-30
6285352

OKI JAPAN
M518222-30
6285352

(c)   (d)

Fig. 3.22 (a), (b) Images of prints on ICs with a rectangular ROI overlaid in light gray. (c), (d) Result of thresholding the images in (a), (b) with $g_{min} = 90$ and $g_{max} = 255$.

background can be separated easily in both cases. Therefore, ideally, we would like to have a method that is able to determine the thresholds automatically. This can be done based on the gray value histogram of the image. Figures 3.23(c) and (d) show the histograms of the images in Figures 3.23(a) and (b). It is obvious that there are two relevant peaks (maxima) in the histograms in both images. The one with the smaller gray value corresponds to the background, while the one with the higher gray value corresponds to the print. Intuitively, a good threshold corresponds to the minimum between the two peaks in the histogram. Unfortunately, neither the two maxima nor the minimum is well defined because of random fluctuations in the gray value histogram. Therefore, to robustly select the threshold that corresponds to the minimum, the histogram must be smoothed, e.g., by convolving it with a 1D Gaussian filter. Since it is not clear which σ to use, a good strategy is to smooth the histogram with progressively larger values of σ until two unique maxima with a unique minimum in between are obtained. The result of using this approach to select the threshold automatically is shown in Figures 3.23(e) and (f). As can be seen, for both images suitable thresholds have been selected. This approach to select the thresholds is not the only approach. Further approaches are described, for example, in [68, 69]. All these approaches have

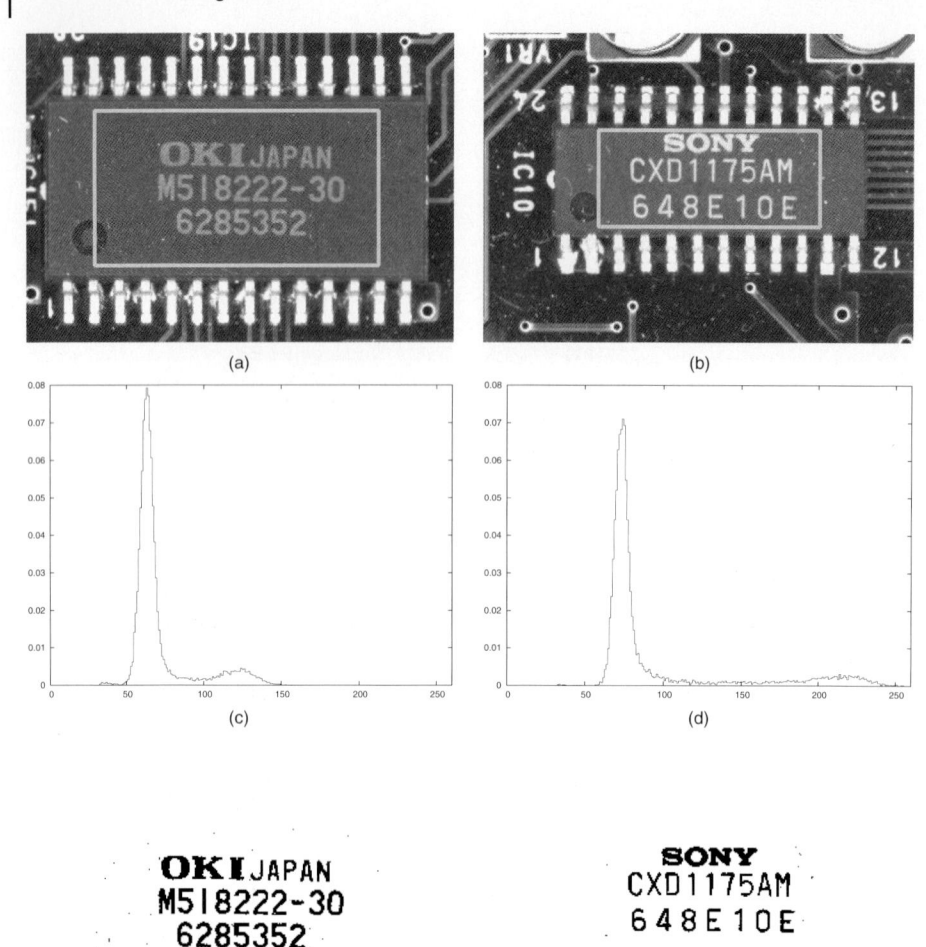

(a) (b)

(c) (d)

(e) (f)

Fig. 3.23 (a), (b) Images of prints on ICs with a rectangular ROI overlaid in light gray. (c), (d) Gray value histogram of the images in (a), (b) within the respective ROI. (e), (f) Result of thresholding the images in (a) and (b) with a threshold selected automatically based on the gray value histogram.

in common that they are based on the gray value histogram of the image. One example of such a different approach is to assume that the gray values in the foreground and background each have a normal (Gaussian) probability distribution, and to jointly fit two Gaussian densities to the histogram. The threshold is then defined as the gray value for which the two Gaussian densities have equal probabilities.

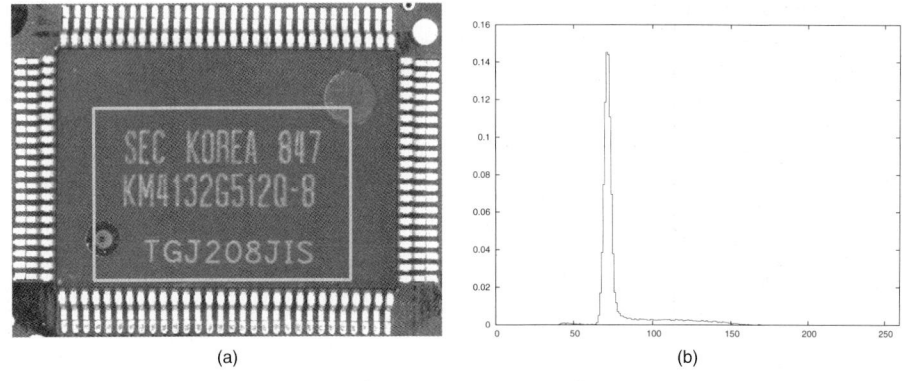

Fig. 3.24 (a) Image of a print on an IC with a rectangular ROI overlaid in light gray. (b) Gray value histogram of the image in (a) within the ROI. Note that there are no significant minima and only one single significant maximum in the histogram.

While calculating the thresholds from the histogram often works extremely well, it fails whenever the assumption that there are two peaks in the histogram is violated. One such example is shown in Figure 3.24. Here, the print is so noisy that the gray values of the print are extremely spread out, and consequently there is no discernible peak for the print in the histogram. Another reason for the failure of the desired peak to appear is an inhomogeneous illumination. This typically destroys the relevant peaks or moves them so that they are in the wrong location. An uneven illumination often even prevents us from using a threshold operation altogether because there are no fixed thresholds that work throughout the entire image. Fortunately, often the objects of interest can be characterized by being locally brighter or darker than their local background. The prints on the ICs we have examined so far are a good example of this. Therefore, instead of specifying global thresholds, we would like to specify by how much a pixel must be brighter or darker than its local background. The only problem we have is how to determine the gray value of the local background. Since a smoothing operation, e.g., the mean, Gaussian, or median filter (see Section 3.2.3), calculates an average gray value in a window around the current pixel, we can simply use the filter output as an estimate of the gray value of the local background. The operation of comparing the image to its local background is called a dynamic thresholding operation. Let the image be denoted by $f_{r,c}$ and the smoothed image be denoted by $g_{r,c}$. Then, the dynamic thresholding operation for bright objects is given by

$$S = \{(r,c) \in R \mid f_{r,c} - g_{r,c} \geq g_{\text{diff}}\} \tag{3.44}$$

while the dynamic thresholding operation for dark objects is given by

$$S = \{(r,c) \in R \mid f_{r,c} - g_{r,c} \leq -g_{\text{diff}}\} \tag{3.45}$$

Figure 3.25 gives an example of how the dynamic thresholding works. In Figure 3.25(a), a small part of a print on an IC with a one-pixel-wide horizontal ROI is shown. Figure 3.25(b) displays the gray value profiles of the image and the image smoothed with a 9 × 9 mean filter. It can be seen that the text is substantially brighter than the local background estimated by the mean filter. Therefore, the characters can be segmented easily with the dynamic thresholding operation.

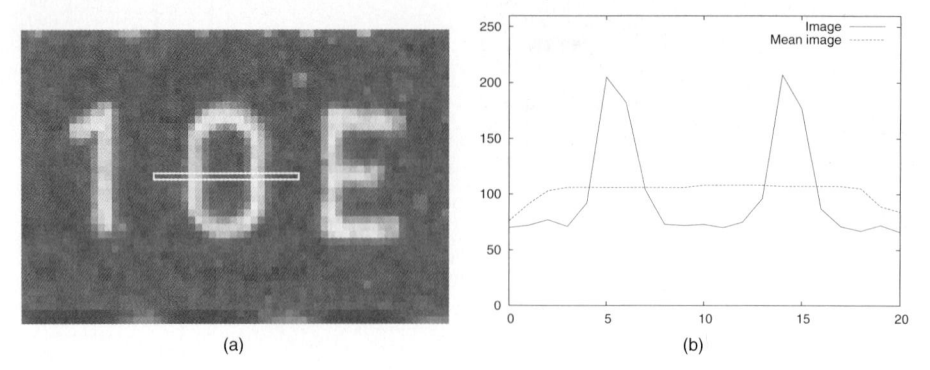

Fig. 3.25 (a) Image showing a small part of a print on an IC with a one-pixel-wide horizontal ROI. (b) Gray value profiles of the image and the image smoothed with a 9 × 9 mean filter. Note that the text is substantially brighter than the local background estimated by the mean filter.

In the dynamic thresholding operation, the size of the smoothing filter determines the size of the objects that can be segmented. If the filter size is too small, the local background will not be estimated well in the center of the objects. As a rule of thumb, the diameter of the mean filter must be larger than the diameter of the objects to be recognized. The same holds for the median filter. An analogous relation exists for the Gaussian filter. Furthermore, in general, if larger filter sizes are chosen for the mean and Gaussian filters, the filter output will be more representative of the local background. For example, for light objects the filter output will become darker within the light objects. For the median filter, this is not true since it will completely eliminate the objects if the filter mask is larger than the diameter of the objects. Hence, the gray values will be representative of the local background if the filter is sufficiently large. If the gray values in the smoothed image are more representative of the local background, we can typically select a larger threshold g_{diff}, and hence can suppress noise in the segmentation better. However, the filter mask cannot be chosen arbitrarily large because neighboring objects might adversely influence the filter output. Finally, it should be noted that the dynamic thresholding operation not only returns a segmentation result for objects that are brighter or darker than their local background. It also returns a segmentation result at the bright or dark region around edges.

Figure 3.26(a) again shows the image of Figure 3.24(a), which could not be segmented with an automatic threshold. In Figure 3.26(b), the result of segmenting the image with a dynamic thresholding operation with $g_{\text{diff}} = 5$ is shown. The local

Fig. 3.26 (a) Image of a print on an IC with a rectangular ROI overlaid in light gray. (b) Result of segmenting the image in (a) with a dynamic thresholding operation with $g_{\text{diff}} = 5$ and a 31 × 31 mean filter.

background was obtained with a 31 × 31 mean filter. Note that the difficult print is segmented very well with the dynamic thresholding.

As described so far, the dynamic thresholding operation can be used to compare the image with its local background, which is obtained by smoothing the image. With a slight modification, the dynamic thresholding operation can also be used to detect errors in an object, e.g., for print inspection. Here, the image $g_{r,c}$ is an image of the ideal object, i.e., the object without errors; $g_{r,c}$ is called the reference image. To detect deviations from the ideal object, we can simply look for too bright or too dark pixels in the image $f_{r,c}$ by using Eq. (3.44) or Eq. (3.45). Often, we are not interested in whether the pixels are too bright or too dark, but simply in whether they deviate too much from the reference image, i.e., the union of Eqs. (3.44) and (3.45), which is given by

$$S = \{(r,c) \in R \mid |f_{r,c} - g_{r,c}| > g_{\text{abs}}\} \tag{3.46}$$

Note that this pixel-by-pixel comparison requires that the image $f_{r,c}$ of the object to check and the reference image $g_{r,c}$ are aligned very accurately to avoid spurious gray value differences that would be interpreted as errors. This can be ensured either by the mechanical setup or by finding the pose of the object in the current image, e.g., using template matching (see Section 3.11), and then transforming the image to the pose of the object in the ideal image (see Section 3.3).

This kind of dynamic thresholding operation is very strict on the shape of the objects. For example, if the size of the object increases by half a pixel and the gray value difference between the object and the background is 200, the gray value difference between the current image and the model image will be 100 at the object's edges. This is a significant gray value difference, which would surely be larger than any reasonable g_{abs}. In real applications, however, small variations of the object's shape typically should be tolerated. On the other hand, small gray value changes in the areas where the object's shape does not change should still be recognized as an error. To achieve this behavior, we can introduce a thresholding operation that takes the expected gray value variations in the image into account. Let us denote the permissible variations in the image by $v_{r,c}$. Ideally, we would like to segment the pixels that differ from the reference image by more than the permissible variations:

$$S = \{(r,c) \in R \mid |f_{r,c} - g_{r,c}| > v_{r,c}\} \tag{3.47}$$

The permissible variations can be determined by learning them from a set of training images. For example, if we use n training images of objects with permissible variations, the standard deviation of the gray values of each pixel can be used to derive $v_{r,c}$. If we use n images to define the variations of the ideal object, we might as well use the mean of each pixel to define the reference image $g_{r,c}$ to reduce noise. Of course, the n training images must be aligned with sufficient accuracy. The mean and standard deviation of the n training images are given by

$$m_{r,c} = \frac{1}{n} \sum_{i=1}^{n} g_{r,c;i}$$

$$s_{r,c} = \sqrt{\frac{1}{n} \sum_{i=1}^{n} (g_{r,c;i} - m_{r,c})^2} \tag{3.48}$$

The images $m_{r,c}$ and $s_{r,c}$ model the reference image and the allowed variations of the reference image. Hence, we can call this approach a variation model. Note that $m_{r,c}$ is identical to the temporal average of Eq. (3.12). To define $v_{r,c}$, ideally we could simply set $v_{r,c}$ to a small multiple c of the standard deviation, i.e., $v_{r,c} = cs_{r,c}$, where, for example, $c = 3$. Unfortunately, this approach does not work well if the variations in the training images are extremely small, e.g., because the noise in the training images is significantly smaller than in the test images, or because parts of the object are near the saturation limit of the camera. In these cases, it is useful to introduce an absolute threshold a for the variation images, which is used whenever the variations in the training images are very small: $v_{r,c} = \max(a, cs_{r,c})$. As a further generalization, it is sometimes useful to have different thresholds for too bright and too dark pixels. With this, the variation threshold is no longer symmetric with respect to $m_{r,c}$, and we need to introduce two threshold images for the too bright and too dark pixels. If we denote the threshold images by $u_{r,c}$ and $l_{r,c}$, the absolute thresholds by a and b, and the factors for the standard deviations by c and d, the variation model segmentation is given by

$$S = \{(r,c) \in R \mid f_{r,c} < l_{r,c} \vee f_{r,c} > u_{r,c}\} \tag{3.49}$$

where

$$u_{r,c} = m_{r,c} + \max(a, cs_{r,c})$$
$$l_{r,c} = m_{r,c} - \max(b, ds_{r,c}) \tag{3.50}$$

3.4 Image Segmentation | 109

(a)　　　　　　　　　　　　(b)

(c)　　　　　　　　　　　　(d)

(e)　　　　　　　　　　　　(f)

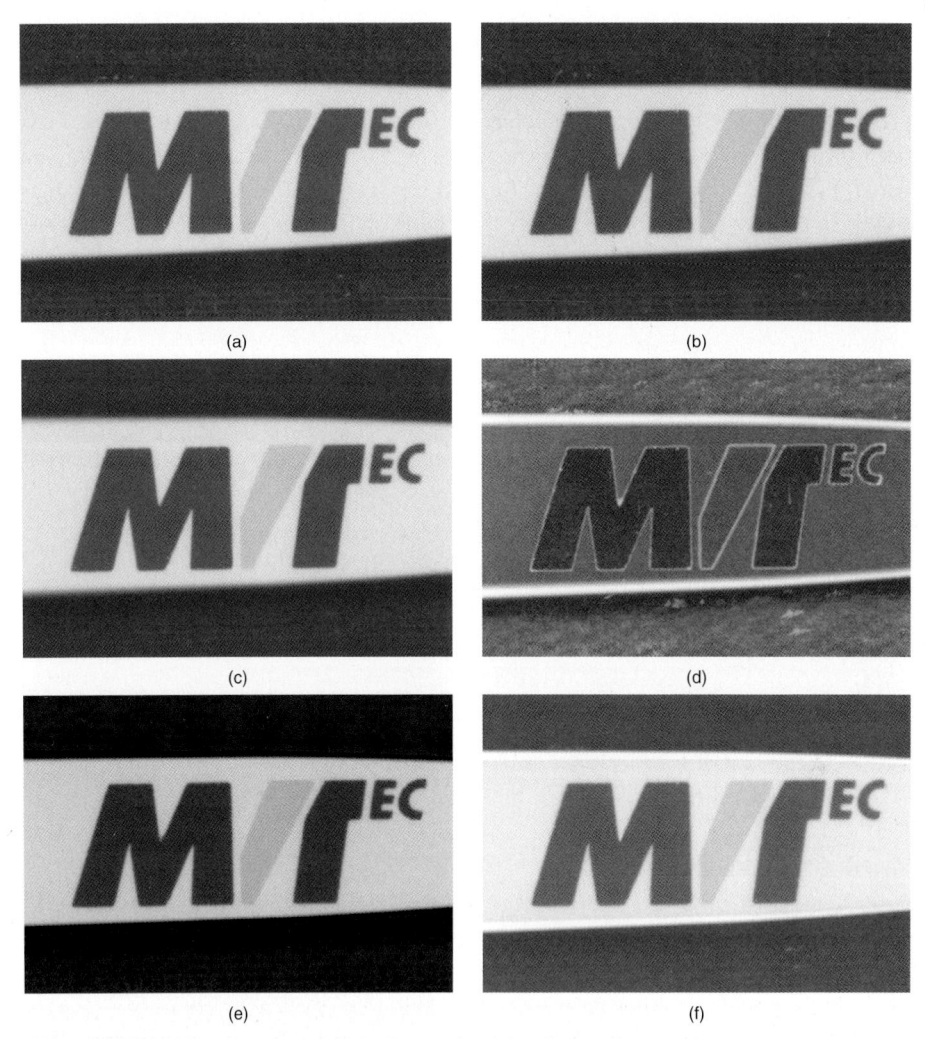

Fig. 3.27 (a), (b) Two images of a sequence of 15 showing a print on the clip of a pen. Note that the letter V in the MVTec logo moves slightly with respect to the rest of the logo. (c) Reference image $m_{r,c}$ of the variation model computed from the 15 training images. (d) Standard deviation image $s_{r,c}$. For better visibility, $s_{r,c}^{1/4}$ is displayed. (e), (f) Minimum and maximum threshold images $u_{r,c}$ and $l_{r,c}$ computed with $a = b = 20$ and $c = d = 3$.

Figures 3.27(a) and (b) display two images of a sequence of 15 showing a print on the clip of a pen. All images are aligned such that the MVTec logo is in the center of the image. Note that the letter V in the MVTec logo moves with respect to the rest of the logo and that the corners of the letters may change their shape slightly. This happens because of the pad printing technology used to print the logo. The two colors of the logo are printed with two different pads, which can move with respect to each other. Furthermore, the size of the letters may vary because of slightly different pressures with which the pads are pressed onto the clip. To ensure that the logo has been printed correctly, the variation model can be used to determine the mean and variation images shown in Figures 3.27(c) and (d), and from them the threshold images shown in Figures 3.27(e) and (f). Note that the variation is large at the letter V of the logo because this letter may move with respect to the rest of the logo. Also note the large variation at the edges of the clip, which occurs because the logo's position varies on the clip.

Figure 3.28(a) shows a logo with errors in the letters T (small hole) and C (too little ink). From Figure 3.28(b), it can be seen that the two errors can be detected reliably. Figure 3.28(c) shows a different kind of error: the letter V has moved too high and to the right. This kind of error can also be detected easily, as shown in Figure 3.28(d).

As described so far, the variation model requires n training images to construct the reference and variation images. In some applications, however, it is only possible to acquire a single reference image. In these cases, there are two options to create the variation model. The first option is to create artificial variations of the model, e.g.,

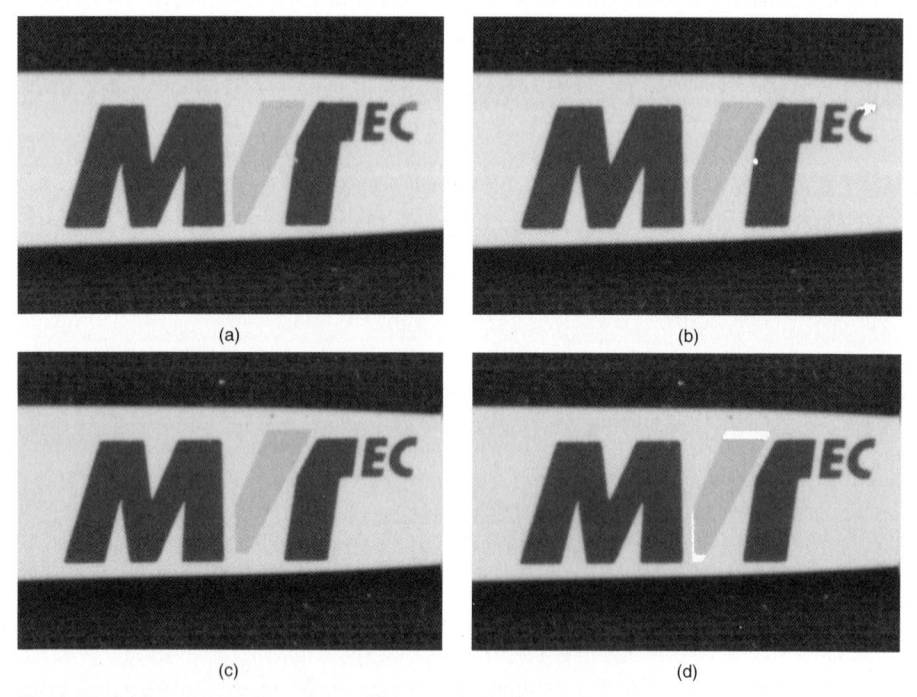

Fig. 3.28 (a) Image showing a logo with errors in the letters T (small hole) and C (too little ink). (b) Errors displayed in white, segmented with the variation model of Figure 3.27. (c) Image showing a logo in which the letter V has moved too high and to the right. (d) Segmented errors.

by creating translated versions of the reference image. Another option can be derived by noting that the variations are necessarily large at the edges of the object if we allow small size and position tolerances. This can be clearly seen in Figure 3.27(d). Consequently, in the absence of training images that show the real variations of the object, a reasonable approximation for $s_{r,c}$ is given by computing the edge amplitude image of the reference image using one of the edge filters described in Section 3.7.3.

3.4.2
Extraction of Connected Components

The segmentation algorithms in the previous section return one region as the segmentation result (recall the definitions in Eqs. (3.43)–(3.45)). Typically, the segmented region contains multiple objects that should be returned individually. For example, in the examples in Figures 3.22–3.26 we are interested in obtaining each character as a separate region. Typically, the objects we are interested in are characterized by forming a connected set of pixels. Hence, to obtain the individual regions we must compute the connected components of the segmented region.

To be able to compute the connected components, we must define when two pixels should be considered connected. On a rectangular pixel grid, there are only two natural options to define the connectivity. The first possibility is to define two pixels as being connected if they have an edge in common, i.e., if the pixel is directly above, below, left, or right of the current pixel, as shown in Figure 3.29(a). Since each pixel has four connected pixels, this definition is called the 4-connectivity or 4-neighborhood. Alternatively, the definition can be extended to also include the diagonally adjacent pixels, as shown in Figure 3.29(b). This definition is called the 8-connectivity or 8-neighborhood.

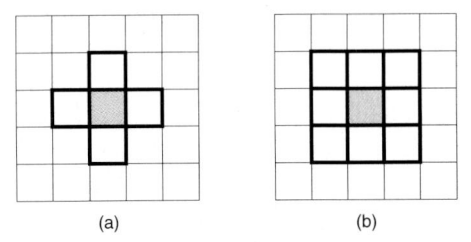

Fig. 3.29 The two possible definitions of connectivity on rectangular pixel grids: (a) 4-connectivity, and (b) 8-connectivity.

While these definitions are easy to understand, they cause problematic behavior if the same definition is used on both the foreground and background. Figure 3.30 shows some of the problems that occur if 8-connectivity is used for the foreground and background. In Figure 3.30(a), there is clearly a single line in the foreground, which divides the background into two connected components. This is what we would intuitively expect. However, as Figure 3.30(b) shows, if the line is slightly rotated we still obtain a single connected component in the foreground. However, now the background is also a single component. This is quite counterintuitive. Figure 3.30(c) shows another peculiarity. Again, the foreground region consists of a single connected component. Intuitively, we would say that the region contains a hole. However, the background also is a single connected component, indicating that the region contains no hole. The only remedy for this problem is to use opposite connectivities on the foreground and background. If, for example, 4-connectivity is used for the background in the examples in Figure 3.30, all of the above problems are solved. Likewise, if 4-connectivity is used for the foreground and 8-connectivity for the background, the inconsistencies are avoided.

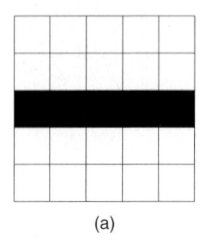

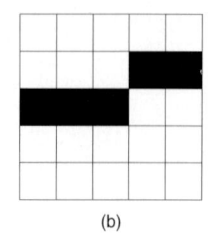

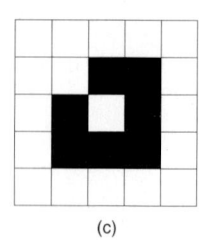

Fig. 3.30 Some peculiarities occur when the same connectivity, in this case 8-connectivity, is used for the foreground and background. (a) The single line in the foreground clearly divides the background into two connected components. (b) If the line is very slightly rotated, there is still a single line, but now the background is a single component, which is counterintuitive. (c) The single region in the foreground intuitively contains one hole. However, the background is also a single connected component, indicating that the region has no hole, which is also counterintuitive.

To compute the connected components on the run-length representation of a region, a classical depth-first search can be performed [70]. We can repeatedly search for the first unprocessed run and then search for overlapping runs in the adjacent rows of the image. The used connectivity determines whether two runs overlap. For 4-connectivity, the runs must at least have one pixel in the same column, while for the 8-connectivity the runs must at least touch diagonally. An example of this procedure is shown in Figure 3.31. The run-length representation of the input region is shown in Figure 3.31(a), the search tree for the depth-first search using the 8-connectivity

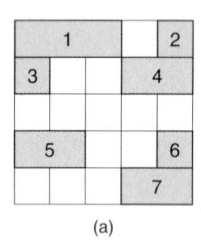

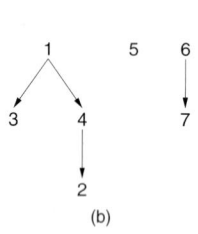

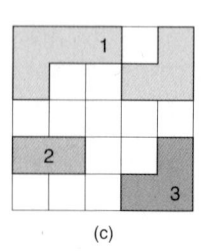

Fig. 3.31 (a) Run-length representation of a region containing seven runs. (b) Search tree when performing a depth-first search for the connected components of the region in (a) using 8-connectivity. The numbers indicate the runs. (c) Resulting connected components.

is shown in Figure 3.31(b), and the resulting connected components are shown in Figure 3.31(c). For the 8-connectivity, three connected components result. If the 4-connectivity had been used, four connected components would result.

It should be noted that the connected components can also be computed from the representation of a region as a binary image. The output of this operation is a label image. Therefore, this operation is also called labeling or component labeling. For a description of algorithms that compute the connected components from a binary image, see [68, 69].

To conclude this section, Figures 3.32(a) and (b) show the result of computing the connected components of the regions in Figures 3.23(e) and (f). As can be seen, each character is a connected component. Furthermore, the noisy segmentation results are also returned as separate components. Thus, it is easy to remove them from the segmentation, e.g., based on their area.

OK JAPAN
M5182 2-30
6 85352

SONY
XD117 AM
648 E 1 E

(a)   (b)

Fig. 3.32 (a), (b) Result of computing the connected components of the regions in Figures 3.23(e) and (f). The connected components are visualized by using eight different gray values cyclically.

3.4.3
Subpixel-Precise Thresholding

All the thresholding operations we have discussed so far have been pixel-precise. In most cases, this precision is sufficient. However, some applications require a higher accuracy than the pixel grid. Therefore, an algorithm that returns a result with subpixel precision is sometimes required. Obviously, the result of this subpixel-precise thresholding operation cannot be a region, which is only pixel-precise. The appropriate data structure for this purpose therefore is a subpixel-precise contour (see Section 3.1.3). This contour will represent the boundary between regions in the image that have gray values above the gray value threshold g_{sub} and regions that have gray values below g_{sub}. To obtain this boundary, we must convert the discrete representation of the image into a continuous function. This can be done, for example, with bilinear interpolation (see Eq. (3.40) in Section 3.3.3). Once we have obtained a continuous representation of the image, the subpixel-precise thresholding operation conceptually consists of intersecting the image function $f(r,c)$ with the constant function $g(r,c) = g_{sub}$. Figure 3.33 shows the bilinearly interpolated image $f(r,c)$ in a 2×2 block of the four closest pixel centers. The closest pixel centers lie at the corners of the graph. The bottom of the graph shows the intersection curve of the image $f(r,c)$ in this 2×2 block with the constant gray value $g_{sub} = 100$. Note that this curve is part of a hyperbola. Since this hyperbolic curve would be quite cumbersome to represent, we can simply substitute it with a straight line segment between the two points where the hyperbola leaves the 2×2 block. This line segment constitutes one segment of the subpixel contour we are interested in. Each 2×2 block in the image typically contains between zero and two of these line segments. If the 2×2 block contains an intersection of two contours, four line segments may occur. To obtain meaningful contours, these segments need to be linked. This can be done by repeatedly selecting the first unprocessed line segment in the image as the first segment of the contour and then tracing the adjacent line segments until the contour closes, reaches the image border, or reaches an intersection point. The result of this linking step typically are closed contours that enclose a region in the image in which the gray values are either larger or smaller than the threshold. Note that, if such a region contains holes, one contour will be created for the outer boundary of the region and one for each hole.

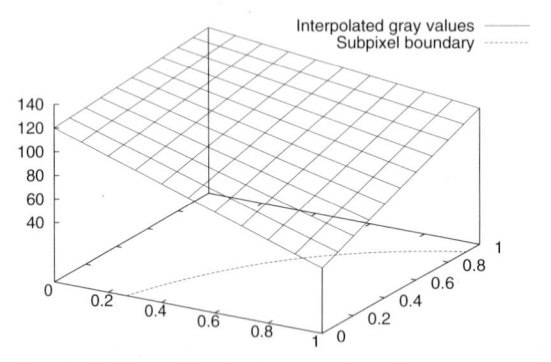

Fig. 3.33 The graph shows gray values that are interpolated bilinearly between four pixel centers, lying at the corners of the graph, and the intersection curve with the gray value $g_{sub} = 100$ at the bottom of the graph. This curve (part of a hyperbola) is the boundary between the region with gray values > 100 and gray values < 100.

Figure 3.34(a) shows an image of a PCB that contains a ball grid array (BGA) of solder pads. To ensure good electrical contact, it must be ensured that the pads have the correct shape and position. This requires high accuracy, and, since in this application typically the resolution of the image is small compared to the size of the balls and pads, the segmentation must be performed with subpixel accuracy. Figure 3.34(b) shows the result of performing a subpixel-precise thresholding operation on the image in Figure 3.34(a). To see enough details of the results, the part that corresponds to the white rectangle in Figure 3.34(a) is displayed. The boundary of the pads is extracted with very good accuracy. Figure 3.34(c) shows even more detail: the left pad in the center row of Figure 3.34(b), which contains an error that must be detected. As can be seen, the subpixel-precise contour correctly captures the erroneous region of the pad. We can also easily see the individual line segments in the subpixel-precise contour and how they are contained in the 2×2 pixel blocks. Note that each block lies between four pixel centers. Therefore, the contour's line segments end at the lines that connect the pixel centers. Note also that in this part of the image there is only one block in which two line segments are contained: at the position where the contour enters the error on the pad. All the other blocks contain one or no line segments.

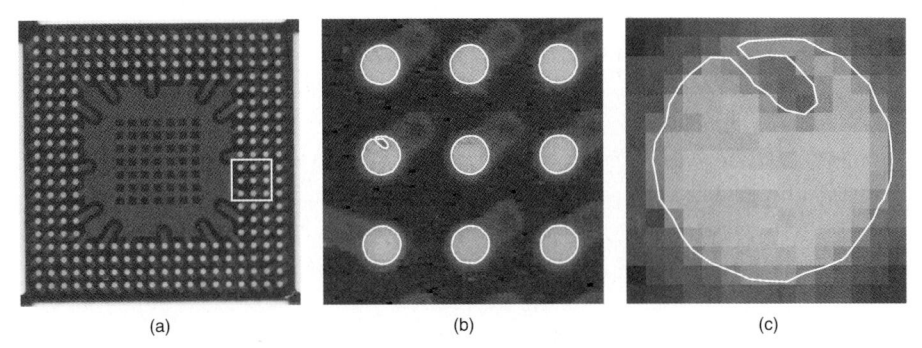

Fig. 3.34 (a) Image of a PCB with BGA solder pads. (b) Result of applying a subpixel-precise threshold to the image in (a). The part that is being displayed corresponds to the white rectangle in (a). (c) Detail of the left pad in the center row of (b).

3.5
Feature Extraction

In the previous sections, we have seen how to extract regions or subpixel-precise contours from an image. While the regions and contours are very useful, they may not be sufficient because they contain the raw description of the segmented data. Often, we must select certain regions or contours from the segmentation result, e.g., to remove unwanted parts of the segmentation. Furthermore, often we are interested in gauging the objects. In other applications, we might want to classify the objects, e.g., in the OCR, to determine the type of the object. All these applications require that we determine one or more characteristic quantities from the regions or contours. The quantities we determine are called features. Typically they are real numbers. The process of determining the features is called feature extraction. There are different kinds of features. Region features are features that can be extracted from the regions themselves. In contrast, gray value features also use the gray values in the image within the region. Finally, contour features are based on the coordinates of the contour.

3.5.1
Region Features

By far the simplest region feature is the area of the region:

$$a = |R| = \sum_{(r,c) \in R} 1 = \sum_{i=1}^{n} ce_i - cs_i + 1 \tag{3.51}$$

Hence, the area a of the region is simply the number of points $|R|$ in the region. If the region is represented as a binary image, the first sum has to be used to compute the area; whereas if a run-length representation is used, the second sum can be used. Recall from Eq. (3.2) that a region can be regarded as the union of its runs, and the area of a run is extremely simple to compute. Note that the second sum contains many fewer terms than the first sum, as discussed in Section 3.1.2. Hence, the run-length representation of a region will lead to a much faster computation of the area. This is true for almost all region features.

Figure 3.35 shows the result of selecting all regions with an area ≥ 20 from the regions in Figures 3.32(a) and (b). Note that all the characters have been selected, while all of the noisy segmentation results have been removed. These regions could now be used as input for the OCR.

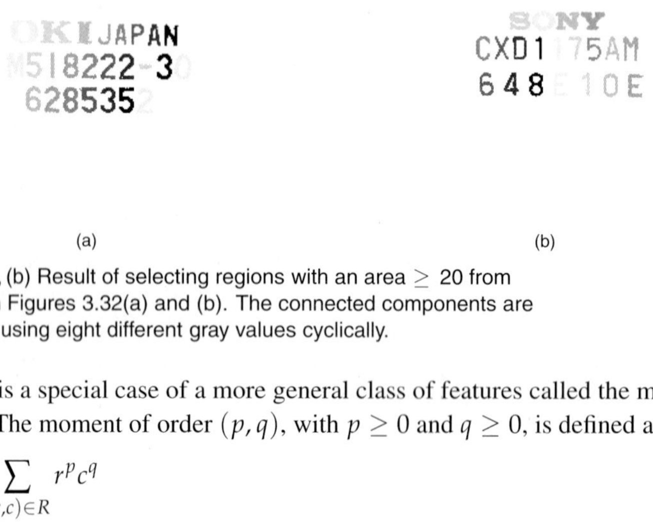

```
OKI JAPAN                        SONY
M5I8222-3                        CXD1175AM
628535                           648E 10E
```

(a)   (b)

Fig. 3.35 (a), (b) Result of selecting regions with an area ≥ 20 from the regions in Figures 3.32(a) and (b). The connected components are visualized by using eight different gray values cyclically.

The area is a special case of a more general class of features called the moments of the region. The moment of order (p, q), with $p \geq 0$ and $q \geq 0$, is defined as

$$m_{p,q} = \sum_{(r,c) \in R} r^p c^q \tag{3.52}$$

Note that $m_{0,0}$ is the area of the region. As for the area, simple formulas to compute the moments solely based on the runs can be derived. Hence, the moments can be computed very efficiently in the run-length representation.

The moments in Eq. (3.52) depend on the size of the region. Often, it is desirable to have features that are invariant to the size of the objects. To obtain such features, we can simply divide the moments by the area of the region if $p+q \geq 1$ to get normalized moments:

$$n_{p,q} = \frac{1}{a} \sum_{(r,c) \in R} r^p c^q \qquad (3.53)$$

The most interesting feature that can be derived from the normalized moments is the center of gravity of the region, which is given by $(n_{1,0}, n_{0,1})$. It can be used to describe the position of the region. Note that the center of gravity is a subpixel-precise feature, even though it is computed from pixel-precise data.

The normalized moments depend on the position in the image. Often, it is useful to make the features invariant to the position of the region in the image. This can be done by calculating the moments relative to the center of gravity of the region. These central moments are given by ($p+q \geq 2$):

$$\mu_{p,q} = \frac{1}{a} \sum_{(r,c) \in R} (r - n_{1,0})^p (c - n_{0,1})^q \qquad (3.54)$$

Note that they are also normalized. The second central moments ($p+q=2$) are particularly interesting. They enable us to define an orientation and an extent for the region. This is done by assuming that the moments of order 1 and 2 of the region were obtained from an ellipse. Then, from these five moments, the five geometric parameters of the ellipse can be derived. Figure 3.36 displays the ellipse parameters graphically. The center of the ellipse is identical to the center of gravity of the region. The major and minor axes r_1 and r_2 and the angle of the ellipse with respect to the column axis are given by

$$\begin{aligned} r_1 &= \sqrt{2 \left(\mu_{2,0} + \mu_{0,2} + \sqrt{(\mu_{2,0} - \mu_{0,2})^2 + 4\mu_{1,1}^2} \right)} \\ r_2 &= \sqrt{2 \left(\mu_{2,0} + \mu_{0,2} - \sqrt{(\mu_{2,0} - \mu_{0,2})^2 + 4\mu_{1,1}^2} \right)} \\ \theta &= -\frac{1}{2} \arctan \frac{2\mu_{1,1}}{\mu_{0,2} - \mu_{2,0}} \end{aligned} \qquad (3.55)$$

For a derivation of these results, see [68] (note that, there, the diameters are used instead of the radii). From the ellipse parameters, we can derive another very useful

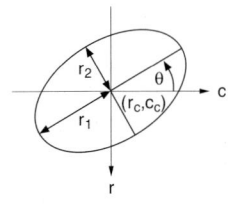

Fig. 3.36 The geometric parameters of an ellipse.

feature: the anisometry r_1/r_2. This is scale-invariant and describes how elongated a region is.

The ellipse parameters are extremely useful to determine orientations and sizes of regions. For example, the angle θ can be used to rectify rotated text. Figure 3.37(a) shows the result of thresholding the image in Figure 3.18(a). The segmentation result is treated as a single region, i.e., the connected components have not been computed. Figure 3.37(a) also displays the ellipse parameters by overlaying the major and minor axes of the equivalent ellipse. Note that the major axis is slightly longer than the region because the equivalent ellipse does not need to have the same area as the region. It only needs to have the same moments of order 1 and 2. The angle of the major axis is a very good estimate for the rotation of the text. In fact, it has been used to rectify the images in Figures 3.18(b) and (c). Figure 3.37(a) shows the axes of the characters after the connected components have been computed. Note how well the orientation of the regions corresponds with our intuition.

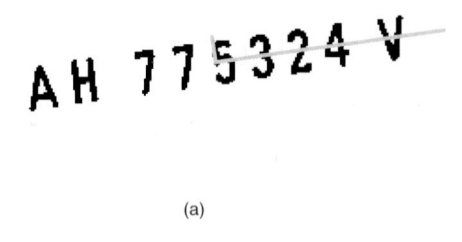

(a)

(b)

Fig. 3.37 Result of thresholding the image in Figure 3.18(a) overlaid with a visualization of the ellipse parameters. The light gray lines represent the major and minor axes of the regions. Their intersection is the center of gravity of the regions. (a) The segmentation is treated as a single region. (b) The connected components of the region are used. The angle of the major axis in (a) has been used to rotate the images in Figures 3.18(b) and (c).

While the ellipse parameters are extremely useful, they have two minor shortcomings. First of all, the orientation can only be determined if $r_1 \neq r_2$. Our first thought might be that this only applies to circles, which have no meaningful orientation anyway. Unfortunately, this is not true. There is a much larger class of objects for which $r_1 = r_2$. All objects that have a fourfold rotational symmetry like squares have $r_1 = r_2$. Hence, their orientation cannot be determined with the ellipse parameters. The second slight problem is that, since the underlying model is an ellipse, the orientation θ can only be determined modulo π (180°). This problem can be solved by determining the point in the region that has the largest distance from the center of gravity and use it to select θ or $\theta + \pi$ as the correct orientation.

In the above discussion, we have used various transformations to make the moment-based features invariant to certain transformations, e.g., translation and scaling. Several approaches have been proposed to create moment-based features that are invariant to a larger class of transformations, e.g., translation, rotation, and scaling [71] or even general affine transformations [72, 73]. They are primarily used to classify objects.

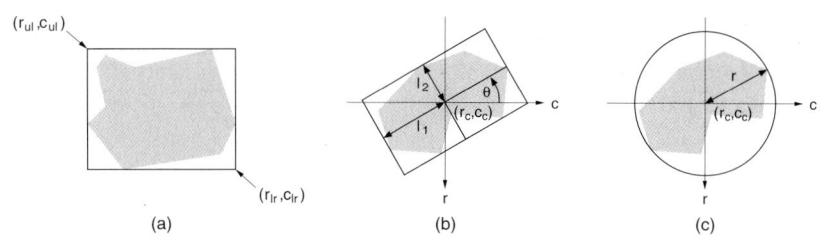

Fig. 3.38 (a) The smallest axis-parallel enclosing rectangle of a region. (b) The smallest enclosing rectangle of arbitrary orientation. (c) The smallest enclosing circle.

Apart from the moment-based features, there are several other useful features that are based on the idea of finding an enclosing geometric primitive for the region. Figure 3.38(a) displays the smallest axis-parallel enclosing rectangle of a region. This rectangle is often also called the bounding box of the region. It can be calculated very easily based on the minimum and maximum row and column coordinates of the region. Based on the parameters of the rectangle, other useful quantities like the width and height of the region and their ratio can be calculated. The parameters of the bounding box are particularly useful if we want to quickly find out whether two regions can intersect. Since the smallest axis-parallel enclosing rectangle sometimes is not very tight, we can also define a smallest enclosing rectangle of arbitrary orientation, as shown in Figure 3.38(b). Its computation is much more complicated than the computation of the bounding box, however, so we cannot give details here. An efficient implementation can be found in [74]. Note that an arbitrarily oriented rectangle has the same parameters as an ellipse. Hence, it also enables us to define the position, size, and orientation of a region. Note that, in contrast to the ellipse parameters, a useful orientation for squares is returned. The final useful enclosing primitive is an enclosing circle, as shown in Figure 3.38(c). Its computation is also quite complex [75]. It also enables us to define the position and size of a region.

The computation of the smallest enclosing rectangle of arbitrary orientation and the smallest enclosing circle is based on first computing the convex hull of the region. The convex hull of a set of points, and in particular a region, is the smallest convex set that contains all the points. A set is convex if, for any two points in the set, the straight line between them is completely contained in the set. The convex hull of a set of points can be computed efficiently [76, 77]. The convex hull of a region is often useful to construct ROIs from regions that have been extracted from the image. Based on the convex hull of the region, another useful feature can be defined: the convexity, which is defined as the ratio of the area of the region to the area of its convex hull. It is a feature between 0 and 1 that measures how compact the region is. A convex region has a convexity of 1. The convexity can, for example, be used to remove unwanted segmentation results, which often are highly non-convex.

Another useful feature of a region is its contour length. To compute it, we need to trace the boundary of the region to get a linked contour of the boundary pixels [68]. Once the contour has been computed, we simply need to sum the Euclidean distances of the contour segments, which are 1 for horizontal and vertical segments and $\sqrt{2}$ for diagonal segments. Based on the contour length l and the area a of the region, we can define another measure for the compactness of a region: $c = l^2/(4\pi a)$. For circular regions, this feature is 1, while all other regions have larger values. The compactness has similar uses to the convexity.

3.5.2
Gray Value Features

We have already seen some gray value features in Section 3.2.1, namely the minimum and maximum gray values within the region:

$$g_{\min} = \min_{(r,c) \in R} g_{r,c} \qquad g_{\max} = \max_{(r,c) \in R} g_{r,c} \qquad (3.56)$$

They are used for the gray value normalization in Section 3.2.1. Another obvious feature is the mean gray value within the region:

$$\bar{g} = \frac{1}{a} \sum_{(r,c) \in R} g_{r,c} \qquad (3.57)$$

Here, a is the area of the region, given by Eq. (3.51). The mean gray value is a measure of the brightness of the region. A single measurement within a reference region can be used to measure additive brightness changes with respect to the conditions when the system was set up. Two measurements within different reference regions can be used to measure linear brightness changes, and hence to compute a linear gray value transformation (see Section 3.2.1) that compensates the brightness change, or to adapt segmentation thresholds. The mean gray value is a statistical feature. Another statistical feature is the variance of the gray values:

$$s^2 = \frac{1}{a-1} \sum_{(r,c) \in R} (g_{r,c} - \bar{g})^2 \qquad (3.58)$$

and the standard deviation $s = \sqrt{s^2}$. Measuring the mean and standard deviation within a reference region can also be used to construct a linear gray value transformation that compensates brightness changes. The standard deviation can be used to adapt segmentation thresholds. Furthermore, the standard deviation is a measure of the amount of texture that is present within the region.

The gray value histogram (3.3) and the cumulative histogram (3.4), which we have already encountered in Section 3.2.1, are also gray value features. From the histogram, we have already used a feature for the robust contrast normalization: the α-quantile

$$g_\alpha = \min\{g : c_g \geq \alpha\} \tag{3.59}$$

where c_g is defined in Eq. (3.4). It was used to obtain the robust minimum and maximum gray values in Section 3.2.1. The quantiles were called p_l and p_u there. Note that for $\alpha = 0.5$ we obtain the median gray value. It has similar uses to the mean gray value.

In the previous section, we have seen that the region's moments are extremely useful features. They can be extended to gray value features in a natural manner. The gray value moment of order (p, q), with $p \geq 0$ and $q \geq 0$, is defined as

$$m_{p,q} = \sum_{(r,c) \in R} g_{r,c} r^p c^q \tag{3.60}$$

This is the natural generalization of the region moments because we obtain the region moments from the gray value moments by using the characteristic function χ_R (Eq. (3.1)) of the region as the gray values. Like for the region moments, the moment $a = m_{0,0}$ can be regarded as the gray value area of the region. It is actually the "volume" of the gray value function $g_{r,c}$ within the region. Like for the region moments, normalized moments can be defined by

$$n_{p,q} = \frac{1}{a} \sum_{(r,c) \in R} g_{r,c} r^p c^q \tag{3.61}$$

The moments $(n_{1,0}, n_{0,1})$ define the gray value center of gravity of the region. With this, central gray value moments can be defined by

$$\mu_{p,q} = \frac{1}{a} \sum_{(r,c) \in R} g_{r,c} (r - n_{1,0})^p (c - n_{0,1})^q \tag{3.62}$$

Like for the region moments, based on the second central moments we can define the ellipse parameters, major and minor axes and the orientation. The formulas are identical to Eqs. (3.55). Furthermore, the anisometry can also be defined identically as for the regions.

All the moment-based gray value features are very similar to their region-based counterparts. Therefore, it is interesting to look at their differences. Like we saw, the gray value moments reduce to the region moments if the characteristic function of the region is used as the gray values. The characteristic function can be interpreted as the membership of a pixel to the region. A membership of 1 means that the pixel belongs to the region, while 0 means that the pixel does not belong to the region. This notion of belonging to the region is crisp, i.e., for every pixel a hard decision must be made. Suppose now that, instead of making a hard decision for every pixel, we could make a "soft" or "fuzzy" decision about whether a pixel belongs to the region, and that we encode the degree of belonging to the region by a number $\in [0, 1]$. We can interpret the degree of belonging as a fuzzy membership value, as opposed to the crisp binary membership value. With this, the gray value image can be regarded as a fuzzy set [78]. The advantage of regarding the image as a fuzzy set is that we do not have to make a hard decision about whether a pixel belongs to the object or not. Instead, the fuzzy membership value determines what percentage of the pixel belongs to the object. This enables us to measure the position and size of the objects much more accurately, especially for small objects, because in the transition zone between the foreground and background there will be some mixed pixels that allow us to capture the geometry of the object more accurately. An example of this is shown in Figure 3.39. Here, a synthetically generated subpixel-precise ideal circle of radius 3 is shifted in subpixel increments. The gray values represent a fuzzy membership, scaled to values between 0 and 200 for display purposes. The figure displays a pixel-precise region, thresholded with a value of 100, which corresponds to a membership above 0.5, as well as two circles that have a center of gravity and area that were obtained from the region and gray value moments. The gray value moments were computed in the entire image. It can be seen that the area and center of gravity are computed

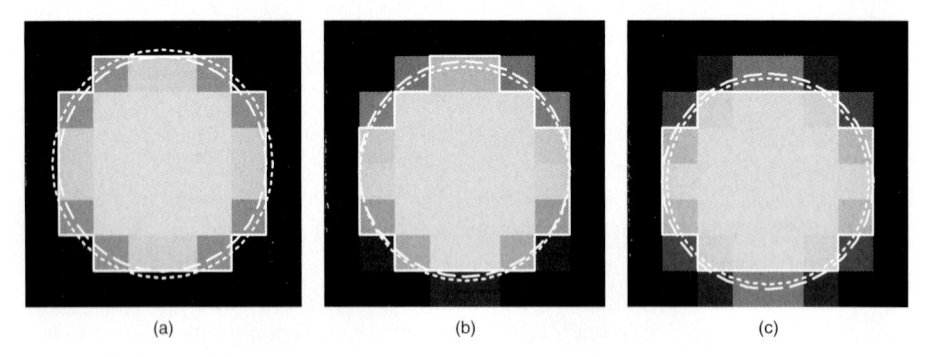

Fig. 3.39 Subpixel-precise circle position and area using the gray value and region moments. The image represents a fuzzy membership, scaled to values between 0 and 200. The solid line is the result of segmenting with a membership of 100. The dotted line is a circle that has the same center of gravity and area as the segmented region. The dashed line is a circle that has the same gray value center of gravity and gray value area as the image. (a) Shift: 0; error in the area for the region moments: 13.2%; for the gray value moments: −0.05%. (b) Shift: 5/32 pixel; error in the row coordinate for the region moments: −0.129; for the gray value moments: 0.003. (c) Shift: 1/2 pixel; error in the area for the region moments: −8.0%; for the gray value moments: −0.015%. Note that the gray value moments yield a significantly better accuracy for this small object.

much more accurately by the gray value moments because the decision about whether a pixel belongs to the foreground or not has been avoided. In this example, the gray value moments result in an area error that is always smaller than 0.25% and a position error smaller than 1/200 pixel. In contrast, the area error for the region moments can be up to 13.2% and the position error can be up to 1/6 pixel. Note that both types of moments yield subpixel-accurate measurements. We can see that on ideal data it is possible to obtain an extremely high accuracy with the gray value moments, even for very small objects. On real data the accuracy will necessarily be somewhat lower. It should also be noted that the accuracy advantage of the gray value moments primarily occurs for small objects. Because the gray value moments must access every pixel within the region, whereas the region moments can be computed solely based on the run-length representation of the region, the region moments can be computed much faster. Hence, the gray moments are typically only used for relatively small regions.

The only question we need to answer is how to define the fuzzy membership value of a pixel. If we assume that the camera has a fill factor of 100% and the gray value response of the image acquisition device and camera are linear, the gray value difference of a pixel from the background is proportional to the portion of the object that is covered by the pixel. Consequently, we can define a fuzzy membership relation as follows: every pixel that has a gray value below the background gray value g_{min} has a membership value of 0. Conversely, every pixel that has a gray value above the foreground gray value g_{max} has a membership value of 1. In between, the membership values are interpolated linearly. Since this procedure would require floating-point images, the membership is scaled to an integer image with b bits, typically 8 bits. Consequently, the fuzzy membership relation is a simple linear gray value scaling, as defined in Section 3.2.1. If we scale the fuzzy membership image in this manner, the gray value area needs to be divided by the maximum gray value, e.g., 255, to obtain the true area. The normalized and central gray value moments do not need to be modified in this manner since they are, by definition, invariant to a scaling of the gray values.

Figure 3.40 displays a real application where the above principles are used. In Figure 3.40(a) a ball grid array (BGA) device with solder balls is displayed, along with two rectangles that indicate the image parts shown in Figures 3.40(b) and (c). The image in Figure 3.40(a) is first transformed into a fuzzy membership image using $g_{min} = 40$ and $g_{max} = 120$ with 8 bit resolution. The individual balls are segmented and then inspected for correct size and shape by using the gray value area and the gray value anisometry. The erroneous balls are displayed with dashed lines. To aid the visual interpretation, the ellipses representing the segmented balls are scaled such that they have the same area as the gray value area. This is done because the gray value ellipse parameters typically return an ellipse with a different area than the gray value area, analogously to the region ellipse parameters (see the discussion following Eqs. (3.55) in Section 3.5.1). As can be seen, all the balls that have an erroneous size or shape, indicating partially missing solder, have been correctly detected.

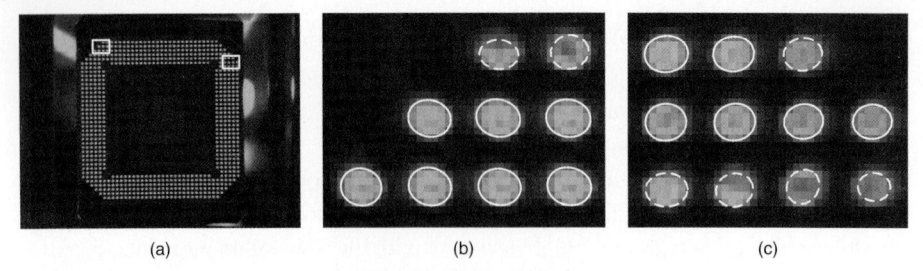

Fig. 3.40 (a) Image of a BGA device. The two rectangles correspond to the image parts shown in (b) and (c). The results of inspecting the balls for correct size (gray value area $\geq$ 20) and correct gray value anisometry ($\leq$ 1.25) are visualized in (b) and (c). Correct balls are displayed as solid ellipses, while defective balls are displayed as dashed ellipses.

3.5.3
Contour Features

Many of the region features we have discussed in Section 3.5.1 can be transferred to subpixel-precise contour features in a straightforward manner. For example, the length of the subpixel-precise contour is even easier to compute because the contour is already represented explicitly by its control points (r_i, c_i), for $i = 1, \ldots, n$. It is also simple to compute the smallest enclosing axis-parallel rectangle (the bounding box) of the contour. Furthermore, the convex hull of the contour can be computed like for regions [76, 77]. From the convex hull, we can also derive the smallest enclosing circles [75] and smallest enclosing rectangles of arbitrary orientation [74].

In the previous two sections, we have seen that the moments are extremely useful features. An interesting question, therefore, is whether they can be defined for contours. In particular, it is interesting whether a contour has an area. Obviously, for this to be true, the contour must enclose a region, i.e., it must be closed and must not intersect itself. To simplify the formulas, let us assume that a closed contour is specified by $(r_1, c_1) = (r_n, c_n)$. Let the subpixel-precise region that the contour encloses be denoted by R. Then, the moment of order (p, q) is defined as

$$m_{p,q} = \iint\limits_{(r,c) \in R} r^p c^q \, dr \, dc \tag{3.63}$$

Like for regions, we can define normalized and central moments. The formulas are identical to Eqs. (3.53) and (3.54) with the sums being replaced by integrals. It can be shown that these moments can be computed solely based on the control points of the contour [79]. For example, the area and center of gravity of the contour are given by

$$a = \frac{1}{2} \sum_{i=1}^{n} r_{i-1}c_i - r_i c_{i-1}$$

$$n_{1,0} = \frac{1}{6a} \sum_{i=1}^{n} (r_{i-1}c_i - r_i c_{i-1})(r_{i-1} + r_i) \qquad (3.64)$$

$$n_{0,1} = \frac{1}{6a} \sum_{i=1}^{n} (r_{i-1}c_i - r_i c_{i-1})(c_{i-1} + c_i)$$

Analogous formulas can be derived for the second-order moments. Based on them, we can again compute the ellipse parameters, major axis, minor axis, and orientation. The formulas are identical to Eqs. (3.55). The moment-based contour features can be used for the same purposes as the corresponding region and gray value features. By performing an evaluation similar to that in Figure 3.39, it can be seen that the contour center of gravity and the ellipse parameters are equally as accurate as the gray value center of gravity. The accuracy of the contour area is slightly worse than the gray value area because we have approximated the hyperbolic segments with line segments. Since the true contour is a circle, the line segments always lie inside the true circle. Nevertheless, subpixel-thresholding and the contour moments could also have been used to detect the erroneous balls in Figure 3.40.

3.6
Morphology

In Section 3.4 we discussed how to segment regions. We have already seen that segmentation results often contain unwanted noisy parts. Furthermore, sometimes the segmentation will contain parts in which the shape of the object we are interested in has been disturbed, e.g., because of reflections. Therefore, often we need to modify the shape of the segmented regions to obtain the desired results. This is the subject of the field of mathematical morphology, which can be defined as a theory for the analysis of spatial structures [80]. For our purposes, mathematical morphology provides a set of extremely useful operations that enable us to modify or describe the shape of objects. Morphological operations can be defined on regions and gray value images. We will discuss both types of operations in this section.

3.6.1
Region Morphology

All region morphology operations can be defined in terms of six very simple operations: union, intersection, difference, complement, translation, and transposition. We will take a brief look at these operations first.

The union of two regions R and S is the set of points that lie in R or in S:

$$R \cup S = \{p \mid p \in R \vee p \in S\} \tag{3.65}$$

One important property of the union is that it is commutative: $R \cup S = S \cup R$. Furthermore, it is associative: $(R \cup S) \cup T = R \cup (S \cup T)$. While this may seem like a trivial observation, it will enable us to derive very efficient implementations for the morphological operations below. The algorithm to compute the union of two binary images is obvious: we simply need to compute the logical *or* of the two images. The runtime complexity of this algorithm obviously is $O(wh)$, where w and h are the width and height of the binary image. In the run-length representation, the union can be computed with a lower complexity: $O(n + m)$, where n and m are the number of runs in R and S. The principle of the algorithm is to merge the runs of the two regions while observing the order of the runs (see Section 3.1.2) and then to pack overlapping runs into single runs.

The intersection of two regions R and S is the set of points that lie in R and in S:

$$R \cap S = \{p \mid p \in R \wedge p \in S\} \tag{3.66}$$

Like the union, the intersection is commutative and associative. Again, the algorithm on binary images is obvious: we compute the logical *and* of the two images. For the run-length representation, again an algorithm that has complexity $O(n + m)$ can be found.

The difference of two regions R and S is the set of points that lie in R but not in S:

$$R \setminus S = \{p \mid p \in R \wedge p \notin S\} = R \cap \overline{S} \tag{3.67}$$

The difference is not commutative and not associative. Note that it can be defined in terms of the intersection and the complement of a region R, which is defined as all the points that do not lie in R:

$$\overline{R} = \{p \mid p \notin R\} \tag{3.68}$$

Since the complement of a finite region is infinite, it is impossible to represent it as a binary image. Therefore, for the representation of regions as binary images, it is important to define the operations without the complement. It is, however, possible to represent it as a run-length-encoded region by adding a flag that indicates whether the region or its complement is being stored. This can be used to define a more general set of morphological operations. There is an interesting relation between the number of connected components of the background $|C(\overline{R})|$ and the number of holes of the foreground $|H(R)|$: thus $|C(\overline{R})| = 1 + |H(R)|$. As discussed in Section 3.4.2, complementary connectivities must be used for the foreground and the background for this relation to hold.

Apart from the set operations, two basic geometric transformations are used in morphological operations. The translation of a region by a vector t is defined as

$$R_t = \{p \mid p - t \in R\} = \{q \mid q = p + t \text{ for } p \in R\} \tag{3.69}$$

Finally, the transposition of a region is defined as a mirroring about the origin:

$$\check{R} = \{-p \mid p \in R\} \tag{3.70}$$

Note that this is the only operation where a special point (the origin) is singled out. All the other operations do not depend on the origin of the coordinate system, i.e., they are translation-invariant.

With these building blocks, we can now take a look at the morphological operations. They typically involve two regions. One of these is the region we want to process, which will be denoted by R below. The other region has a special meaning. It is called the structuring element, and will be denoted by S. The structuring element is the means by which we can describe the shapes we are interested in.

The first morphological operation we consider is the Minkowski addition, which is defined by

$$R \oplus S = \{r + s \mid r \in R, s \in S\} = \bigcup_{s \in S} R_s$$
$$= \bigcup_{r \in R} S_r = \{t \mid R \cap (\check{S})_t \neq \emptyset\} \tag{3.71}$$

It is interesting to interpret the formulas. The first formula says that, to get the Minkowski addition of R with S, we take every point in R and every point in S and compute the vector sum of the points. The result of the Minkowski addition is the set of all points thus obtained. If we single out S, this can also be interpreted as taking all points in S, translating the region R by the vector corresponding to the point s from S, and computing the union of all the translated regions. Thus, we obtain the second formula. By symmetry, we can also translate S by all points in R to obtain the third formula. Another way to look at the Minkowski addition is the fourth formula. It tells us that we move the transposed structuring element around in the plane. Whenever the translated transposed structuring element and the region have at least one point in common, we copy the translated reference point into the output. Figure 3.41 shows an example of the Minkowski addition.

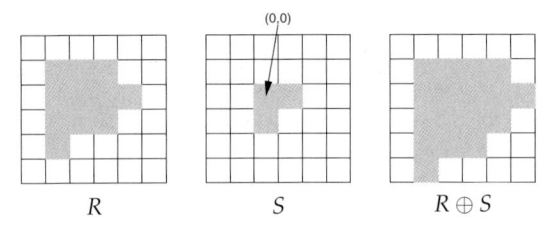

Fig. 3.41 Example of the Minkowski addition $R \oplus S$.

While the Minkowski addition has a simple formula, it has one small drawback. Its geometric criterion is that the transposed structuring element has at least one point in common with the region. Ideally, we would like to have an operation that returns all translated reference points for which the structuring element itself has at least one point in common with the region. To achieve this, we only need to use the transposed structuring element in the Minkowski addition. This operation is called a dilation, and is defined by

$$R \oplus \check{S} = \{t \mid R \cap S_t \neq \emptyset\} = \bigcup_{s \in S} R_{-s} \qquad (3.72)$$

Figure 3.42 shows an example of the dilation. Note that the result of the Minkowski addition and dilation are different. This is true whenever the structuring element is not symmetric with respect to the origin. If the structuring element is symmetric, the Minkowski addition and dilation are identical. Please be aware that this is assumed in many discussions about and implementations of the morphology. Therefore, the dilation is often defined without the transposition, which is technically incorrect.

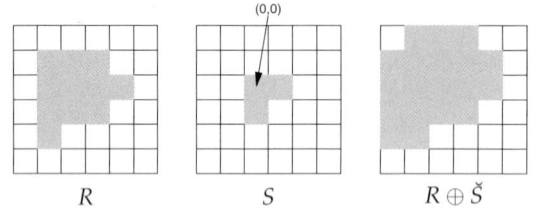

Fig. 3.42 Example of the dilation $R \oplus \check{S}$.

The implementation of the Minkowski addition for binary images is straightforward. As suggested by the second formula in Eq. (3.71), it can be implemented as a nonlinear filter with logical *or* operations. The runtime complexity is proportional to the size of the image times the number of pixels in the structuring element. The second factor can be reduced to roughly the number of boundary pixels in the structuring element [62]. Also, for binary images represented with one bit per pixel, very efficient algorithms can be developed for special structuring elements [81]. Nevertheless, in both cases the runtime complexity is proportional to the number of pixels in the image. To derive an implementation for the run-length representation of the regions, we first need to examine some algebraic properties of the Minkowski addition. It is commutative: $R \oplus S = S \oplus R$. Furthermore, it is distributive with respect to the union: $(R \cup S) \oplus T = R \oplus T \cup S \oplus T$. Since a region can be regarded as the union of its runs, we can use the commutativity and distributivity to transform the Minkowski addition as follows:

$$R \oplus S = \left(\bigcup_{i=1}^{n} \mathbf{r}_i\right) \oplus \left(\bigcup_{j=1}^{m} \mathbf{s}_j\right) = \bigcup_{j=1}^{m}\left(\left(\bigcup_{i=1}^{n} \mathbf{r}_i\right) \oplus \mathbf{s}_j\right) = \bigcup_{i=1}^{n} \bigcup_{j=1}^{m} \mathbf{r}_i \oplus \mathbf{s}_j \qquad (3.73)$$

Thus, the Minkowski addition can be implemented as the union of nm dilations of single runs, which are trivial to compute. Since the union of the runs can be computed easily, the runtime complexity is $O(mn)$, which is better than for binary images.

As we have seen above, the dilation and Minkowski addition enlarge the input region. This can be used, for example, to merge separate parts of a region into a single part, and thus to obtain the correct connected components of objects. One example of this is shown in Figure 3.43. Here, we want to segment each character as a separate connected component. If we compute the connected components of the thresholded region in Figure 3.43(b), we can see that the characters and their dots are separate components (Figure 3.43(c)). To solve this problem, we first need to connect the dots with their characters. This can be achieved using a dilation with a circle of diameter 5 (Figure 3.43(d)). With this, the correct connected components are obtained (Figure 3.43(e)). Unfortunately, they have the wrong shape because of the dilation. This can be corrected by intersecting the components with the originally segmented region. Figure 3.43(f) shows that, with these simple steps, we have obtained one component with the correct shape for each character.

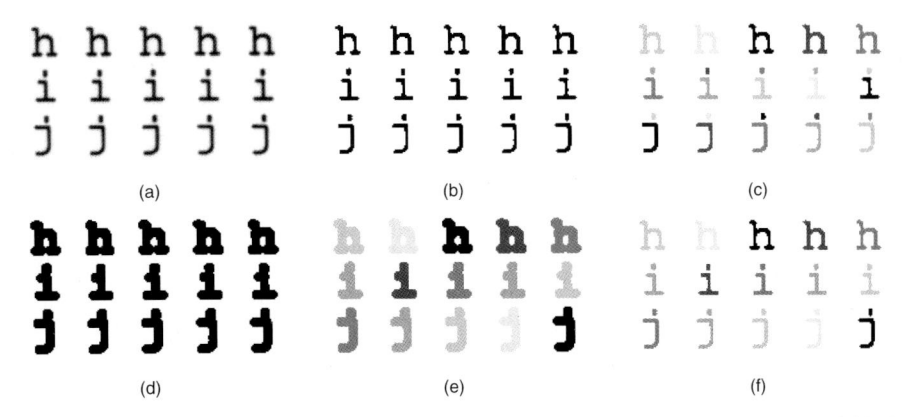

Fig. 3.43 (a) Image of a print of several characters. (b) Result of thresholding (a). (c) Connected components of (b) displayed with six different gray values. Note that the characters and their dots are separate connected components, which is undesirable. (d) Result of dilating the region in (b) with a circle of diameter 5. (e) Connected components of (d). Note that each character is now a single connected component. (f) Result of intersecting the connected components in (e) with the original segmentation in (b). This transforms the connected components into the correct shape.

The dilation is also very useful for constructing ROIs based on regions that were extracted from the image. We will see an example of this in Section 3.7.3.

The second type of morphological operation is the Minkowski subtraction. It is defined by

$$R \ominus S = \bigcap_{s \in S} R_s = \{r \mid \forall s \in S : r - s \in R\} = \{t \mid (\check{S})_t \subseteq R\} \tag{3.74}$$

The first formula is similar to the second formula in Eq. (3.71) with the union having been replaced by an intersection. Hence, we can still think about moving the region R by all vectors s from S. However, now the points must be contained in all translated regions (instead of at least one translated region). This is what the second formula in Eq. (3.74) expresses. Finally, if we look at the third formula, we see that we can also move the transposed structuring element around in the plane. If it is completely contained in the region R, we add its reference point to the output. Again, note the similarity to the Minkowski addition, where the structuring element had to have at least one point in common with the region. For the Minkowski subtraction it must lie completely within the region. Figure 3.44 shows an example of the Minkowski subtraction.

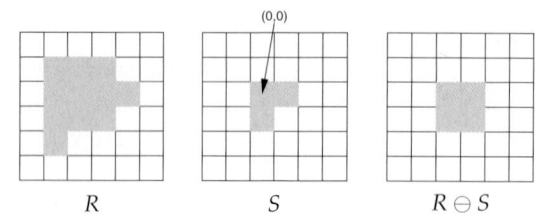

Fig. 3.44 Example of the Minkowski subtraction $R \ominus S$.

The Minkowski subtraction has the same small drawback as the Minkowski addition: its geometric criterion is that the transposed structuring element must completely lie within the region. As for the dilation, we can use the transposed structuring element. This operation is called an erosion and is defined by

$$R \ominus \check{S} = \bigcap_{s \in S} R_{-s} = \{t \mid S_t \subseteq R\} \qquad (3.75)$$

Figure 3.45 shows an example of the erosion. Again, note that the Minkowski subtraction and erosion only produce identical results if the structuring element is symmetric with respect to the origin. Please be aware that this is often silently assumed and the erosion is defined as a Minkowski subtraction, which is technically incorrect.

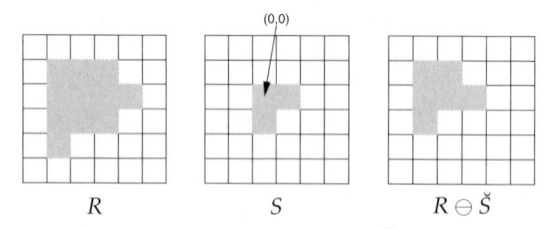

Fig. 3.45 Example of the erosion $R \ominus \check{S}$.

As we have seen from the small examples, the Minkowski subtraction and erosion shrink the input region. This can, for example, be used to separate objects that are attached to each other. Figure 3.46 shows an example of this. Here, the goal is to segment the individual globular objects. The result of thresholding the image is shown in Figure 3.46(b). If we compute the connected components of this region, an incorrect result is obtained because several objects touch each other (Figure 3.46(c)). The solution is to erode the region with a circle of diameter 15 (Figure 3.46(d)) before computing the connected components (Figure 3.46(e)). Unfortunately, the connected components have the wrong shape. Here, we cannot use the same strategy that we used for the dilation (intersecting the connected components with the original segmentation) because the erosion has shrunk the region. To approximately get the original shape back, we can dilate the connected components with the same structuring element that we used for the erosion (Figure 3.46(f)).

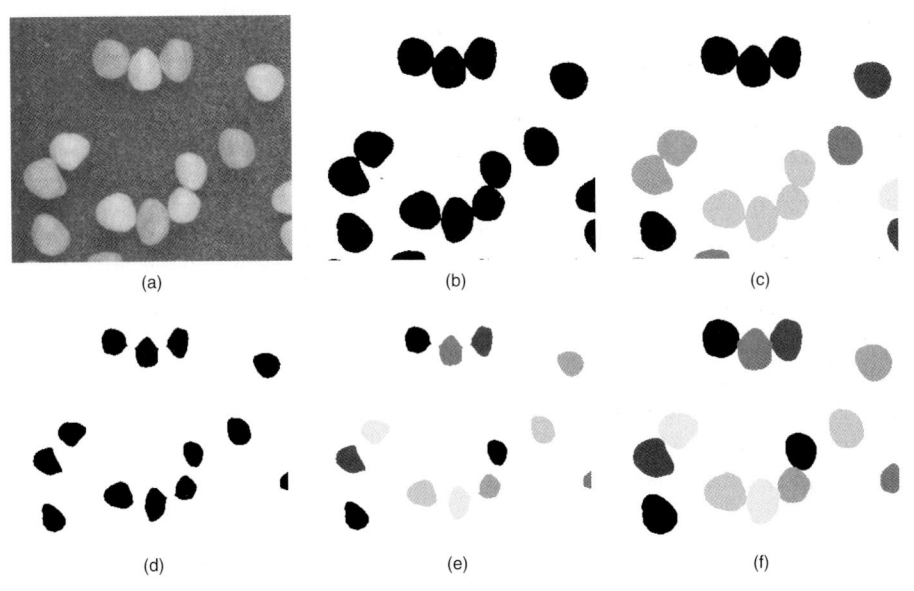

Fig. 3.46 (a) Image of several globular objects. (b) Result of thresholding (a). (c) Connected components of (b) displayed with six different gray values. Note that several objects touch each other and hence are in the same connected component. (d) Result of eroding the region in (b) with a circle of diameter 15. (e) Connected components of (d). Note that each object is now a single connected component. (f) Result of dilating the connected components in (e) with a circle of diameter 15. This transforms the correct connected components into approximately the correct shape.

We can see another use of the erosion if we remember its definition: it returns the translated reference point of the structuring element S for every translation for which S_t completely fits into the region R. Hence, the erosion acts like a template matching operation. An example of this use of the erosion is shown in Figure 3.47. In Figure 3.47(a), we can see an image of a print of several letters with the structuring element used for the erosion overlaid in white. The structuring element corresponds to the center line of the letter "e." The reference point of the structuring element is its center of gravity. The result of eroding the thresholded letters (Figure 3.47(b)) with the structuring element is shown in Figure 3.47(c). Note that all letters "e" have been correctly identified. In Figures 3.47(d)–(f), the experiment is repeated with another set of letters. The structuring element is the center line of the letter "o." Note that the

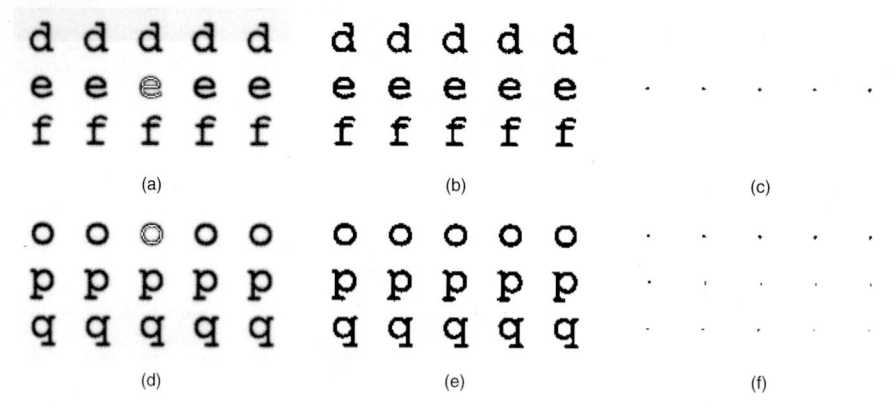

```
d  d  d  d  d       d  d  d  d  d
e  e  ⊜  e  e       e  e  e  e  e       ·    ·    ·    ·    ·
f  f  f  f  f       f  f  f  f  f
      (a)                 (b)                   (c)

o  o  ⊚  o  o       o  o  o  o  o
p  p  p  p  p       p  p  p  p  p       ·    ·    ·    ·    ·
q  q  q  q  q       q  q  q  q  q
      (d)                 (e)                   (f)
```

Fig. 3.47 (a) Image of a print of several characters with the structuring element used for the erosion overlaid in white. (b) Result of thresholding (a). (c) Result of the erosion of (b) with the structuring element in (a). Note that the reference point of all letters "e" has been found. (d) A different set of characters with the structuring element used for the erosion overlaid in white. (e) Result of thresholding (d). (f) Result of the erosion of (e) with the structuring element in (d). Note that the reference point of the letter "o" has been identified correctly. In addition, the circular parts of the letters "p" and "q" have been extracted.

erosion correctly finds the letters "o." However, additionally the circular parts of the letters "p" and "q" are found, since the structuring element completely fits into them.

An interesting property of the Minkowski addition and subtraction as well as the dilation and erosion is that they are dual to each other with respect to the complement operation. For the Minkowski addition and subtraction we have:

$$R \oplus S = \overline{\overline{R} \ominus S} \quad \text{and} \quad R \ominus S = \overline{\overline{R} \oplus S}$$

The same identities hold for the dilation and erosion. Hence, a dilation of the foreground is identical to an erosion of the background and vice versa. We can make use of the duality whenever we want to avoid computing the complement explicitly, and hence to speed up some operations. Note that the duality only holds if the complement can be infinite. Hence, it does not hold for binary images, where the complemented region needs to be clipped to a certain image size.

One extremely useful application of the erosion and dilation is the calculation of the boundary of a region. The algorithm to compute the true boundary as a linked list of contour points is quite complicated [68]. However, an approximation to the boundary can be computed very easily. If we want to compute the inner boundary, we simply need to erode the region appropriately and to subtract the eroded region from the original region:

$$\partial R = R \setminus (R \ominus S) \tag{3.76}$$

By duality, the outer boundary (the inner boundary of the background) can be computed with a dilation:

$$\partial R = (R \oplus S) \setminus R \tag{3.77}$$

To get a suitable boundary, the structuring element S must be chosen appropriately. If we want to obtain an 8-connected boundary, we must use the structuring element S_8 in Figure 3.48. If we want a 4-connected boundary, we must use S_4.

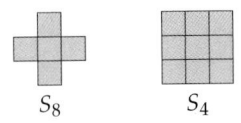

Fig. 3.48 The structuring elements for computing the boundary of a region with 8-connectivity (S_8) and 4-connectivity (S_4).

Figure 3.49 displays an example of the computation of the inner boundary of a region. A small part of the input region is shown in Figure 3.49(a). The boundary of the region computed by Eq. (3.76) with S_8 is shown in Figure 3.49(b), while the result with S_4 is shown in Figure 3.49(c). Note that the boundary is only approximately 8- or 4-connected. For example, in the 8-connected boundary there are occasional 4-connected pixels. Finally, the boundary of the region as computed by an algorithm that traces around the boundary of the region and links the boundary points into contours is shown in Figure 3.49(d). Note that this is the true boundary of the region. Also note that, since only part of the region is displayed, there is no boundary at the bottom of the displayed part.

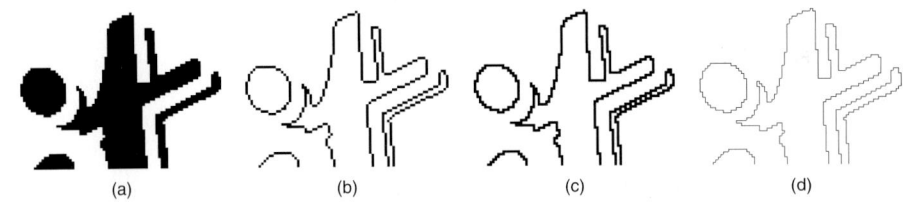

Fig. 3.49 (a) Detail of a larger region. (b) The 8-connected boundary of (a) computed by Eq. (3.76). (c) The 4-connected boundary of (a). (d) Linked contour of the boundary of (a).

As we have seen above, the erosion can be used as a template matching operation. However, sometimes it is not selective enough and returns too many matches. The reason for this is that the erosion does not take into account the background. For this reason, an operation that explicitly models the background is needed. This operation is called the hit-or-miss transform. Since the foreground and background should be taken into account, it uses a structuring element that consists of two parts: $S = (S^f, S^b)$ with $S^f \cap S^b = \emptyset$. With this, the hit-or-miss transform is defined as

$$R \otimes S = (R \ominus \check{S}^f) \cap (\overline{R} \ominus \check{S}^b) = (R \ominus \check{S}^f) \setminus (R \oplus \check{S}^b) \tag{3.78}$$

Hence the hit-or-miss transform returns those translated reference points for which the foreground structuring element S^f completely lies within the foreground and the background structuring element S^b completely lies within the background. The second equation is especially useful from an implementation point of view since it avoids having to compute the complement. The hit-or-miss transform is dual to itself if the foreground and background structuring elements are exchanged: $R \otimes S = \overline{R} \otimes S'$, where $S' = (S^b, S^f)$.

Figure 3.50 shows the same image as Figure 3.47(d). The goal here is to match only the letters "o" in the image. To do so, we can define a structuring element that crosses the vertical strokes of the letters "p" and "q" (and also "b" and "d"). One possible structuring element for this purpose is shown in Figure 3.50(b). With the hit-or-miss transform, we are able to remove the found matches for the letters "p" and "q" from the result, as can be seen from Figure 3.50(c).

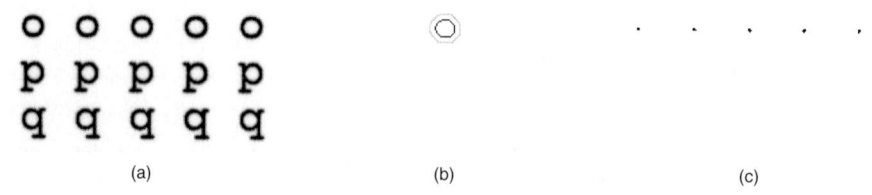

Fig. 3.50 (a) Image of a print of several characters. (b) The structuring element used for the hit-or-miss transform. The black part is the foreground structuring element and the light gray part is the background structuring element. (c) Result of the hit-or-miss transform of the thresholded image (see Figure 3.47(e)) with the structuring element in (b). Note that only the reference point of the letter "o" has been identified, in contrast to the erosion (see Figure 3.47(f)).

We now turn our attention to operations in which the basic operations we have discussed so far are executed in succession. The first such operation is the opening:

$$R \circ S = (R \ominus \check{S}) \oplus S = \bigcup_{S_t \subseteq R} S_t \qquad (3.79)$$

Hence, the opening is an erosion followed by a Minkowski addition with the same structuring element. The second equation tells us that we can visualize the opening by moving the structuring element around the plane. Whenever the structuring element completely lies within the region, we add the entire translated structuring element to the output region (and not just the translated reference point as in the erosion). The opening's definition causes the location of the reference point to cancel out, which can be seen from the second equation. Therefore, the opening is translation-invariant with respect to the structuring element. In contrast to the erosion and dilation, the opening is idempotent, i.e., applying it multiple times has the same effect as applying it once: $(R \circ S) \circ S = R \circ S$.

Like the erosion, the opening can be used as a template matching operation. In contrast to the erosion and hit-or-miss transform, it returns all points of the input region into which the structuring element fits. Hence it preserves the shape of the object to find. An example of this is shown in Figure 3.51, where the same input images and structuring elements as in Figure 3.47 are used. Note that the opening has found the same instances of the structuring elements as the erosion but has preserved the shape of the matched structuring elements. Hence, in this example it also finds the letters "p" and "q." To find only the letters "o", we could combine the hit-or-miss transformation with a Minkowski addition to get a hit-or-miss opening: $R \odot S = (R \otimes S) \oplus S^f$.

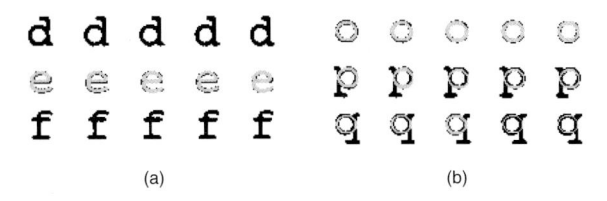

Fig. 3.51 (a) Result of applying an opening with the structuring element in Figure 3.47(a) to the segmented region in Figure 3.47(b). (b) Result of applying an opening with the structuring element in Figure 3.47(d) to the segmented region in Figure 3.47(e). The result of the opening is overlaid in light gray onto the input region, displayed in black. Note that the opening finds the same instances of the structuring elements as the erosion but preserves the shape of the matched structuring elements.

Another very useful property of the opening results if structuring elements like circles or rectangles are used. If an opening with these structuring elements is performed, parts of the region that are smaller than the structuring element are removed from the region. This can be used to remove unwanted appendages from the region and to smooth the boundary of the region by removing small protrusions. Furthermore, small bridges between object parts can be removed, which can be used to separate objects. Finally, the opening can be used to suppress small objects. Figure 3.52 shows an example of using the opening to remove unwanted appendages and small objects from the segmentation. In Figure 3.52(a), an image of a ball-bonded die is shown. The goal is to segment the balls on the pads. If the image is thresholded (Figure 3.52(b)), the wires that are attached to the balls are also extracted. Furthermore, there are extraneous small objects in the segmentation. By performing an opening with a circle of diameter 31, the wires and small objects are removed, and only smooth region parts that correspond to the balls are retained.

The second interesting operation in which the basic morphological operations are executed in succession is the closing:

$$R \bullet S = (R \oplus \check{S}) \ominus S = \bigcup_{S_t \subseteq R} S_t \tag{3.80}$$

Hence, the closing is a dilation followed by a Minkowski subtraction with the same structuring element. There is, unfortunately, no simple formula that tells us how the closing can be visualized. The second formula is actually defined by the duality of the opening and the closing, namely a closing on the foreground is identical to an opening on the background and vice versa:

$$R \bullet S = \overline{\overline{R} \circ S} \quad \text{and} \quad R \circ S = \overline{\overline{R} \bullet S}$$

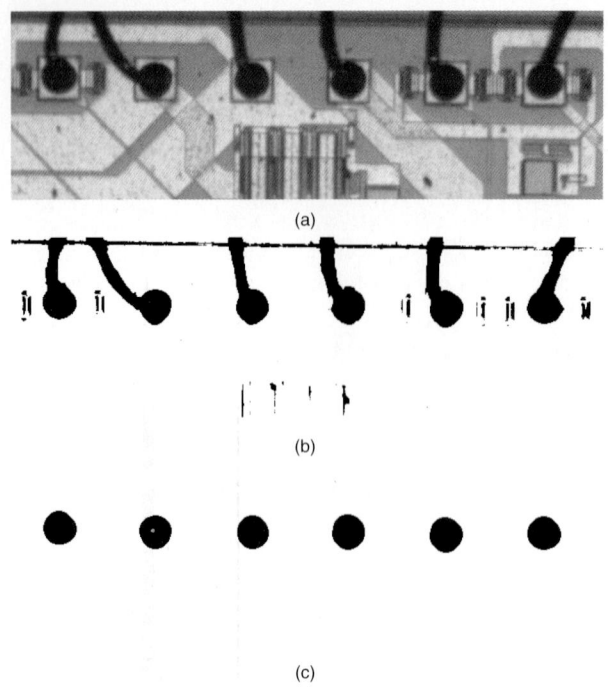

Fig. 3.52 (a) Image of a ball-bonded die. The goal is to segment the balls. (b) Result of thresholding (a). The segmentation includes the wires that are bonded to the pads. (c) Result of performing an opening with a circle of diameter 31. The wires and the other extraneous segmentation results have been removed by the opening and only the balls remain.

Like the opening, the closing is translation-invariant with respect to the structuring element. Furthermore, it is also idempotent.

Since the closing is dual to the opening, it can be used to merge objects that are separated by gaps that are smaller than the structuring element. If structuring elements like circles or rectangles are used, the closing can be used to close holes and to remove indentations that are smaller than the structuring element. The second property enables us to smooth the boundary of the region.

Figure 3.53 shows how the closing can be used to remove indentations in a region. In Figure 3.53(a), a molded plastic part with a protrusion is shown. The goal is to detect the protrusion because it is a production error. Since the actual object is circular, if the entire part were visible the protrusion could be detected by performing an opening with a circle that is almost as large as the object and then subtracting the opened region from the original segmentation. However, only a part of the object is visible, so the erosion in the opening would create artifacts or remove the object entirely. Therefore, by duality we can pursue the opposite approach: we can segment the background and perform a closing on it. Figure 3.53(b) shows the result of thresholding the background. The protrusion is now an indentation in the background. The result of performing a closing with a circle of diameter 801 is shown in Figure 3.53(c). The diameter of the circle was set to 801 because it is large enough to completely fill the indentation and to recover the circular shape of the object. If much smaller circles were used, e.g., with a diameter of 401, the indentation would not be filled completely. To detect the error itself, we can compute the difference between the closing and the original segmentation. To remove some noisy pixels that result because the boundary of the original segmentation is not as smooth as the closed region, the difference can be post-processed with an opening, e.g., with a 5×5 rectangle, to remove the noisy pixels. The resulting error region is shown in Figure 3.53(d).

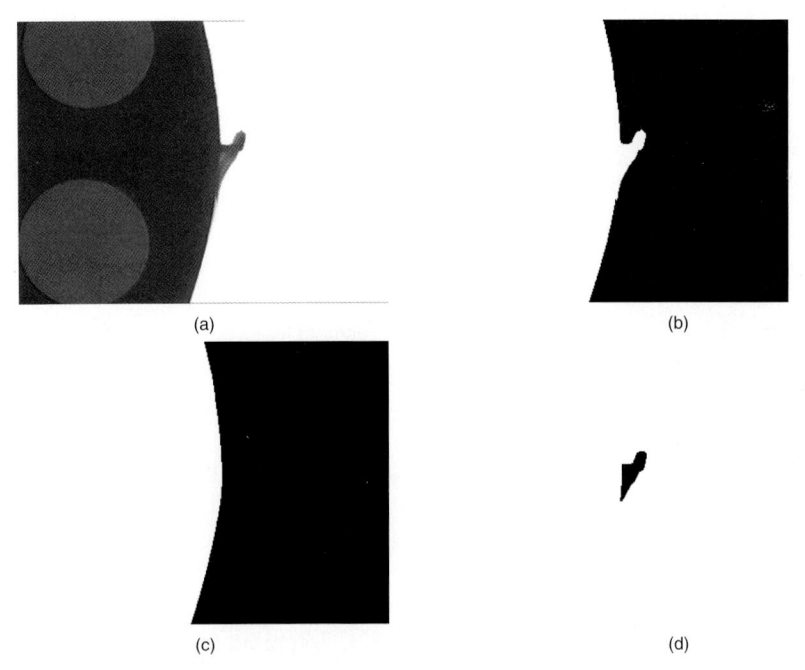

Fig. 3.53 (a) Image of size 768×576 showing a molded plastic part with a protrusion. (b) Result of thresholding the background of (a). (c) Result of a closing on (b) with a circle of diameter 801. Note that the protrusion (the indentation in the background) has been filled in and the circular shape of the plastic part has been recovered. (d) Result of computing the difference between (c) and (b) and performing an opening with a 5×5 rectangle on the difference to remove small parts. The result is the erroneous protrusion of the mould.

The operations we have discussed so far have been mostly concerned with the region as a 2D object. The only exception has been the calculation of the boundary of a region, which reduces a region to its 1D outline, and hence gives a more condensed description of the region. If the objects are mostly linear, i.e., are regions that have a much greater length than width, a more salient description of the object would be obtained if we could somehow capture its one-pixel-wide center line. This center line is called the skeleton or medial axis of the region. Several definitions of a skeleton can be given [80]. One intuitive definition can be obtained if we imagine that we try to fit circles that are as large as possible into the region. More precisely, a circle C is maximal in the region R if there is no other circle in R that is a superset of C. The skeleton then is defined as the set of the centers of the maximal circles. Consequently, a point on the skeleton has at least two different points on the boundary of the region to which it has the same shortest distance. Algorithms to compute the skeleton are given in [80, 82]. They basically can be regarded as sequential hit-or-miss transforms that find points on the boundary of the region that cannot belong to the skeleton and delete them. The goal of the skeletonization is to preserve the homotopy of the region, i.e., the number of connected components and holes. One set of structuring elements for computing an 8-connected skeleton is shown in Figure 3.54 [80]. These structuring elements are used sequentially in all four possible orientations to find pixels with the hit-or-miss transform that can be deleted from the region. The iteration is continued until no changes occur. It should be noted that the skeletonization is an example of an algorithm that can be implemented more efficiently on binary images than on the run-length representation.

Fig. 3.54 The structuring elements for computing an 8-connected skeleton of a region. These structuring elements are used sequentially in all four possible orientations to find pixels that can be deleted.

Figure 3.55(a) shows a part of an image of a printed circuit board with several tracks. The image is threshold (Figure 3.55(b)), and the skeleton of the thresholded region is computed with the above algorithm (Figure 3.55(c)). Note that the skeleton contains several undesirable branches on the upper two tracks. For this reason, many different skeletonization algorithms have been proposed. One algorithm that produces relatively few unwanted branches is described in [83]. The result of this algorithm is shown in Figure 3.55(d). Note that there are no undesirable branches in this case.

The final region morphology operation that we will discuss is the distance transform, which returns an image instead of a region. This image contains, for each point in the region R, the shortest distance to a point outside the region (i.e., to $\overline{R}$). Consequently, all points on the inner boundary of the region have a distance of 1. Typically, the distance of the other points is obtained by considering paths that must be contained in the pixel grid. Thus, the chosen connectivity defines which paths are allowed. If the 4-connectivity is used, the corresponding distance is called the city-block distance. Let (r_1, c_1) and (r_2, c_2) be two points. Then the city-block distance is given by

$$d_4 = |r_2 - r_1| + |c_2 - c_1|$$

Figure 3.56(a) shows the city-block distance between two points. In the example, the city-block distance is 5. On the other hand, if 8-connectivity is used, the corresponding

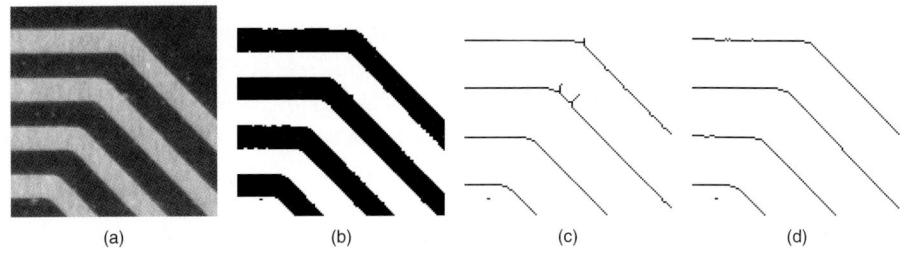

Fig. 3.55 (a) Image showing a part of a PCB with several tracks. (b) Result of thresholding (a). (c) The 8-connected skeleton computed with an algorithm that uses the structuring elements in Figure 3.54 [80]. (d) Result of computing the skeleton with an algorithm that produces fewer skeleton branches [83].

distance is called the chessboard distance. It is given by

$$d_8 = \max\{|r_2 - r_1|, |c_2 - c_1|\}$$

In the example in Figure 3.56(b), the chessboard distance between the two points is 3. Both of these distances are approximations to the Euclidean distance, given by

$$d_e = \sqrt{(r_2 - r_1)^2 + (c_2 - c_1)^2}$$

For the example in Figure 3.56(c), the Euclidean distance is $\sqrt{13}$.

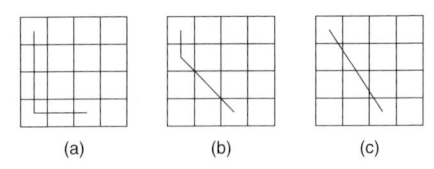

Fig. 3.56 (a) City-block distance between two points. (b) Chessboard distance. (c) Euclidean distance.

Algorithms to compute the distance transform are described in [84]. They work by initializing the distance image outside the region with 0 and within the region with a suitably chosen maximum distance, i.e., $2^b - 1$, where b is the number of bits in the distance image, e.g., $2^{16} - 1$. Then, two sequential line-by-line scans through the image are performed, one from the top left to the bottom right corner, and the second in the opposite direction. In each case, a small mask is placed at the current pixel and the minimum over the elements in the mask of the already computed distances plus the elements in the mask is computed. The two masks are shown in Figure 3.57. If $d_1 = 1$ and $d_2 = \infty$ are used (i.e., d_2 is ignored), the city-block distance is computed. For $d_1 = 1$ and $d_2 = 1$, the chessboard distance results. Interestingly, if $d_1 = 3$ and $d_2 = 4$ is used and the distance image is divided by 3, a very good approximation to the Euclidean distance results, which can be computed solely with integer operations.

This distance is called the chamfer-3-4 distance [84]. With slight modifications, the true Euclidean distance can be computed [85]. The principle is to compute the number of horizontal and vertical steps to reach the boundary using masks similar to the ones in Figure 3.57, and then to compute the Euclidean distance from the number of steps.

d_2	d_1	d_2
d_1	0	

0	d_1	
d_2	d_1	d_2

Fig. 3.57 Masks used in the two sequential scans to compute the distance transform. The left mask is used in the left-to-right, top-to-bottom scan. The right mask is used in the scan in the opposite direction.

The skeleton and the distance transform can be combined to compute the width of linear objects very efficiently. In Figure 3.58(a), a PCB with tracks that have several errors is shown. The protrusions on the tracks are called spurs, while the indentations are called mouse bites [86]. They are deviations from the correct track width. Figure 3.58(b) shows the result of computing the distance transform with the chamfer-3-4 distance on the segmented tracks. The errors are clearly visible in the distance transform. To extract the width of the tracks, we need to calculate the skeleton of the segmented tracks (Figure 3.58(c)). If the skeleton is used as the ROI for the distance image, each point on the skeleton will have the corresponding distance to the border of the track. Since the skeleton is the center line of the track, this distance is the width of the track. Hence, to detect errors, we simply need to threshold the distance image within the skeleton. Note that in this example it is extremely useful that we have defined that images can have an arbitrary region of interest. Figure 3.58(d) shows the result of drawing circles at the centers of gravity of the connected components of the error region. All major errors have been detected correctly.

3.6.2
Gray Value Morphology

Because morphological operations are very versatile and useful, the question of whether they can be extended to gray value images arises quite naturally. This can indeed be done. In analogy to the region morphology, let $g(r, c)$ denote the image that should be processed and let $s(r, c)$ be an image with ROI S. Like in the region morphology, the image s is called the structuring element. The gray value Minkowski addition is then defined as

$$g \oplus s = (g \oplus s)_{r,c} = \max_{(i,j) \in S} \{g_{r-i, c-j} + s_{i,j}\} \quad (3.81)$$

This is a natural generalization because the Minkowski addition for regions is obtained as a special case if the characteristic function of the region is used as the gray value image. If, additionally, an image with gray value 0 within the ROI S is used as the

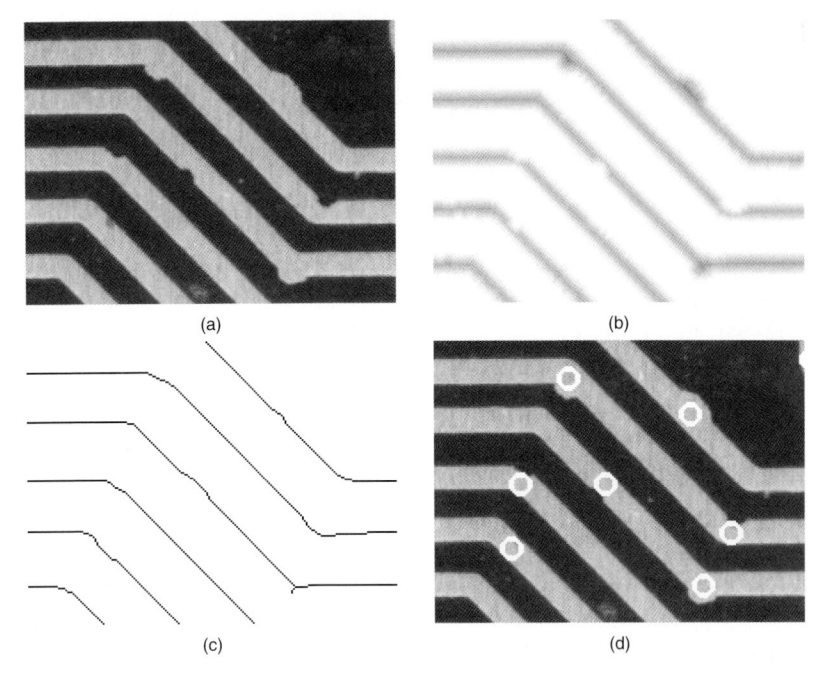

Fig. 3.58 (a) Image showing a part of a PCB with several tracks that have spurs and mouse bites. (b) Distance transform of the result of thresholding (a). The distance image is visualized inverted (dark gray values correspond to large distances). (c) Skeleton of the segmented region. (d) Result of extracting too narrow or too wide parts of the tracks by using (c) as the ROI for (b) and thresholding the distances. The errors are visualized by drawing circles at the centers of gravity of the connected components of the error region.

structuring element, the Minkowski addition becomes

$$g \oplus s = \max_{(i,j) \in S} \{g_{r-i,c-j}\} \quad (3.82)$$

For characteristic functions, the maximum operation corresponds to the union. Furthermore, $g_{r-i,c-j}$ corresponds to the translation of the image by the vector (i, j). Hence, Eq. (3.82) is equivalent to the second formula in Eq. (3.71).

Like in the region morphology, the dilation can be obtained by transposing the structuring element. This results in the following definition:

$$g \oplus \check{s} = (g \oplus \check{s})_{r,c} = \max_{(i,j) \in S} \{g_{r+i,c+j} + s_{i,j}\} \quad (3.83)$$

The typical choice for the structuring element in the gray value morphology is the flat structuring element that was already used above: $s(r, c) = 0$ for $(r, c) \in S$. With this, the gray value dilation has a similar effect as the region dilation: it enlarges the foreground, i.e., parts in the image that are brighter than their surroundings, and shrinks the background, i.e., parts in the image that are darker than their surroundings. Hence, it can be used to connect disjoint parts of a bright object in the gray value image. This is sometimes useful if the object cannot be segmented easily using region operations alone. Conversely, the dilation can be used to split dark objects.

The Minkowski subtraction for gray value images is given by

$$g \ominus s = (g \ominus s)_{r,c} = \min_{(i,j) \in S} \{g_{r-i,c-j} - s_{i,j}\} \tag{3.84}$$

As above, by transposing the structuring element we obtain the gray value erosion:

$$g \ominus \check{s} = (g \ominus \check{s})_{r,c} = \min_{(i,j) \in S} \{g_{r+i,c+j} - s_{i,j}\} \tag{3.85}$$

Like the region erosion, the gray value erosion shrinks the foreground and enlarges the background. Hence, the erosion can be used to split touching bright objects and to connect disjoint dark objects. In fact, the dilation and erosion, as well as the Minkowski addition and subtraction, are dual to each other, like for regions. For the duality, we need to define what the complement of an image should be. If the images are stored with b bits, the natural definition for the complement operation is $\overline{g}_{r,c} = 2^b - 1 - g_{r,c}$. With this, it can be easily shown that the erosion and dilation are dual:

$$g \oplus s = \overline{\overline{g} \ominus s} \quad \text{and} \quad g \ominus s = \overline{\overline{g} \oplus s}$$

Therefore, all the properties that hold for one operation for bright objects hold for the other operation for dark objects and vice versa.

Note that the dilation and erosion can also be regarded as two special rank filters (see Section 3.2.3) if flat structuring elements are used. They select the minimum and maximum gray values within the domain of the structuring element, which can be regarded as the filter mask. Therefore, the dilation and erosion are sometimes referred to as the maximum and minimum filters (or max and min filters).

Efficient algorithms to compute the dilation and erosion are given in [62]. Their runtime complexity is $O(whn)$, where w and h are the dimensions of the image, while n is roughly the number of points on the boundary of the domain of the structuring element for flat structuring elements. For rectangular structuring elements, algorithms with a runtime complexity of $O(wh)$, i.e., with a constant number of operations per pixel, can be found [87]. This is similar to the recursive implementation of a linear filter.

With these building blocks, we can define a gray value opening like for regions as an erosion followed by a Minkowski addition, i.e.

$$g \circ s = (g \ominus \check{s}) \oplus s \tag{3.86}$$

and the closing as a dilation followed by a Minkowski subtraction, i.e.

$$g \bullet s = (g \oplus \check{s}) \ominus s \tag{3.87}$$

The gray value opening and closing have similar properties to their region counterparts. In particular, with the above definition of the complement for images, they are dual to each other:

$$g \circ s = \overline{\overline{g} \bullet s} \quad \text{and} \quad g \bullet s = \overline{\overline{g} \circ s}$$

Like the region operations, they can be used to fill in small holes or, by duality, to remove small objects. Furthermore, they can be used to join or separate objects and to smooth the inner and outer boundaries of objects in the gray value image.

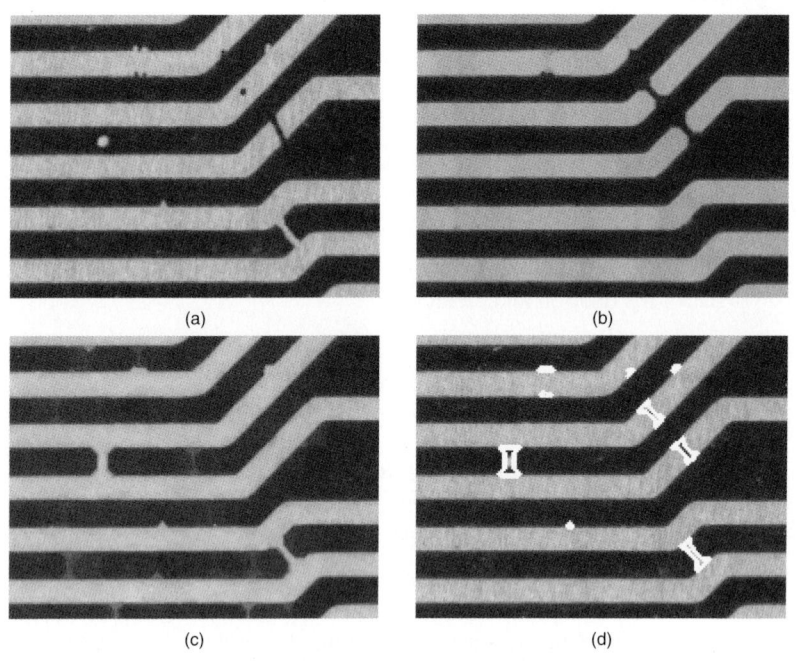

Fig. 3.59 (a) Image showing a part of a PCB with several tracks that have spurs, mouse bites, pinholes, spurious copper, and open and short circuits. (b) Result of performing a gray value opening with an octagon of diameter 11 on (a). (c) Result of performing a gray value closing with an octagon of diameter 11 on (a). (d) Result of segmenting the errors in (a) by using a dynamic threshold operation with the images of (b) and (c).

Figure 3.59 shows how the gray value opening and closing can be used to detect errors in the tracks on a PCB. We have already seen in Figure 3.58 that some of these errors can be detected by looking at the width of the tracks with the distance transform and the skeleton. This technique is very useful because it enables us to detect relatively large areas with errors. However, small errors are harder to detect with this technique because the distance transform and skeleton are only pixel-precise, and consequently the width of the track can only be determined reliably with a precision of two pixels. Smaller errors can be detected more reliably with the gray value morphology. Figure 3.59(a) shows a part of a PCB with several tracks that have spurs, mouse bites, pinholes, spurious copper, and open and short circuits [86]. The result of performing a gray value opening and closing with an octagon of diameter 11 are shown in Figures 3.59(b) and (c). Because of the horizontal, vertical, and diagonal layout of the tracks, using an octagon as the structuring element is preferable. It can be seen that the opening smooths out the spurs, while the closing smooths out the mouse bites. Furthermore, the short circuit and spurious copper are removed by the opening, while the pinhole and open circuit are removed by the closing. To detect these errors, we can require that the opened and closed images should not differ too much. If there were no errors, the differences would solely be caused by the texture on the tracks. Since the gray values of the opened image are always smaller than those of the closed image, we can use the dynamic threshold operation for bright objects (Eq. (3.44)) to perform the required segmentation. Every pixel that has a gray value difference greater than g_{diff} can be considered as an error. Figure 3.59(d) shows the result of segmenting the errors using a dynamic threshold $g_{\text{diff}} = 60$. This detects all the errors on the board.

We conclude this section with an operator that computes the range of gray values that occur within the structuring element. This can be obtained easily by calculating the difference between the dilation and erosion:

$$g \diamond s = (g \oplus š) - (g \ominus š) \tag{3.88}$$

Since this operator produces similar results to a gradient filter (see Section 3.7), it is sometimes called the morphological gradient.

Figure 3.60 shows how the gray range operator can be used to segment punched serial numbers. Because of the scratches, texture, and illumination, it is difficult to segment the characters in Figure 3.60(a) directly. In particular, the scratch next to the upper left part of the "2" cannot be separated from the "2" without splitting several of the other numbers. The result of computing the gray range within a 9×9 rectangle is shown in Figure 3.60(b). With this, it is easy to segment the numbers (Figure 3.60(c)) and to separate them from other segmentation results (Figure 3.60(d)).

3.7
Edge Extraction

In Section 3.4 we discussed several segmentation algorithms. They have in common that they are based on thresholding the image, with either pixel or subpixel accuracy. It is possible to achieve very good accuracies with these approaches, as we saw in Section 3.5. However, in most cases the accuracy of the measurements that we can derive from the segmentation result critically depends on choosing the correct threshold for the segmentation. If the threshold is chosen incorrectly, the extracted objects typically become larger or smaller because of the smooth transition from the foreground to the background gray value. This problem is especially grave if the illumination can change, since in this case the adaptation of the thresholds to the changed illumina-

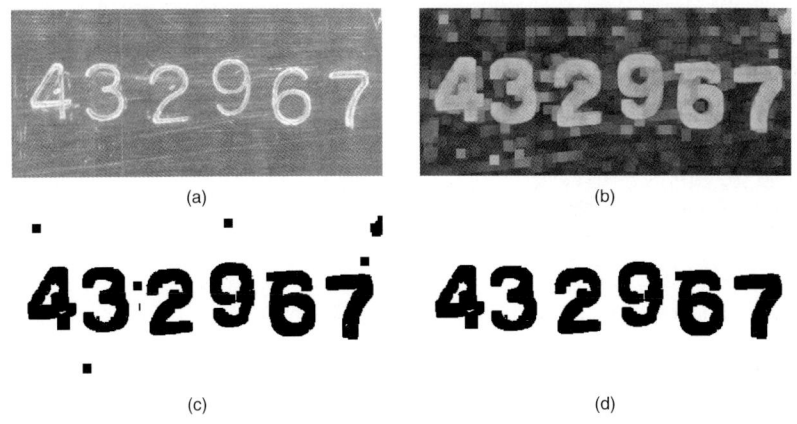

Fig. 3.60 (a) Image showing a punched serial number. Because of the scratches, texture, and illumination, it is difficult to segment the characters directly. (b) Result of computing the gray range within a 9 × 9 rectangle. (c) Result of thresholding (b). (d) Result of computing the connected components of (c) and selecting the characters based on their size.

tion must be very accurate. Therefore, a segmentation algorithm that is robust with respect to illumination changes is extremely desirable. From the above discussion, we see that the boundary of the segmented region or subpixel-precise contour moves if the illumination changes or the thresholds are chosen inappropriately. Therefore, the goal of a robust segmentation algorithm must be to find the boundary of the objects as robustly and accurately as possible. The best way to describe the boundaries of the objects robustly is by regarding them as edges in the image. Therefore, in this section we will examine methods to extract edges.

3.7.1
Definition of Edges in 1D and 2D

To derive an edge extraction algorithm, we need to define what edges actually are. For the moment, let us make the simplifying assumption that the gray values in the object and in the background are constant. In particular, we assume that the image contains no noise. Furthermore, let us assume that the image is not discretized, i.e., is continuous. To illustrate this, Figure 3.61(b) shows an idealized gray value profile across the part of a workpiece that is indicated in Figure 3.61(a).

From the above example, we can see that edges are areas in the image in which the gray values change significantly. To formalize this, let us regard the image for the moment as a 1D function $f(x)$. From elementary calculus we know that the gray values change significantly if the first derivative of $f(x)$ differs significantly from 0: thus $|f'(x)| \gg 0$. Unfortunately, this alone is insufficient to define a unique edge location because there are typically many connected points for which this condition

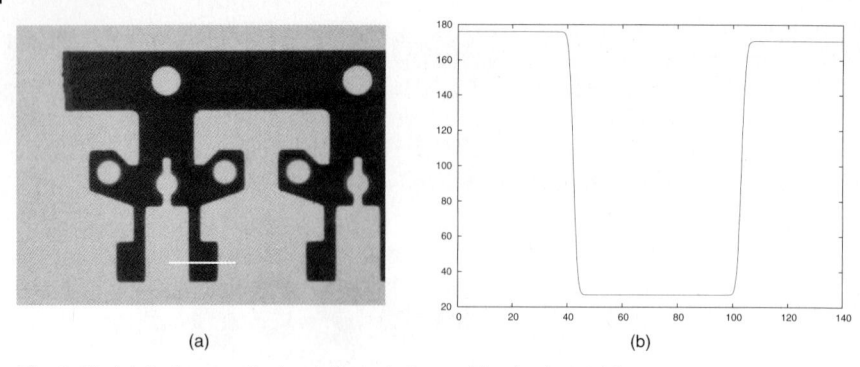

Fig. 3.61 (a) An image of a back-lit workpiece with a horizontal line that indicates the location of the idealized gray value profile in (b).

is true since the transition between the background and foreground gray values is smooth. This can be seen in Figure 3.62(a), where the first derivative $f'(x)$ of the ideal gray value profile in Figure 3.61(b) is displayed. Note, for example, that there is an extended range of points for which $|f'(x)| \geq 20$. Therefore, to obtain a unique edge position, we must additionally require that the absolute value of the first derivative $|f'(x)|$ is locally maximal. This is called non-maximum suppression.

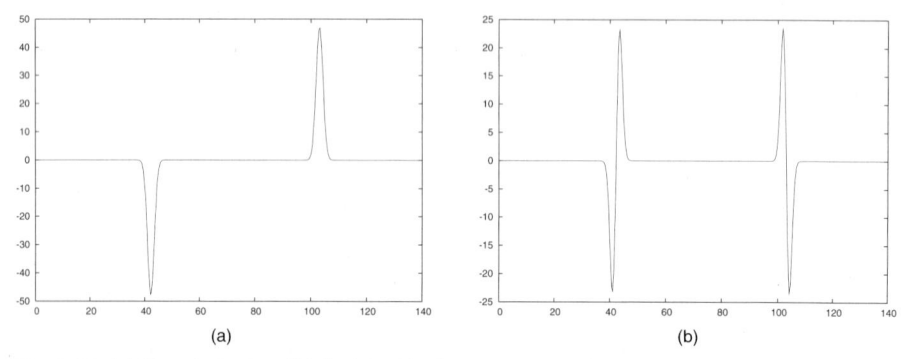

Fig. 3.62 (a) First derivative $f'(x)$ of the ideal gray value profile in Figure 3.61(b). (b) Second derivative $f''(x)$.

From elementary calculus we know that, at the points where $|f'(x)|$ is locally maximal, the second derivative vanishes: $f''(x) = 0$. Hence, edges are given by the locations of inflection points of $f(x)$. To remove flat inflection points, we would additionally have to require that $f'(x)f'''(x) < 0$. However, this restriction is seldom observed. Therefore, in 1D an alternative and equivalent definition to the maxima of the absolute value of the first derivative is to define edges as the locations of the zero crossings of the second derivative. Figure 3.62(b) displays the second derivative $f''(x)$ of the ideal gray value profile in Figure 3.61(b). Clearly, the zero crossings are in the same positions as the maxima of the absolute value of the first derivative in Figure 3.62(a).

From Figure 3.62(a), we can also see that in 1D we can easily associate a polarity with an edge based on the sign of $f'(x)$. We speak of a positive edge if $f'(x) > 0$ and of a negative edge if $f'(x) < 0$.

We now turn to edges in continuous 2D images. Here, the edge itself is a curve $s(t) = (r(t), c(t))$, which is parameterized by a parameter t, e.g., its arc length. At each point of the edge curve, the gray value profile perpendicular to the curve is a 1D edge profile. With this, we can adapt the first 1D edge definition above for the 2D case: we define an edge as the points in the image where the directional derivative in the direction perpendicular to the edge is locally maximal. From differential geometry we know that the direction $n(t)$ perpendicular to the edge curve $s(t)$ is given by $n(t) = s'(t)^\perp \parallel s''(t)$. Unfortunately, the edge definition seemingly requires us to know the edge position $s(t)$ already to obtain the direction perpendicular to the edge, and hence looks like a circular definition. Fortunately, the direction $n(t)$ perpendicular to the edge can be determined easily from the image itself. It is given by the gradient vector of the image, which points into the direction of steepest ascent of the image function $f(r, c)$. The gradient of the image is given by the vector of its first partial derivatives:

$$\nabla f = \nabla f(r,c) = \left(\frac{\partial f(r,c)}{\partial r}, \frac{\partial f(r,c)}{\partial c} \right) = (f_r, f_c) \qquad (3.89)$$

In the last equation, we have used a subscript to denote the partial derivative with respect to the subscripted variable. We will use this convention throughout this section. The Euclidean length

$$\|\nabla f\|_2 = \sqrt{f_r^2 + f_c^2}$$

of the gradient vector is the equivalent of the absolute value of the first derivative $|f'(x)|$ in 1D. We will also call the length of the gradient vector its magnitude. It is also often called the amplitude. The gradient direction is, of course, directly given by the gradient vector. We can also convert it to an angle by calculating $\phi = -\arctan(f_r/f_c)$. Note that ϕ increases in the mathematically positive direction (counterclockwise) starting at the column axis. This is the usual convention. With the above definitions, we can define edges in 2D as the points in the image where the gradient magnitude is locally maximal in the direction of the gradient. To illustrate this definition, Figure 3.63(a) shows a plot of the gray values of an idealized corner. The corresponding gradient magnitude is shown in Figure 3.63(b). The edges are the points at the top of the ridge in the gradient magnitude.

In 1D, we have seen that the second edge definition (the zero crossings of the second derivative) is equivalent to the first definition. Therefore, it is natural to ask whether this definition can be adapted for the 2D case. Unfortunately, there is no direct equivalent for the second derivative in 2D, since there are three partial derivatives of order two. A suitable definition for the second derivative in 2D is the Laplacian operator

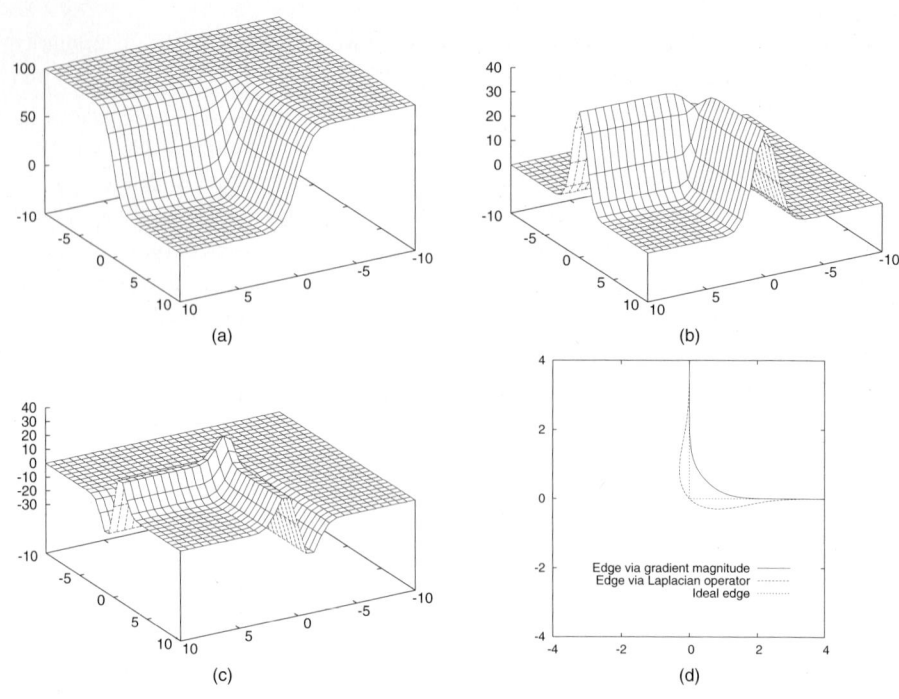

Fig. 3.63 (a) Image of an idealized corner, e.g., one of the corners at the bottom of the workpiece in Figure 3.61(a). (b) Gradient magnitude of (a). (c) Laplacian of (a). (d) Comparison of the edges that result from the two definitions in 2D.

(Laplacian for short), defined by

$$\Delta f = \Delta f(r,c) = \frac{\partial^2 f(r,c)}{\partial r^2} + \frac{\partial^2 f(r,c)}{\partial c^2} = f_{rr} + f_{cc} \qquad (3.90)$$

With this, the edges can be defined as the zero crossings of the Laplacian: $\Delta f(r,c) = 0$. Figure 3.63(c) shows the Laplacian of the idealized corner in Figure 3.63(a). The results of the two edge definitions are shown in Figure 3.63(d). It can be seen that, unlike for the 1D edges, the two definitions do not result in the same edge positions. The edge positions are only identical for straight edges. Whenever the edge is significantly curved, the two definitions return different results. It can be seen that the definition via the maxima of the gradient magnitude always lies inside the ideal corner, whereas the definition via the zero crossings of the Laplacian always lies outside the corner and passes directly through the ideal corner. The Laplacian edge also is in a different position from the true edge for a larger part of the edge. Therefore, in 2D the definition via the maxima of the gradient magnitude is usually preferred. However, in some applications the fact that the Laplacian edge passes through the corner can be used to measure objects with sharp corners more accurately.

3.7.2
1D Edge Extraction

We now turn our attention to edges in real images, which are discrete and contain noise. In this section, we will discuss how to extract edges from 1D gray value profiles. This is a very useful operation that is used frequently in machine vision applications because it is extremely fast. It is typically used to determine the position or diameter of an object.

The first problem we have to address is how to compute the derivatives of the discrete 1D gray value profile. Our first idea might be to use the differences of consecutive gray values on the profile: $f'_i = f_i - f_{i-1}$. Unfortunately, this definition is not symmetric. It would compute the derivative at the "half-pixel" positions $f_{i-\frac{1}{2}}$. A symmetric way to compute the first derivative is given by

$$f'_i = \tfrac{1}{2}(f_{i+1} - f_{i-1}) \tag{3.91}$$

This formula is obtained by fitting a parabola through three consecutive points of the profile and computing the derivative of the parabola at the center point. The parabola is uniquely defined by the three points. With the same mechanism, we can also derive a formula for the second derivative:

$$f''_i = \tfrac{1}{2}(f_{i+1} - 2f_i + f_{i-1}) \tag{3.92}$$

Note that the above methods to compute the first and second derivatives are linear filters, and hence can be regarded as the following two convolution masks:

$$\tfrac{1}{2} \cdot (1 \quad 0 \quad -1) \quad \text{and} \quad \tfrac{1}{2} \cdot (1 \quad -2 \quad 1)$$

Note that the -1 is the last element in the first derivative mask because the elements of the mask are mirrored in the convolution (see Eq. (3.16)).

Figure 3.64(a) displays the true gray value profile taken from the horizontal line in the image in Figure 3.61(a). Its first derivative, computed with Eq. (3.91), is shown in Figure 3.64(b). We can see that the noise in the image causes a very large number of local maxima in the absolute value of the first derivative, and consequently also a large number of zero crossings in the second derivative. The salient edges can easily be selected by thresholding the absolute value of the first derivative: $|f'_i| \geq t$. For the second derivative, the edges cannot be selected as easily. In fact, we have to resort to calculating the first derivative as well to be able to select the relevant edges. Hence, the edge definition via the first derivative is preferable because it can be done with one filter operation instead of two, and consequently the edges can be extracted much faster.

The gray value profile in Figure 3.64(a) already contains relatively little noise. Nevertheless, in most cases it is desirable to suppress the noise even further. If the object we are measuring has straight edges in the part in which we are performing the measurement, we can use the gray values perpendicular to the line along which we are

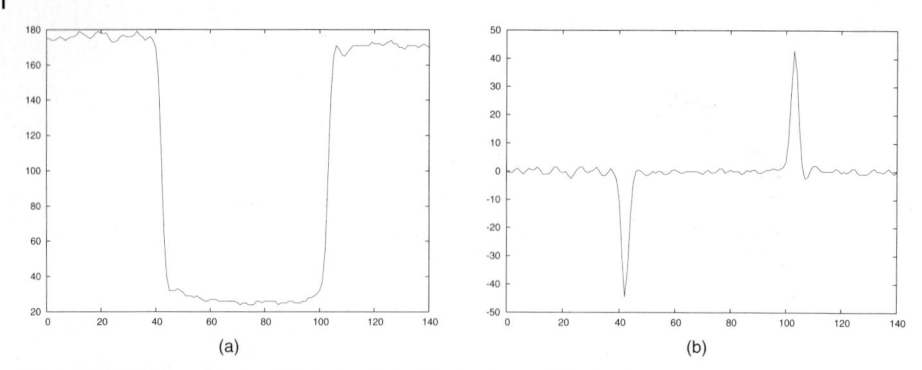

Fig. 3.64 (a) Gray value profile taken from the horizontal line in the image in Figure 3.61(a). (b) First derivative f'_i of the gray value profile.

extracting the gray value profile and average them in a suitable manner. The simplest way to do this is to compute the mean of the gray values perpendicular to the line. If, for example, the line along which we are extracting the gray value profile is horizontal, we can calculate the mean in the vertical direction as follows:

$$f_i = \frac{1}{2m+1} \sum_{j=-m}^{m} f_{r+j,c+i} \tag{3.93}$$

This acts like a mean filter in one direction. Hence, the noise variance is reduced by a factor of $2m+1$. Of course, we could also use a 1D Gaussian filter to average the gray values. However, since this would require larger filter masks for the same noise reduction, and consequently would lead to longer execution times, in this case the mean filter is preferable.

If the line along which we want to extract the gray value profile is horizontal or vertical, the calculation of the profile is simple. If we want to extract the profile from inclined lines or from circles or ellipses, the computation is slightly more difficult. To enable meaningful measurements for distances, we must sample the line with a fixed distance, typically one pixel. Then, we need to generate lines perpendicular to the curve along which we want to extract the profile. This procedure is shown for an inclined line in Figure 3.65. Because of this, the points from which we must extract the gray values typically do not lie on pixel centers. Therefore, we will have to interpolate them. This can be done with the techniques discussed in Section 3.3.3, i.e., with nearest-neighbor or bilinear interpolation.

Figure 3.66 shows a gray value profile and its first derivative obtained by vertically averaging the gray values along the line shown in Figure 3.61(a). The size of the 1D mean filter was 21 pixels in this case. If we compare this with Figure 3.64, which shows the profile obtained from the same line without averaging, we can see that the noise in the profile has been reduced significantly. Because of this, the salient edges are even easier to select than without the averaging.

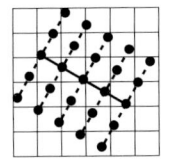

Fig. 3.65 Creation of the gray value profile from an inclined line. The line is visualized by the heavy solid line. The circles indicate the points that are used to compute the profile. Note that they do not lie on the pixel centers. The direction in which the 1D mean is computed is visualized by the dashed lines.

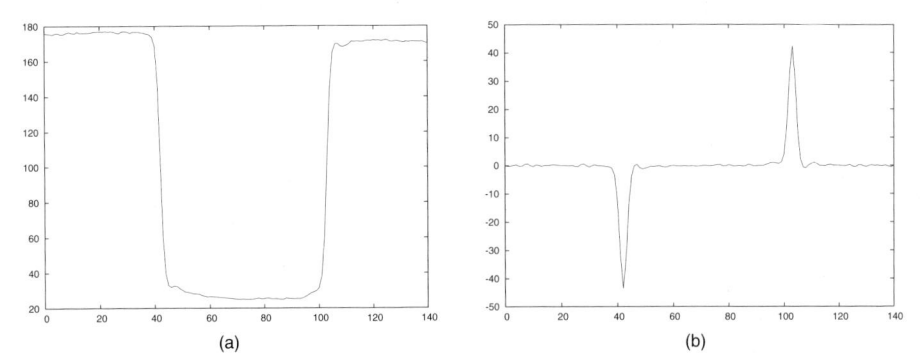

Fig. 3.66 (a) Gray value profile taken from the horizontal line in the image in Figure 3.61(a) and averaged vertically over 21 pixels. (b) First derivative f'_i of the gray value profile.

Unfortunately, the averaging perpendicular to the curve along which the gray value profile is extracted is sometimes insufficient to smooth the profiles enough to enable us to extract the relevant edges easily. One example is shown in Figure 3.67. Here, the object to be measured has a significant amount of texture, which is not as random as noise and consequently does not average out completely. Note that, on the right side of the profile, there is a negative edge with an amplitude almost as large as the edges we want to extract. Another reason for the noise not to cancel out completely may be that we cannot choose the size of the averaging large enough, e.g., because the object's boundary is curved.

To solve these problems, we must smooth the gray value profile itself to suppress the noise even further. This is done by convolving the profile with a smoothing filter: $f_s = f * h$. We can then extract the edges from the smoothed profile via its first derivative. This would involve two convolutions: one for the smoothing filter, and one for the derivative filter. Fortunately, the convolution has a very interesting property that we can use to save one convolution. The derivative of the smoothed function is identical to the convolution of the function with the derivative of the smoothing filter: $(f * h)' = f * h'$. We can regard h' as an edge filter.

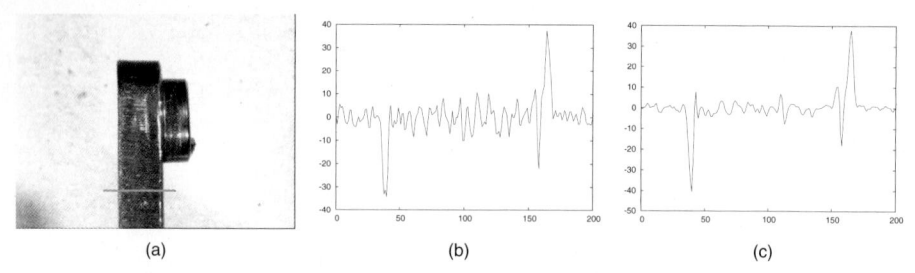

Fig. 3.67 (a) An image of a relay with a horizontal line that indicates the location of the gray value profile. (b) First derivative of the gray value profile without averaging. (c) First derivative of the gray value profile with vertical averaging over 21 pixels.

Like for the smoothing filters, the natural question to ask is which edge filter is optimal. This problem was addressed by Canny [88]. He proposed three criteria that an edge detector should fulfill. First of all, it should have a good detection quality, i.e., it should have a low probability of falsely detecting an edge point and also a low probability of erroneously missing an edge point. This criterion can be formalized as maximizing the signal-to-noise ratio of the output of the edge filter. Second, the edge detector should have good localization quality, i.e., the extracted edges should be as close as possible to the true edges. This can be formalized by minimizing the variance of the extracted edge positions. Finally, the edge detector should return only a single edge for each true edge, i.e., it should avoid multiple responses. This criterion can be formalized by maximizing the distance between the extracted edge positions. Canny then combined these three criteria into one optimization criterion and solved it using the calculus of variations. To do so, he assumed that the edge filter has a finite extent (mask size). Since adapting the filter to a particular mask size involves solving a relatively complex optimization problem, Canny looked for a simple filter that could be written in closed form. He found that the optimal edge filter can be approximated very well with the first derivative of the Gaussian filter:

$$g'_\sigma(x) = \frac{-x}{\sqrt{2\pi}\,\sigma^3} e^{-x^2/(2\sigma^2)} \tag{3.94}$$

One drawback of using the true derivative of the Gaussian filter is that the edge amplitudes become progressively smaller as σ is increased. Ideally, the edge filter should return the true edge amplitude independent of the smoothing. To achieve this for an idealized step edge, the output of the filter must be multiplied with $\sqrt{2\pi}\,\sigma$.

Note that the optimal smoothing filter would be the integral of the optimal edge filter, i.e., the Gaussian smoothing filter. It is interesting to note that, like the criteria in Section 3.2.3, Canny's formulation indicates that the Gaussian filter is the optimal smoothing filter.

Since the Gaussian filter and its derivatives cannot be implemented recursively (see Section 3.2.3), Deriche used Canny's approach to find optimal edge filters that can be implemented recursively [89]. He derived the following two filters:

$$d'_\alpha(x) = -\alpha^2 x e^{-\alpha|x|}$$
$$e'_\alpha(x) = -2\alpha \sin(\alpha x) e^{-\alpha|x|} \qquad (3.95)$$

The corresponding smoothing filters are:

$$d_\alpha(x) = \tfrac{1}{4}\alpha(\alpha|x|+1) e^{-\alpha|x|}$$
$$e_\alpha(x) = \tfrac{1}{2}\alpha[\sin(\alpha|x|) + \cos(\alpha|x|)] e^{-\alpha|x|} \qquad (3.96)$$

Note that, in contrast to the Gaussian filter, where larger values for σ indicate more smoothing, smaller values for α indicate more smoothing in the Deriche filters. The Gaussian filter has similar effects to the first Deriche filter for $\sigma = \sqrt{\pi}/\alpha$. For the second Deriche filter, the relation is $\sigma = \sqrt{\pi}/(2\alpha)$. Note that the Deriche filters are significantly different from the Canny filter. This can also be seen from Figure 3.68, which compares the Canny and Deriche smoothing and edge filters with equivalent filter parameters.

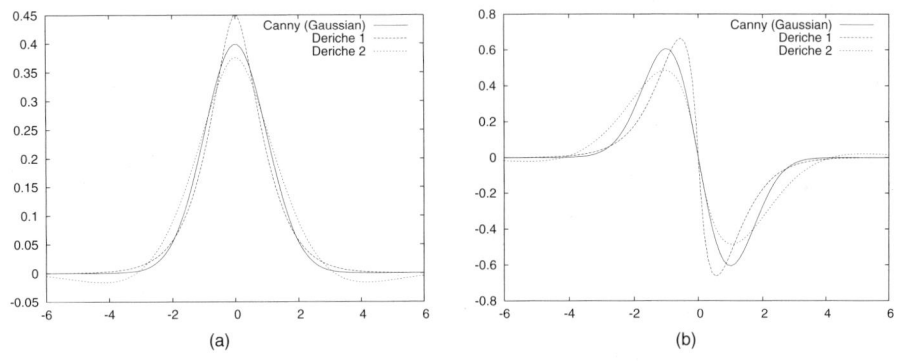

Fig. 3.68 Comparison of the Canny and Deriche filters. (a) Smoothing filters. (b) Edge filters.

Figure 3.69 shows the result of using the Canny edge detector with $\sigma = 1.5$ to compute the smoothed first derivative of the gray value profile in Figure 3.67(a). Like in Figure 3.67(c), the profile was obtained by averaging over 21 pixels vertically. Note that the amplitude of the unwanted edge on the right side of the profile has been reduced significantly. This enables us to select the salient edges more easily.

To extract the edge position, we need to perform the non-maximum suppression. If we are only interested in the edge positions with pixel accuracy, we can proceed as follows. Let the output of the edge filter be denoted by $e_i = |f * h'|_i$, where h' denotes one of the above edge filters. Then, the local maxima of the edge amplitude are given by the points for which $e_i > e_{i-1} \wedge e_i > e_{i+1} \wedge e_i \geq t$, where t is the threshold to select the relevant edges.

Unfortunately, extracting the edges with pixel accuracy is often not accurate enough. To extract edges with subpixel accuracy, we can note that around the max-

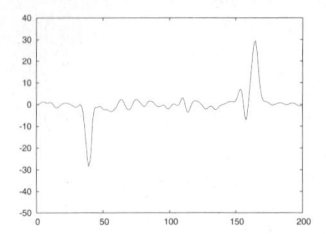

Fig. 3.69 Result of applying the Canny edge filter with $\sigma = 1.5$ to the gray value profile in Figure 3.67(a) with vertical averaging over 21 pixels.

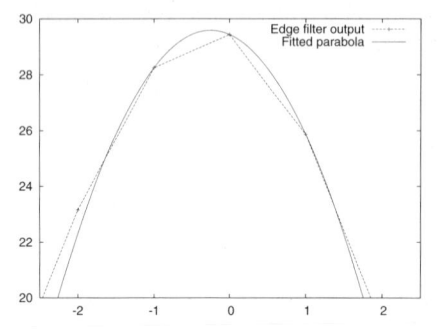

Fig. 3.70 Principle of extracting edge points with subpixel accuracy. The local maximum of the edge amplitude is detected. Then, a parabola is fitted through the three points around the maximum. The maximum of the parabola is the subpixel-accurate edge location. The edge amplitude was taken from the right edge in Figure 3.69.

imum the edge amplitude can be approximated well with a parabola. Figure 3.70 illustrates this by showing a zoomed part of the edge amplitude around the right edge in Figure 3.69. If we fit a parabola through three points around the maximum edge amplitude and calculate the maximum of the parabola, we can obtain the edge position with subpixel accuracy. If an ideal camera system is assumed, this algorithm is as accurate as the precision with which the floating-point numbers are stored in the computer [90].

We conclude the discussion of the 1D edge extraction by showing the results of the edge extraction on the two examples we have used so far. Figure 3.71(a) shows the edges that have been extracted along the line shown in Figure 3.61(a) with the Canny filter with $\sigma = 1.0$. From the two zoomed parts around the extracted edge positions, we can see that by coincidence both edges lie very close to the pixel centers. Figure 3.71(b) displays the result of extracting edges along the line shown in Figure 3.67(a) with the Canny filter with $\sigma = 1.5$. In this case, the left edge is almost exactly in the middle of two pixel centers. Hence, we can see that the algorithm is successful in extracting the edges with subpixel precision.

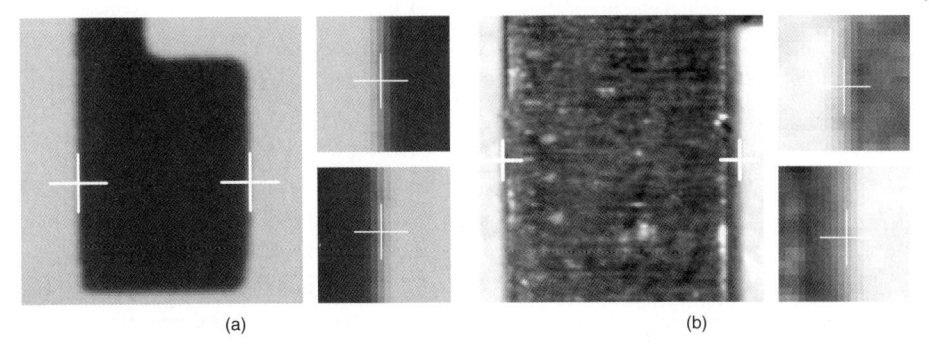

Fig. 3.71 (a) Result of extracting 1D edges along the line shown in Figure 3.61(a). The two small images show a zoomed part around the edge positions. In this case, they both lie very close to the pixel centers. The distance between the two edges is 60.95 pix- els. (b) Result of extracting 1D edges along the line shown in Figure 3.67(a). Note that the left edge, shown in detail in the upper right image, is almost exactly in the middle between two pixel centers. The distance between the two edges is 125.37 pixels.

3.7.3
2D Edge Extraction

As discussed in Section 3.7.1, there are two possible definitions for edges in 2D, which are not equivalent. Like in the 1D case, the selection of salient edges will require us to perform a thresholding based on the gradient magnitude. Therefore, the definition via the zero crossings of the Laplacian requires us to compute more partial derivatives than the definition via the maxima of the gradient magnitude. Consequently, we will concentrate on the maxima of the gradient magnitude for the 2D case. We will add some comments on the zero crossings of the Laplacian at the end of this section.

As in the 1D case, the first question we need to answer is how to compute the partial derivatives of the image that are required to calculate the gradient. Similar to Eq. (3.91), we could use finite differences to calculate the partial derivatives. In 2D, they would be

$$f_{r;i,j} = \tfrac{1}{2}(f_{i+1,j} - f_{i-1,j}) \quad \text{and} \quad f_{c;i,j} = \tfrac{1}{2}(f_{i,j+1} - f_{i,j-1})$$

However, as we have seen above, typically the image must be smoothed to obtain good results. For time-critical applications, the filter masks should be as small as possible, i.e., 3×3. All 3×3 edge filters can be brought into the following form by scaling the coefficients appropriately (note that the filter masks are mirrored in the convolution):

$$\begin{pmatrix} 1 & 0 & -1 \\ a & 0 & -a \\ 1 & 0 & -1 \end{pmatrix} \quad \begin{pmatrix} 1 & a & 1 \\ 0 & 0 & 0 \\ -1 & -a & -1 \end{pmatrix} \quad (3.97)$$

If we use $a = 1$ we obtain the Prewitt filter. Note that it performs a mean filter perpendicular to the derivative direction. For $a = \sqrt{2}$ the Frei filter is obtained; while for $a = 2$ we obtain the Sobel filter, which performs an approximation to a Gaussian smoothing perpendicular to the derivative direction. Of the above three filters, the Sobel filter returns the best results because it uses the best smoothing filter. Interestingly enough, 3 × 3 filters are still actively investigated. For example, Ando [91] has recently proposed an edge filter that tries to minimize the artifacts that invariably are obtained with small filter masks. In our notation, his filter would correspond to $a = 2.435\,101$. Unfortunately, like the Frei filter, it requires floating-point calculations, which makes it unattractive for time-critical applications.

The 3 × 3 edge filters are primarily used to quickly find edges with moderate accuracy in images of relatively good quality. Since speed is important and the calculation of the gradient magnitude via the Euclidean length (the 2-norm) of the gradient vector ($\|\nabla f\|_2 = \sqrt{f_r^2 + f_c^2}$) requires an expensive square root calculation, the gradient magnitude is typically computed by one of the following norms: the 1-norm $\|\nabla f\|_1 = |f_r| + |f_c|$, or the maximum norm $\|\nabla f\|_\infty = \max(|f_r|, |f_c|)$. Note that the first norm corresponds to the city-block distance in the distance transform, while the second norm corresponds to the chessboard distance (see Section 3.6.1). Furthermore, the non-maximum suppression also is relatively expensive and is often omitted. Instead, the gradient magnitude is simply thresholded. Because this results in edges that are wider than one pixel, the thresholded edge regions are skeletonized. Note that this implicitly assumes that the edges are symmetric.

Figure 3.72 shows an example where this simple approach works quite well because the image is of good quality. Figure 3.72(a) displays the edge amplitude around the leftmost hole of the workpiece in Figure 3.61(a) computed with the Sobel filter and the 1-norm. The edge amplitude is thresholded (Figure 3.72(b)) and the skeleton of the resulting region is computed (Figure 3.72(c)). Since the assumption that the edges are symmetric is fulfilled in this example, the resulting edges are in the correct location.

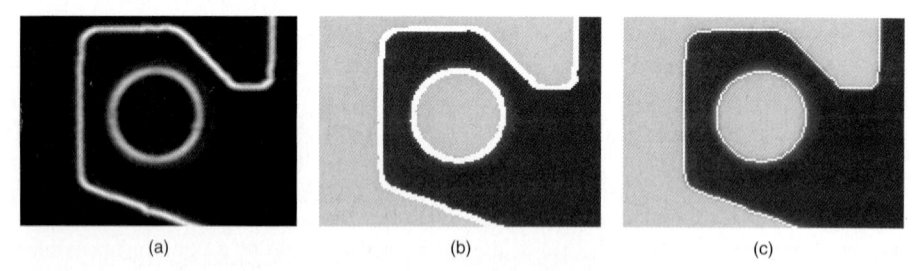

(a)　　　　　　　　　　(b)　　　　　　　　　　(c)

Fig. 3.72 (a) Edge amplitude around the leftmost hole of the workpiece in Figure 3.61(a) computed with the Sobel filter and the 1-norm. (b) Thresholded edge region. (c) Skeleton of (b).

This approach fails to produce good results on the more difficult image of the relay in Figure 3.67(a). As can be seen from Figure 3.73(a), the texture on the relay causes many areas with high gradient magnitude, which are also present in the segmentation (Figure 3.73(b)) and the skeleton (Figure 3.73(c)). Another interesting thing to note is that the vertical edge at the right corner of the top edge of the relay is quite blurred and

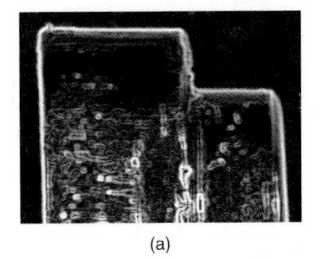

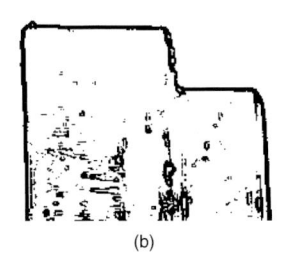

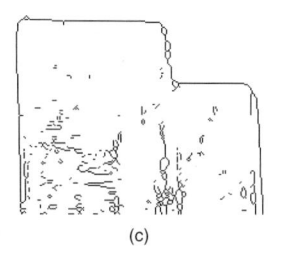

Fig. 3.73 (a) Edge amplitude around the top part of the relay in Figure 3.67(a) computed with the Sobel filter and the 1-norm. (b) Thresholded edge region. (c) Skeleton of (b).

asymmetric. This produces holes in the segmented edge region, which are exacerbated by the skeletonization.

Because the 3×3 filters are not robust against noise and other disturbances, e.g., textures, we need to adapt the approach to optimal 1D edge extraction described in the previous section to the 2D case. In 2D, we can derive the optimal edge filters by calculating the partial derivatives of the optimal smoothing filters, since the properties of the convolution again allow us to move the derivative calculation into the filter. Consequently, Canny's optimal edge filters in 2D are given by the partial derivatives of the Gaussian filter. Because the Gaussian filter is separable, so are its derivatives:

$$g_r = \sqrt{2\pi}\,\sigma g'_\sigma(r)g_\sigma(c) \quad \text{and} \quad g_c = \sqrt{2\pi}\,\sigma g_\sigma(r)g'_\sigma(c)$$

(see the discussion following Eq. (3.94) for the factors of $\sqrt{2\pi}\,\sigma$). To adapt the Deriche filters to the 2D case, the separability of the filters is postulated. Hence, the optimal 2D Deriche filters are given by $d'_\alpha(r)d_\alpha(c)$ and $d_\alpha(r)d'_\alpha(c)$ for the first Deriche filter, and by $e'_\alpha(r)e_\alpha(c)$ and $e_\alpha(r)e'_\alpha(c)$ for the second Deriche filter (see Eq. (3.95)).

The advantage of the Canny filter is that it is isotropic, i.e., rotation-invariant (see Section 3.2.3). Its disadvantage is that it cannot be implemented recursively. Therefore, the execution time depends on the amount of smoothing specified by σ. The Deriche filters, on the other hand, can be implemented recursively, and hence their runtime is independent of the smoothing parameter α. However, they are anisotropic, i.e., the edge amplitude they calculate depends on the angle of the edge in the image. This is undesirable because it makes the selection of the relevant edges harder. Lanser has shown that the anisotropy of the Deriche filters can be corrected [92]. We will refer to the isotropic versions of the Deriche filters as the Lanser filters.

Figure 3.74 displays the result of computing the edge amplitude with the second Lanser filter with $\alpha = 0.5$. Compared to the results with the Sobel filter, the Lanser filter was able to suppress the noise and texture significantly better. This can be seen from the edge amplitude image (Figure 3.74(a)) as well as the thresholded edge region (Figure 3.74(b)). Note, however, that the edge region still contains a hole for the vertical edge that starts at the right corner of the topmost edge of the relay. This happens because the edge amplitude has only been thresholded and the important step of the non-maximum suppression has been omitted to compare the results of the Sobel and Lanser filters.

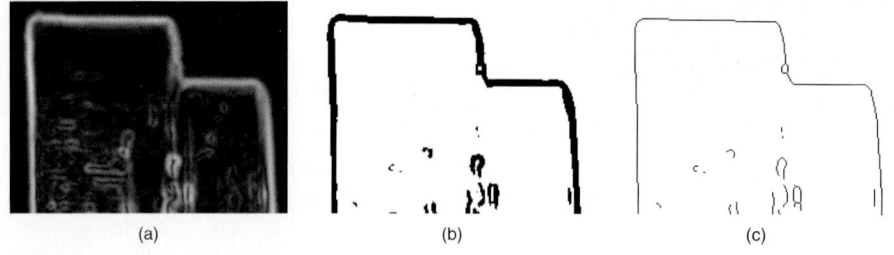

Fig. 3.74 (a) Edge amplitude around the top part of the relay in Figure 3.67(a) computed with the second Lanser filter with $\alpha = 0.5$. (b) Thresholded edge region. (c) Skeleton of (b).

As we saw in the above examples, thresholding the edge amplitude and then skeletonizing the region sometimes does not yield the desired results. To obtain the correct edge locations, we must perform the non-maximum suppression (see Section 3.7.1). In the 2D case, this can be done by examining the two neighboring pixels that lie closest to the gradient direction. Conceptually, we can think of transforming the gradient vector into an angle. Then, we divide the angle range into eight sectors. Figure 3.75 shows two examples of this. Unfortunately, with this approach, diagonal edges often are still two pixels wide. Consequently, the output of the non-maximum suppression must still be skeletonized.

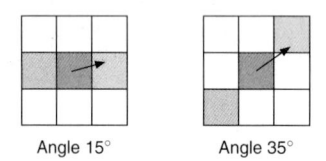

Angle 15°   Angle 35°

Fig. 3.75 Examples of the pixels that are examined in the non-maximum suppression for different gradient directions.

Figure 3.76 shows the result of applying the non-maximum suppression to the edge amplitude image in Figure 3.74(a). From the thresholded edge region in Figure 3.76(b), it can be seen that the edges are now in the correct locations. In particular, the incorrect hole in Figure 3.74 is no longer present. We can also see that the few diagonal edges are sometimes two pixels wide. Therefore, their skeleton is computed and displayed in Figure 3.76(c).

Up to now, we have been using simple thresholding to select the salient edges. This works well as long as the edges we are interested in have roughly the same contrast or have a contrast that is significantly different from the contrast of noise, texture, or other irrelevant objects in the image. In many applications, however, we face the problem that, if we select the threshold so high that only the relevant edges are selected, they are often fragmented. If, on the other hand, we set the threshold so low

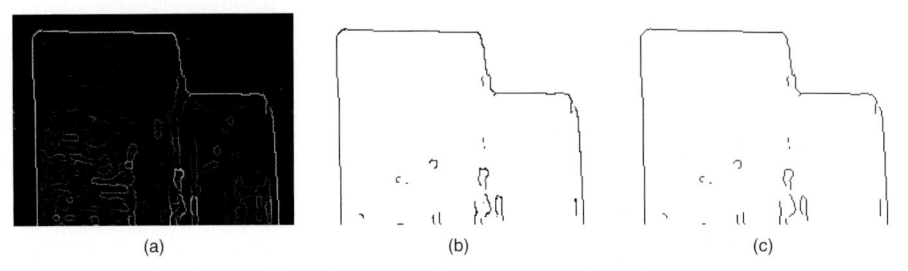

Fig. 3.76 (a) Result of applying the non-maximum suppression to the edge amplitude image in Figure 3.74(a). (b) Thresholded edge region. (c) Skeleton of (b).

that the edges we are interested in are not fragmented, we end up with many irrelevant edges. These two situations are illustrated in Figures 3.77(a) and (b). A solution to this problem was proposed by Canny [88]. He devised a special thresholding algorithm for segmenting edges: the hysteresis thresholding. Instead of a single threshold, it uses two thresholds. Points with an edge amplitude greater than the higher threshold are immediately accepted as safe edge points. Points with an edge amplitude smaller than the lower threshold are immediately rejected. Points with an edge amplitude between the two thresholds are only accepted if they are connected to safe edge points via a path in which all points have an edge amplitude above the lower threshold. We can also think of this operation as first selecting the edge points with an amplitude above the upper threshold, and then extending the edges as far as possible while remaining above the lower threshold. Figure 3.77(c) shows that the hysteresis thresholding enables us to select only the relevant edges without fragmenting them or missing edge points.

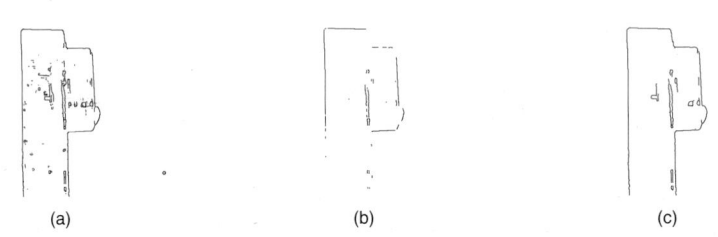

Fig. 3.77 (a) Result of thresholding the edge amplitude for the entire relay image in Figure 3.67(a) with a threshold of 60. This causes many irrelevant texture edges to be selected. (b) Result of thresholding the edge amplitude with a threshold of 140. This only selects relevant edges. However, they are severely fragmented and incomplete. (c) Result of hysteresis thresholding with a low threshold of 60 and a high threshold of 140. Only the relevant edges are selected, and they are complete.

Like in the 1D case, the pixel-accurate edges we have extracted so far are often not accurate enough. We can use a similar approach as for 1D edges to extract edges with subpixel accuracy: we can fit a 2D polynomial to the edge amplitude and extract its maximum in the direction of the gradient vector [90, 93]. The fitting of the polynomial can be done with convolutions with special filter masks (so-called facet model masks) [68, 94]. To illustrate this, Figure 3.78(a) shows a 7×7 part of an edge amplitude image. The fitted 2D polynomial obtained from the central 3×3 amplitudes is shown in Figure 3.78(b), along with an arrow that indicates the gradient direction. Furthermore, contour lines of the polynomial are shown. They indicate that the edge point is offset by approximately a quarter of a pixel in the direction of the arrow.

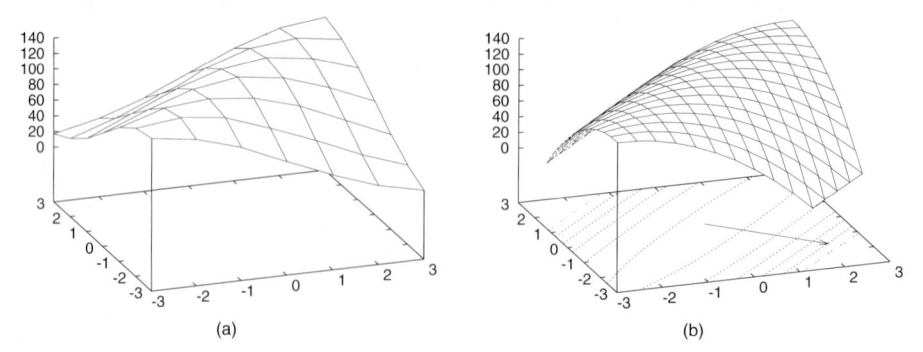

Fig. 3.78 (a) A 7×7 part of an edge amplitude image. (b) Fitted 2D polynomial obtained from the central 3×3 amplitudes in (a). The arrow indicates the gradient direction. The contour lines in the plot indicate that the edge point is offset by approximately a quarter of a pixel in the direction of the arrow.

The above procedure gives us one subpixel-accurate edge point per non-maximum suppressed pixel. These individual edge points must be linked into subpixel-precise contours. This can be done by repeatedly selecting the first unprocessed edge point to start the contour and then successively finding adjacent edge points until the contour closes, reaches the image border, or reaches an intersection point.

Figure 3.79 illustrates the subpixel edge extraction along with a very useful strategy to increase the processing speed. The image in Figure 3.79(a) is the same workpiece as in Figure 3.61(a). Because the subpixel edge extraction is relatively costly, we want to reduce the search space as much as possible. Since the workpiece is back-lit, we can threshold it easily (Figure 3.79(b)). If we calculate the inner boundary of the region with Eq. (3.76), the resulting points are close to the edge points we want to extract. We only need to dilate the boundary slightly, e.g., with a circle of diameter 5 (Figure 3.79(c)), to obtain a region of interest for the edge extraction. Note that the ROI is only a small fraction of the entire image. Consequently, the edge extraction can be done an order of magnitude faster than in the entire image, without any loss of information. The resulting subpixel-accurate edges are shown in Figure 3.79(d) for the part of the image indicated by the rectangle in Figure 3.79(a). Note how well they capture the shape of the hole.

We conclude this section with a look at the second edge definition via the zero crossings of the Laplacian. Since the zero crossings are just a special threshold, we

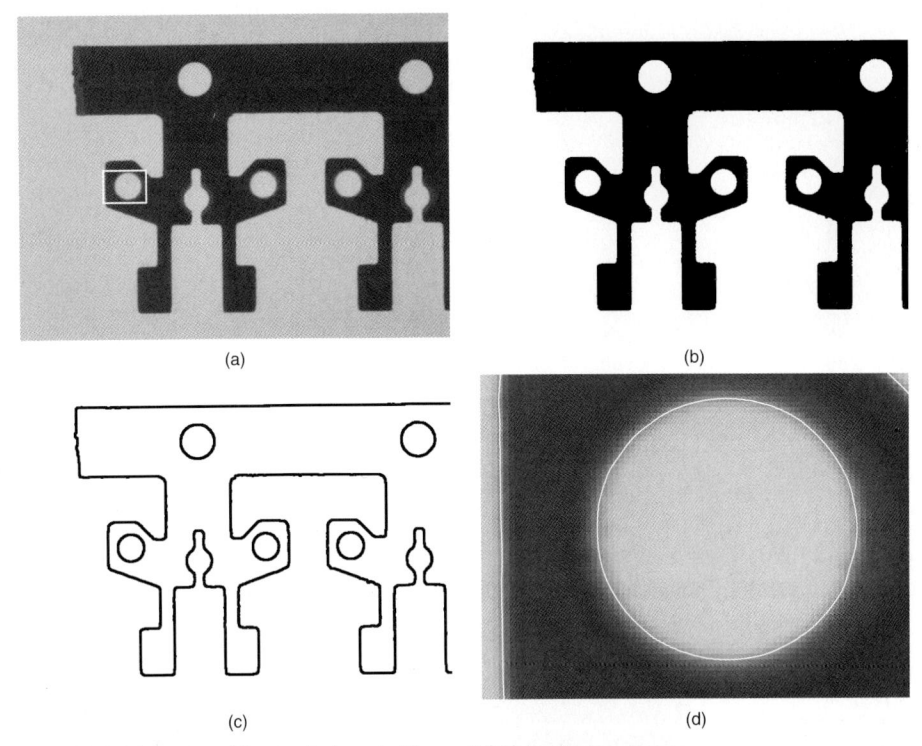

Fig. 3.79 (a) Image of the workpiece in Figure 3.61(a) with a rectangle that indicates the image part shown in (d). (b) Thresholded workpiece. (c) Dilation of the boundary of (b) with a circle of diameter 5. This is used as the ROI for the subpixel edge extraction. (d) Subpixel-accurate edges of the workpiece extracted with the Canny filter with $\sigma = 1$.

can use the subpixel-precise thresholding operation, defined in Section 3.4.3, to extract edges with subpixel accuracy. To make this as efficient as possible, we must first compute the edge amplitude in the entire ROI of the image. Then, we threshold the edge amplitude and use the resulting region as the ROI for the computation of the Laplacian and for the subpixel-precise thresholding. The resulting edges for two parts of the workpiece image are compared to the gradient magnitude edge in Figure 3.80. Note that, since the Laplacian edges must follow the corners, they are much more curved than the gradient magnitude edges, and hence are more difficult to process further. This is another reason why the edge definition via the gradient magnitude is usually preferred.

Despite the above arguments, the property that the Laplacian edge exactly passes through corners in the image can be used advantageously in some applications. Figure 3.81(a) shows an image of a bolt for which the depth of the thread must be measured. Figures 3.81(b)–(d) display the results of extracting the border of the bolt with subpixel-precise thresholding, the gradient magnitude edges with a Canny filter with

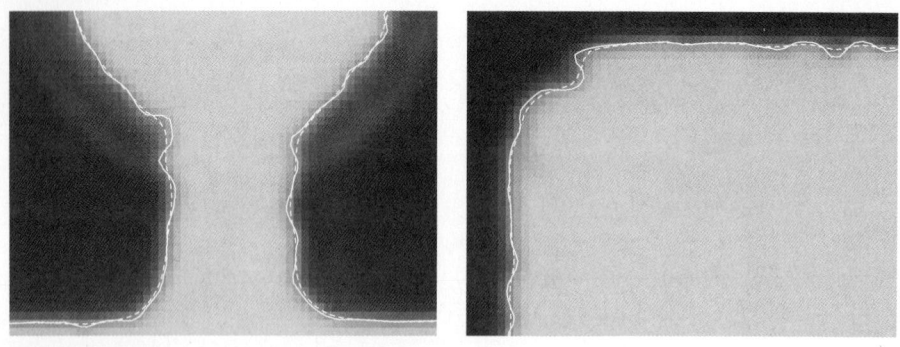

Fig. 3.80 Comparison of the subpixel-accurate edges extracted via the maxima of the gradient magnitude in the gradient direction (dashed lines) and the edges extracted via the subpixel-accurate zero crossings of the Laplacian. In both cases, a Gaussian filter with $\sigma = 1$ was used. Note that, since the Laplacian edges must follow the corners, they are much more curved than the gradient magnitude edges.

$\sigma = 0.7$, and the Laplacian edges with a Gaussian filter with $\sigma = 0.7$. Note that in this case the most suitable results are obtained with the Laplacian edges.

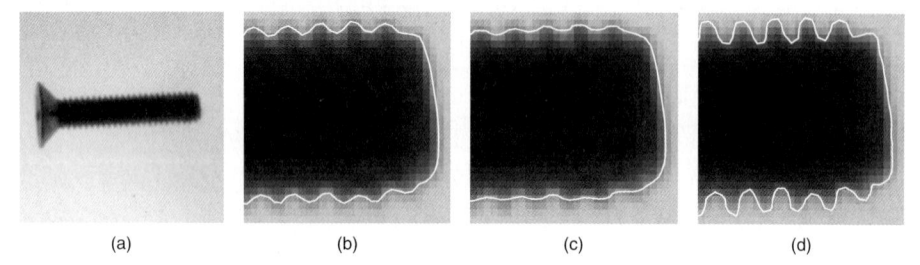

Fig. 3.81 (a) Image of a bolt for which the depth of the thread must be measured. (b) Result of performing a subpixel-precise thresholding operation. (c) Result of extracting the gradient magnitude edges with a Canny filter with $\sigma = 0.7$. (d) Result of extracting the Laplacian edges with a Gaussian filter with $\sigma = 0.7$. Note that for this application the Laplacian edges return the most suitable result.

3.7.4 Accuracy of Edges

In the previous two sections, we have seen that edges can be extracted with subpixel resolution. We have used the terms "subpixel-accurate" and "subpixel-precise" to describe these extraction mechanisms without actually justifying the use of the words "accurate" and "precise." Therefore, in this section we will examine whether the edges we can extract are actually subpixel-accurate and subpixel-precise.

Since the words "accuracy" and "precision" are often confused or used interchangeably, let us first define what we mean by them. By precision, we denote how close on average an extracted value is to its mean value [95]. Hence, precision measures how repeatable we can extract the value. The official name for precision is repeatability [96]. However, we will stick to precision throughout this text to avoid awkward terms such as "subpixel-repeatable." By accuracy, on the other hand, we denote how close on average the extracted value is to its true value [95, 96]. Note that the precision does not tell us anything about the accuracy of the extracted value. The measurements could, for example, be offset by a systematic bias, but still be very precise. Conversely, the accuracy does not necessarily tell us how precise the extracted value is. The measurement could be quite accurate, but not very repeatable. Figure 3.82 shows the different situations that can occur. Also note that the accuracy and precision are statements about the average distribution of the extracted values. From a single value, we cannot tell whether the measurements are accurate or precise.

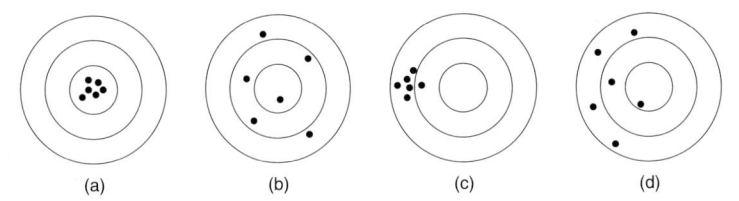

Fig. 3.82 Comparison of accuracy and precision. The center of the circles indicates the true value of the feature. The dots indicate the outcome of the measurements of the feature. (a) Accurate and precise. (b) Accurate but not precise. (c) Not accurate but precise. (d) Neither accurate nor precise.

If we adopt a statistical point of view, the extracted values can be regarded as random variables. With this, the precision of the values is given by the variance of the values: $V[x] = \sigma_x^2$. If the extracted values are precise, they have a small variance. On the other hand, the accuracy can be described by the difference of the expected value $E[x]$ of the values from the true value T: $|E[x] - T|$. Since we typically do not know anything about the true probability distribution of the extracted values, and consequently cannot determine $E[x]$ and $V[x]$, we must estimate them with the empirical mean and variance of the extracted values.

The accuracy and precision of edges is analyzed extensively in [90, 97]. The precision of ideal step edges extracted with the Canny filter is derived analytically. If we denote the true edge amplitude by a and the noise variance in the image by σ_n^2, it can be shown that the variance of the edge positions σ_e^2 is given by

$$\sigma_e^2 = \frac{3}{8} \frac{\sigma_n^2}{a^2} \tag{3.98}$$

Even though this result was derived analytically for continuous images, it also holds in the discrete case. This result has also been verified empirically in [90, 97]. Note that it is quite intuitive: the larger the noise in the image, the less precisely the edges can be located; furthermore, the larger the edge amplitude, the higher is the precision of the edges. Note also that, possibly contrary to our intuition, increasing the smoothing does not increase the precision. This happens because the noise reduction achieved by the larger smoothing cancels out exactly with the weaker edge amplitude that results from the smoothing. From Eq. (3.98), we can see that the Canny filter is subpixel-precise ($\sigma_e \leq 1/2$) if the signal-to-noise ratio $a^2/\sigma_n^2 \geq 3/2$. This can, of course, be easily achieved in practice. Consequently, we see that we were justified in calling the Canny filter subpixel-precise.

The same derivation can also be performed for the Deriche and Lanser filters. For continuous images, the following variances result:

$$\sigma_e^2 = \frac{5}{64} \frac{\sigma_n^2}{a^2} \quad \text{and} \quad \sigma_e^2 = \frac{3}{16} \frac{\sigma_n^2}{a^2} \tag{3.99}$$

Note that the Deriche and Lanser filters are more precise than the Canny filter. Like for the Canny filter, the smoothing parameter α has no influence on the precision. In the discrete case, this is, unfortunately, no longer true because of the discretization of the filter. Here, less smoothing (larger values of α) leads to slightly worse precision than predicted by Eq. (3.99). However, for practical purposes, we can assume that the smoothing for all the edge filters that we have discussed has no influence on the precision of the edges. Consequently, if we want to control the precision of the edges, we must maximize the signal-to-noise ratio by using proper lighting, cameras, and frame grabbers. Furthermore, if analog cameras are used, the frame grabber should have a line jitter that is as small as possible.

For ideal step edges, it is also easy to convince oneself that the expected position of the edge under noise corresponds to its true position. This happens because both the ideal step edge and the above filters are symmetric with respect to the true edge positions. Therefore, the edges that are extracted from noisy ideal step edges must be distributed symmetrically around the true edge position. Consequently, their mean value is the true edge position. This is also verified empirically for the Canny filter in [90, 97]. Of course, it can also be verified for the Deriche and Lanser filters.

While it is easy to show that edges are very accurate for ideal step edges, we must also perform experiments on real images to test the accuracy on real data. This is important because some of the assumptions that are used in the edge extraction algorithms may not hold in practice. Because these assumptions are seldom stated explicitly, we should examine them carefully here. Let us focus on straight edges because, as we have seen from the discussion in Section 3.7.1, especially Figure 3.63, sharply curved edges will necessarily lie in incorrect positions. See also [98] for a thorough discussion on the positional errors of the Laplacian edge detector for ideal corners of two straight edges with varying angles. Because we concentrate on straight edges, we can reduce the edge detection to the 1D case, which is simpler to analyze. From Section 3.7.1 we know that 1D edges are given by the inflection points of the gray value profiles. This implicitly assumes that the gray value profile, and consequently its derivatives, are symmetric with respect to the true edge. Furthermore, to obtain subpixel positions, the edge detection implicitly assumes that the gray values at the edge change smoothly and continuously as the edge moves in subpixel increments through a pixel. For example, if an edge covers 25% of a pixel, we would assume that the gray value in the pixel is a mixture of 25% of the foreground gray value and 75% of the background gray value. We will see whether these assumptions hold in real images below.

To test the accuracy of the edge extraction on real images, it is instructive to repeat the experiments in [90, 97] with a different camera. In [90, 97], a print of an edge is mounted on an xy-stage and shifted in 50 μm increments, which corresponds to approximately 1/10 pixel, for a total of 1 mm. The goals are to determine whether the shifts of 1/10 pixel can be detected reliably and to obtain information about the absolute accuracy of the edges. Figure 3.83(a) shows an image used in this experiment. We are not going to repeat the test to see whether the subpixel shifts can be detected reliably here. The 1/10 pixel shifts can be detected with a very high confidence (more than 99.999 99%). What is more interesting is to look at the absolute accuracy. Since we do not know the true edge position, we must get an estimate for it. Because the edge was shifted in linear increments in the test images, such an estimate can be obtained by fitting a straight line through the extracted edge positions and subtracting the line from the measured edge positions.

Figure 3.83(b) displays the result of extracting the edge in Figure 3.83(a) along a horizontal line with the Canny filter with $\sigma = 1$. The edge position error is shown in Figure 3.83(c). We can see that there are errors of up to approximately 1/22 pixel. What causes these errors? As we discussed above, for ideal cameras, no error occurs, so one of the assumptions must be violated. In this case, the assumption that is violated is that the gray value is a mixture of the foreground and background gray values that is proportional to the area of the pixel covered by the object. This happens because the camera did not have a fill factor of 100%, i.e., the light-sensitive area of a pixel on the sensor was much smaller than the total area of the pixel. Consider what happens when the edge moves across the pixel and the image is perfectly focused. In the light-sensitive area of the pixel, the gray value changes as expected when the edge moves across the pixel because the sensor integrates the incoming light. However, when the edge enters the light-insensitive area, the gray value no longer changes [99]. Consequently, the edge does not move in the image. In the real image, the focus is not perfect. Hence, the light is spread slightly over adjacent sensor elements. Therefore, the edges do not jump as they would in a perfectly focused image, but shift continuously. Nevertheless, the poor fill factor causes errors in the edge positions. This can be seen very clearly from Figure 3.83(c). Recall that the shift of 50 μm corresponds to 1/10 pixel. Consequently, the entire shift of 1 mm corresponds to two pixels. This is why we see a sine wave with two periods in Figure 3.83(c). Each period corresponds exactly to one pixel. That these effects are caused by the fill factor can also be seen if the lens is defocused. In this case, the light is spread over more sensor elements. This helps to create an artificially increased fill factor, which causes smaller errors.

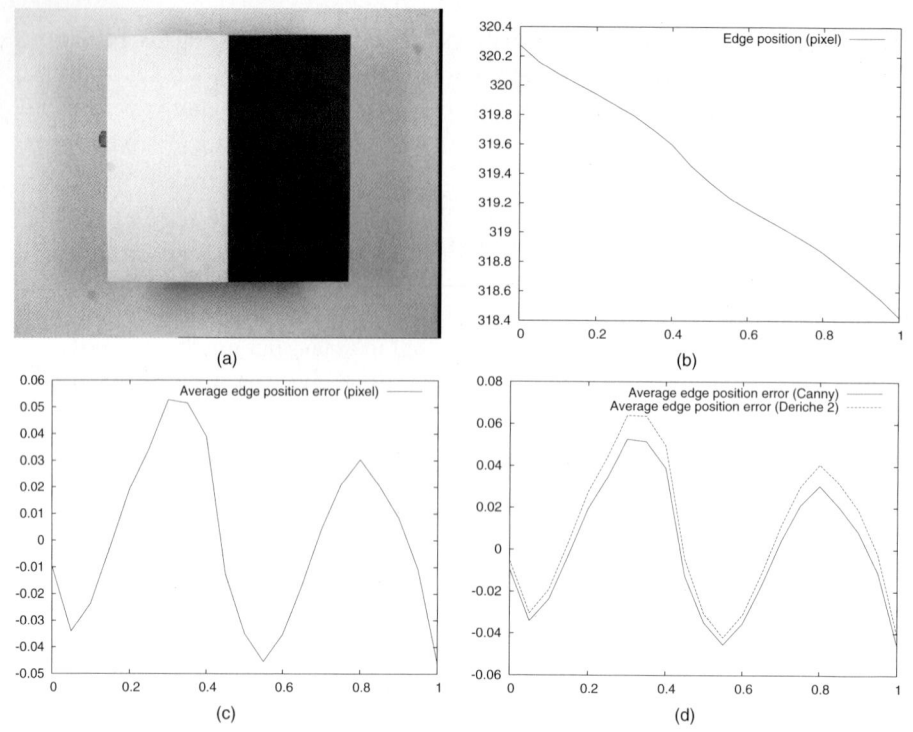

Fig. 3.83 (a) Edge image used in the accuracy experiment. (b) Edge position extracted along a horizontal line in the image with the Canny filter. The edge position is given in pixels as a function of the true shift in millimeters. (c) Error of the edge positions obtained by fitting a line through the edge positions in (b) and subtracting the line from (b). (d) Comparison of the errors obtained with the Canny and second Deriche filters.

From the above discussion, it would appear that the edge position can be extracted with an accuracy of 1/22 pixel. To check whether this is true, let us repeat the experiment with the second Deriche filter. Figure 3.83(d) shows the result of extracting the edges with $\alpha = 1$ and computing the errors with the line fitted through the Canny edge positions. The last part is done to make the errors comparable. We can see to our surprise that the Deriche edge positions are systematically shifted in one direction. Does this mean that the Deriche filter is less accurate than the Canny filter? Of course, it does not, since on ideal data both filters return the same result. It shows us that another assumption must be violated. In this case, it is the assumption that the edge profile is symmetric with respect to the true edge position. This is the only reason why the two filters, which are symmetric themselves, can return a different result.

There are many reasons why edge profiles may become asymmetric. One reason is that the gray value responses of the camera and image acquisition device are nonlinear. Figure 3.84 illustrates that an originally symmetric edge profile becomes asymmetric by a nonlinear gray value response function. It can be seen that the edge position accuracy is severely degraded by the nonlinear response. To correct the nonlinear response of the camera, it must be calibrated radiometrically with the methods described in Section 3.2.2.

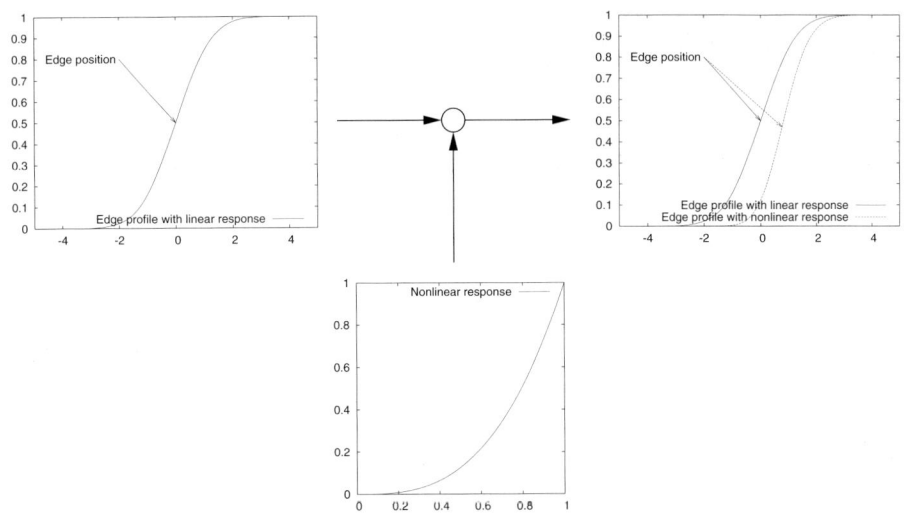

Fig. 3.84 Result of applying a nonlinear gray value response curve to an ideal symmetric edge profile. The ideal edge profile is shown in the upper left graph, the nonlinear response in the bottom graph. The upper right graph shows the modified gray value profile along with the edge positions on the profiles. Note that the edge position is affected substantially by the nonlinear response.

Unfortunately, even if the camera has a linear response or is calibrated radiometrically, other factors may cause the edge profiles to become asymmetric. In particular, lens aberrations like coma, astigmatism, and chromatic aberrations may cause asymmetric profiles (see Section 2.2.5). Since lens aberrations cannot be corrected easily with image processing algorithms, they should be as small as possible.

While all the error sources discussed above influence the edge accuracy, we have so far neglected the largest source of errors. If the camera is not calibrated geometrically, extracting edges with subpixel accuracy is pointless because the lens distortions alone are sufficient to render any subpixel position meaningless. Let us, for example, assume that the lens has a distortion that is smaller than 1% in the entire field of view. At the corners of the image, this means that the edges are offset by 4 pixels for a 640 × 480 image. We can see that extracting edges with subpixel accuracy is an exercise in futility if the lens distortions are not corrected, even for this relatively small distortion. This is illustrated in Figure 3.85, where the result of correcting the lens distortions after calibrating the camera as described in Section 3.9 is shown. Note that, despite the fact that the application used a very high-quality lens, the lens distortions cause an error of approximately 3 pixels.

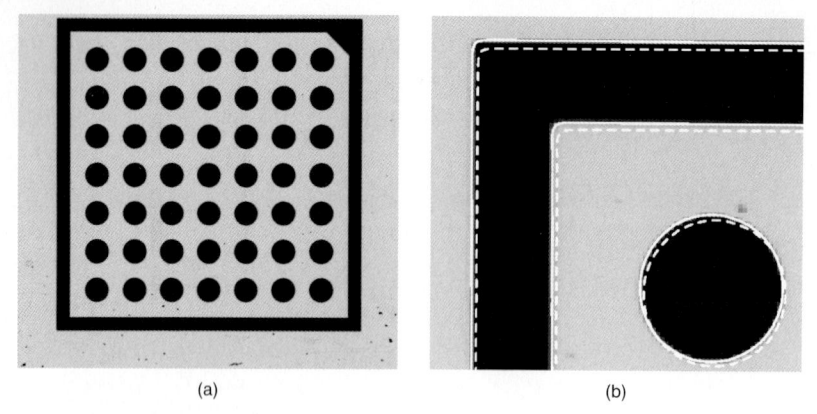

Fig. 3.85 (a) Image of a calibration target. (b) Extracted subpixel-accurate edges (solid lines) and edges after the correction of the lens distortions (dashed lines). Note that the lens distortions cause an error of approximately 3 pixels.

Another detrimental influence on the accuracy of the extracted edges is caused by the perspective distortions in the image. They happen whenever we cannot mount the camera perpendicularly to the objects we want to measure. Figure 3.86(a) shows the result of extracting the 1D edges along the ruler markings on a caliper. Because of the severe perspective distortions, the distances between the ruler markings vary greatly throughout the image. If the camera is calibrated, i.e., its interior orientation and the exterior orientation of the plane in which the objects to measure lie have been determined with the approach described in Section 3.9, the measurements in the image can be converted into measurements in world coordinates in the plane determined by the calibration. This is done by intersecting the optical ray that corresponds to each edge point in the image with the plane in the world. Figure 3.86(b) displays the results of converting the measurements in Figure 3.86(a) into millimeters with this approach. Note that the measurements are extremely accurate even in the presence of severe perspective distortions.

From the above discussion, we can see that extracting edges with subpixel accuracy relies on careful selection of the hardware components. First of all, the gray value responses of the camera and image acquisition device should be linear. To ensure this, the camera should be calibrated radiometrically. Furthermore, lenses with very small aberrations (like coma and astigmatism) should be chosen. Furthermore, monochromatic light should be used to avoid the effects of chromatic aberrations. In addition, the fill factor of the camera should be as large as possible to avoid the effects of "blind spots." Finally, the camera should be calibrated geometrically to obtain meaningful results. All these requirements for the hardware components are, of course, also valid for other subpixel algorithms, e.g., the subpixel-precise thresholding (see Section 3.4.3), the gray value moments (see Section 3.5.2), and the contour features (see Section 3.5.3).

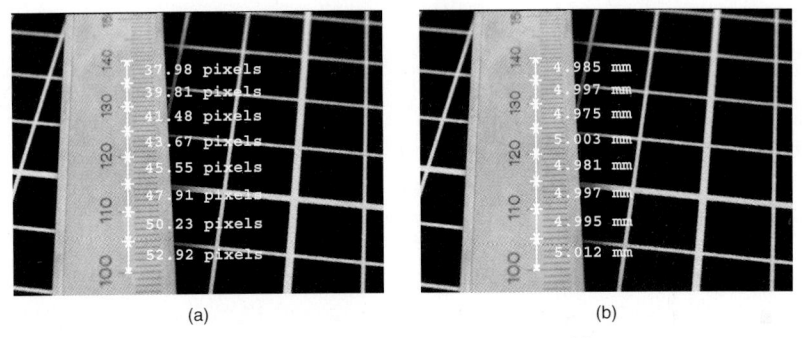

Fig. 3.86 Result of extracting 1D edges along the ruler markings on a caliper. (a) Pixel distances between the markings. (b) Distances converted to world units using camera calibration.

3.8
Segmentation and Fitting of Geometric Primitives

In Sections 3.4 and 3.7, we have seen how to segment images by thresholding and edge extraction. In both cases, the boundary of objects either is returned explicitly or can be derived by some post-processing (see Section 3.6.1). Therefore, for the purposes of this section, we can assume that the result of the segmentation is a contour with the points of the boundary, which may be subpixel-accurate. This approach often creates an enormous amount of data. For example, the subpixel-accurate edge of the hole in the workpiece in Figure 3.79(d) contains 172 contour points. However, we are typically not interested in such a large amount of information. For example, in the application in Figure 3.79(d), we would probably be content with knowing the position and radius of the hole, which can be described with just three parameters. Therefore, in this section we will discuss methods to fit geometric primitives to contour data. We will only examine the most relevant geometric primitives: lines, circles, and ellipses. Furthermore, we will examine how contours can be segmented automatically into parts that correspond to the geometric primitives. This will enable us to substantially reduce the amount of data that needs to be processed, while also providing us with a symbolic description of the data. Furthermore, the fitting of the geometric primitives will enable us to reduce the influence of incorrectly or inaccurately extracted points (so-called outliers). We will start by examining the fitting of the geometric primitives in Sections 3.8.1–3.8.3. In each case, we will assume that the contour or part of the contour we are examining corresponds to the primitive we are trying to fit, i.e., we are assuming that the segmentation into the different primitives has already been performed. The segmentation itself will be discussed in Section 3.8.4.

3.8.1
Fitting Lines

If we want to fit lines, we first need to think about the representation of lines. In images, lines can occur in any orientation. Therefore, we have to use a representation that enables us to represent all lines. For example, the common representation $y = mx + b$ does not allow us to do this. One representation that can be used is the Hessian normal form of the line, given by

$$\alpha r + \beta c + \gamma = 0 \qquad (3.100)$$

This is actually an over-parameterization, since the parameters (α, β, γ) are homogeneous [66, 67]. Therefore, they are only defined up to a scale factor. The scale factor in the Hessian normal form is fixed by requiring that $\alpha^2 + \beta^2 = 1$. This has the advantage that the distance of a point to the line can simply be obtained by substituting its coordinates into Eq. (3.100).

To fit a line through a set of points (r_i, c_i), $i = 1, \ldots, n$, we can minimize the sum of the squared distances of the points to the line:

$$\varepsilon^2 = \sum_{i=1}^{n} (\alpha r_i + \beta c_i + \gamma)^2 \qquad (3.101)$$

While this is correct in principle, it does not work in practice, because we can achieve a zero error if we select $\alpha = \beta = \gamma = 0$. This is caused by the over-parameterization of the line. Therefore, we must add the constraint $\alpha^2 + \beta^2 = 1$ as a Lagrange multiplier, and hence must minimize the following error:

$$\varepsilon^2 = \sum_{i=1}^{n} (\alpha r_i + \beta c_i + \gamma)^2 - \lambda(\alpha^2 + \beta^2 - 1)n \qquad (3.102)$$

The solution to this optimization problem is derived in [68]. It can be shown that (α, β) is the eigenvector corresponding to the smaller eigenvalue of the following matrix:

$$\begin{pmatrix} \mu_{2,0} & \mu_{1,1} \\ \mu_{1,1} & \mu_{0,2} \end{pmatrix} \qquad (3.103)$$

With this, γ is given by $\gamma = -(\alpha n_{1,0} + \beta n_{0,1})$. Here, $\mu_{2,0}$, $\mu_{1,1}$, and $\mu_{0,2}$ are the second-order central moments of the point set (r_i, c_i), while $n_{1,0}$ and $n_{0,1}$ are the normalized first-order moments (the center of gravity) of the point set. If we replace the area a of a region with the number n of points and sum over the points in the point set instead of the points in the region, the formulas to compute these moments are identical to the region moments of Eqs. (3.53) and (3.54) in Section 3.5.1. It is interesting to note that the vector (α, β) thus obtained, which is the normal vector of the line, is the minor axis that would be obtained from the ellipse parameters of the point set. Consequently, the major axis of the ellipse is the direction of the line. This

3.8 Segmentation and Fitting of Geometric Primitives

is a very interesting connection between the ellipse parameters and the line fitting, because the results were derived using different approaches and models.

Figure 3.87(b) illustrates the line fitting procedure for an oblique edge of the workpiece shown in Figure 3.87(a). Note that, by fitting the line, we were able to reduce the effects of the small protrusion on the workpiece. As mentioned above, by inserting the coordinates of the edge points into the line equation (3.100), we can easily calculate the distances of the edge points to the line. Therefore, by thresholding the distances, the protrusion can be detected easily.

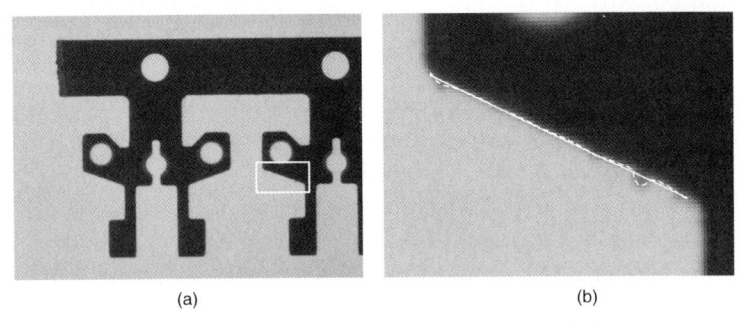

Fig. 3.87 (a) Image of a workpiece with the part shown in (b) indicated by the white rectangle. (b) Extracted edge within a region around the inclined edge of the workpiece (dashed line) and straight line fitted to the edge (solid line).

As can be seen from the above example, the line fit is robust to small deviations from the assumed model (small outliers). However, Figure 3.88 shows that large outliers severely affect the quality of the fitted line. In this example, the line is fitted through the straight edge as well as the large arc caused by the relay contact. Since the line fit must minimize the sum of the squared distances of the contour points, the fitted line has a direction that deviates from that of the straight edge.

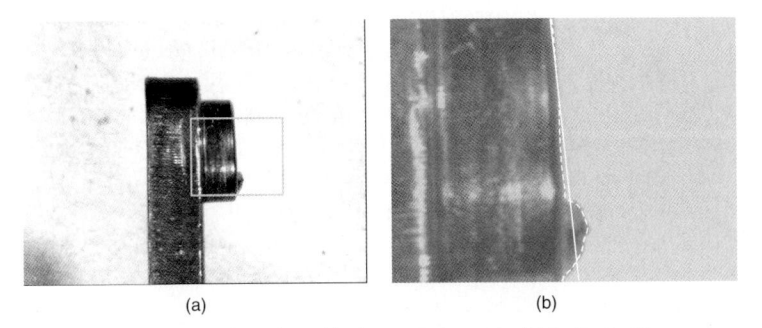

Fig. 3.88 (a) Image of a relay with the part shown in (b) indicated by the light gray rectangle. (b) Extracted edge within a region around the vertical edge of the relay (dashed line) and straight line fitted to the edge (solid line). To provide a better visibility of the edge and line, the contrast of the image has been reduced in (b).

The least-squares line fit is not robust to large outliers since points that lie far from the line have a very large weight in the optimization because of the squared distances. To reduce the influence of distant points, we can introduce a weight w_i for each point. The weight should be $\ll 1$ for distant points. Let us assume for the moment that we have a way to compute these weights. Then, the minimization becomes

$$\varepsilon^2 = \sum_{i=1}^{n} w_i(\alpha r_i + \beta c_i + \gamma)^2 - \lambda(\alpha^2 + \beta^2 - 1)n \tag{3.104}$$

The solution of this optimization problem is again given by the eigenvector corresponding to the smaller eigenvalue of a moment matrix like in Eq. (3.103) [100]. The only difference is that the moments are computed by taking the weights w_i into account. If we interpret the weights as gray values, the moments are identical to the gray value center of gravity and the second-order central gray value moments (see Eqs. (3.61) and (3.62) in Section 3.5.2). As above, the fitted line corresponds to the major axis of the ellipse obtained from the weighted moments of the point set. Hence, there is an interesting connection to the gray value moments.

The only remaining problem is how to define the weights w_i. Since we want to give smaller weights to points with large distances, the weights must be based on the distances $\delta_i = |\alpha r_i + \beta c_i + \gamma|$ of the points to the line. Unfortunately, we do not know the distances without fitting the line, so this seems an impossible requirement. The solution is to fit the line in several iterations. In the first iteration, $w_i = 1$ is used, i.e., a normal line fit is performed to calculate the distances δ_i. They are used to define weights for the following iterations by using a weight function $w(\delta)$ [100]. In practice, one of the following two weight functions can be used. They both work very well. The first weight function was proposed by Huber [100, 101]. It is given by

$$w(\delta) = \begin{cases} 1 & |\delta| \leq \tau \\ \tau/|\delta| & |\delta| > \tau \end{cases} \tag{3.105}$$

The parameter τ is the clipping factor. It defines which points should be regarded as outliers. We will see how it is computed below. For now, please note that all points with a distance $\leq \tau$ receive a weight of 1. This means that, for small distances, the squared distance is used in the minimization. Points with a distance $> \tau$, on the other hand, receive a progressively smaller weight. In fact, the weight function is chosen such that points with large distances use the distance itself and not the squared distance in the optimization. Sometimes, these weights are not small enough to suppress outliers completely. In this case, the Tukey weight function can be used [100, 102]. It is given by

$$w(\delta) = \begin{cases} [1 - (\delta/\tau)^2]^2 & |\delta| \leq \tau \\ 0 & |\delta| > \tau \end{cases} \tag{3.106}$$

Again, τ is the clipping factor. Note that this weight function completely disregards points that have a distance $> \tau$. For distances $\leq \tau$, the weight changes smoothly from 1 to 0.

In the above two weight functions, the clipping factor specifies which points should be regarded as outliers. Since the clipping factor is a distance, it could simply be set manually. However, this would ignore the distribution of the noise and the outliers in the data, and consequently would have to be adapted for each application. It is more convenient to derive the clipping factor from the data itself. This is typically done based on the standard deviation of the distances to the line. Since we expect outliers in the data, we cannot use the normal standard deviation, but must use a standard deviation that is robust to outliers. Typically, the following formula is used to compute the robust standard deviation:

$$\sigma_\delta = \frac{\text{median} |\delta_i|}{0.6745} \qquad (3.107)$$

The constant in the denominator is chosen such that, for normally distributed distances, the standard deviation of the normal distribution is computed. The clipping factor is then set to a small multiple of σ_δ, e.g., $\tau = 2\sigma_\delta$.

In addition to the Huber and Tukey weight functions, other weight functions can be defined. Several other possibilities are discussed in [66].

Figure 3.89 displays the result of fitting a line robustly to the edge of the relay using the Tukey weight function with a clipping factor of $\tau = 2\sigma_\delta$ with five iterations. If we compare this to the standard least-squares line fit in Figure 3.88(b), we see that with the robust fit the line is now fitted to the straight-line part of the edge and the outliers caused by the relay contact have been suppressed.

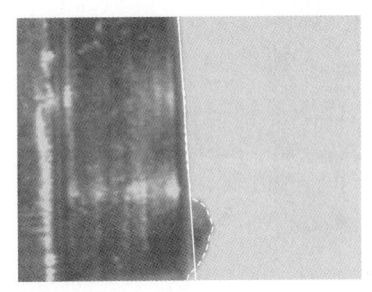

Fig. 3.89 Straight line (solid line) fitted robustly to the vertical edge (dashed line). In this case, the Tukey weight function with a clipping factor of $\tau = 2\sigma_\delta$ with five iterations was used. Compared to Figure 3.88(b), the line is now fitted to the straight-line part of the edge.

It should also be noted that the above approach to outlier suppression by weighting down the influence of points with large distances can sometimes fail because the initial fit, which is a standard least-squares fit, can produce a solution that is dominated by the outliers. Consequently, the weight function will drop inliers. In this case, other robust methods must be used. The most important approach is the random sample consensus (RANSAC) algorithm, proposed by Fischler and Bolles in [103]. Instead of dropping outliers successively, it constructs a solution (e.g., a line fit) from the minimum number of points (e.g., two for lines), which are selected randomly, and then checks how many points are consistent with the solution. The process of randomly selecting points, constructing the solution, and checking the number of consistent points is continued until a certain probability of having found the correct solution, e.g., 99%, is achieved. At the end, the solution with the largest number of consistent points is selected.

3.8.2
Fitting Circles

Fitting circles or circular arcs to a contour uses the same idea as fitting lines, that is, we want to minimize the sum of the squared distances of the contour points to the circle:

$$\varepsilon^2 = \sum_{i=1}^{n} \left(\sqrt{(r_i - \alpha)^2 + (c_i - \beta)^2} - \rho \right)^2 \tag{3.108}$$

Here, (α, β) is the center of the circle and ρ is its radius. Unlike the line fitting, this leads to a nonlinear optimization problem, which can only be solved iteratively using nonlinear optimization techniques. Details can be found in [68, 104, 105].

Figure 3.90(a) shows the result of fitting circles to the edges of the holes of a workpiece, along with the extracted radii in pixels. In Figure 3.90(b), details of the upper right hole are shown. Note how well the circle fits the extracted edges.

(a)   (b)

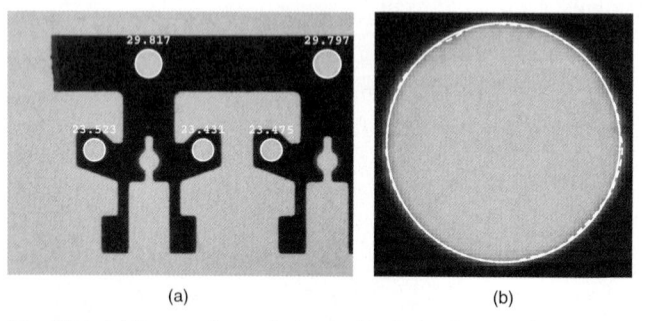

Fig. 3.90 (a) Image of a workpiece with circles fitted to the edges of the holes in the workpiece. (b) Details of the upper right hole with the extracted edge (dashed line) and the fitted circle (solid line).

Like the least-squares line fit, the least-squares circle fit is not robust to outliers. To make the circle fit robust, we can use the same approach that we used for the line fitting: we can introduce a weight that is used to reduce the influence of the outliers. Again, this requires that we perform a normal least-squares fit first and then use the distances that result from it to calculate the weights in later iterations. Since it is possible that large outliers prevent this algorithm from converging to the correct solution, a RANSAC approach might be necessary in extreme cases.

Figure 3.91 compares the standard circle fitting with the robust circle fitting using the BGA example of Figure 3.34. With the standard fitting (Figure 3.91(b)), the circle is affected by the error in the pad, which acts like an outlier. This is corrected with the robust fitting (Figure 3.91(c)).

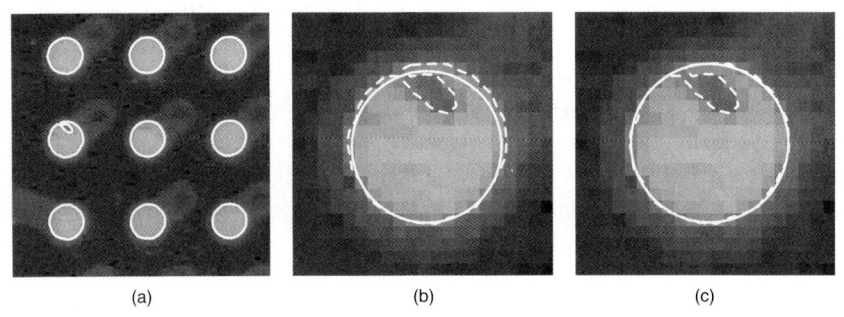

Fig. 3.91 (a) Image of a BGA with pads extracted by subpixel-precise thresholding (see also Figure 3.34). (b) Circle fitted to the left pad in the center row of (a). The fitted circle is shown as a solid line, while the extracted contour is shown as a dashed line. The fitted circle is affected by the error in the pad, which acts like an outlier. (c) Result of robustly fitting a circle. The fitted circle corresponds to the true boundary of the pad.

To conclude this section, we should give some thought to what happens when a circle is fitted to a contour that only represents a part of a circle (a circular arc). In this case, the accuracy of the parameters becomes progressively worse as the angle of the circular arc becomes smaller. An excellent analysis of this effect is performed in [104]. This effect is obvious from the geometry of the problem. Simply think about a contour that only represents a 5° arc. If the contour points are disturbed by noise, we have a very large range of radii and centers that lead to almost the same fitting error. On the other hand, if we fit to a complete circle, the geometry of the circle is much more constrained. This effect is caused by the geometry of the fitting problem and not by a particular fitting algorithm, i.e., it will occur for all fitting algorithms.

3.8.3
Fitting Ellipses

To fit an ellipse to a contour, we would like to use the same principles as for lines and circles: i.e., minimize the distance of the contour points to the ellipse. This requires us to determine the closest point to each contour point on the ellipse. While this can be determined easily for lines and circles, for ellipses it requires finding the roots of a fourth-degree polynomial. Since this is quite complicated and expensive, ellipses are often fitted by minimizing a different kind of distance. The principle is similar to the line fitting approach: we write down an implicit equation for ellipses (for lines, the implicit equation is given by Eq. (3.100)), and then substitute the point coordinates into the implicit equation to get a distance measure for the points to the ellipse. For the line fitting problem, this procedure returns the true distance to the line. For the ellipse fitting, it only returns a value that has the same properties as a distance, but is not the true distance. Therefore, this distance is called the algebraic distance. Ellipses are described by the following implicit equation:

$$\alpha r^2 + \beta rc + \gamma c^2 + \delta r + \zeta c + \eta = 0 \tag{3.109}$$

Like for lines, the set of parameters is a homogeneous quantity, i.e., only defined up to scale. Furthermore, Eq. (3.109) also describes hyperbolas and parabolas. Ellipses require $\beta^2 - 4\alpha\gamma < 0$. We can solve both problems by requiring $\beta^2 - 4\alpha\gamma = -1$. An elegant solution to fitting ellipses by minimizing the algebraic error with a linear method was proposed by Fitzgibbon. The interested reader is referred to [106] for details. Unfortunately, minimizing the algebraic error can result in biased ellipse parameters. Therefore, if the ellipse parameters should be determined with maximum accuracy, the geometric error should be used. A nonlinear approach for fitting ellipses based on the geometric error is proposed in [105]. It is significantly more complicated than the linear approach in [106].

Like the least-squares line and circle fits, fitting ellipses via the algebraic or geometric distance is not robust to outliers. We can again introduce weights to create a robust fitting procedure. If the ellipses are fitted with the algebraic distance, this again results in a linear algorithm in each iteration of the robust fit [100]. In applications with a very large number of outliers or with very large outliers, a RANSAC approach might be necessary.

The ellipse fitting is very useful in camera calibration, where often circular marks are used on the calibration targets (see Section 3.9 and [99, 107, 108]). Since circles project to ellipses, fitting ellipses to the edges in the image is the natural first step in the calibration process. Figure 3.92(a) displays an image of a calibration target. The ellipses fitted to the extracted edges of the calibration marks are shown in Figure 3.92(b). In Figure 3.92(c), a detailed view of the center mark with the fitted ellipse is shown. Since the subpixel edge extraction is very accurate, there is hardly any visible difference between the edge and the ellipse, and consequently the edge is not shown in the figure.

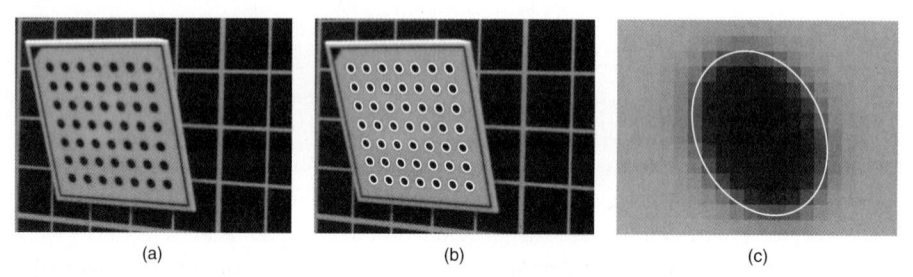

Fig. 3.92 (a) Image of a calibration target. (b) Ellipses fitted to the extracted edges of the circular marks of the calibration target. (c) Detail of the center mark of the calibration target with the fitted ellipse.

To conclude this section, we should note that, if ellipses are fitted to contours that only represent a part of an ellipse, the same comments that were made for circular arcs at the end of the last section apply: the accuracy of the parameters will become worse as the angle that the arc subtends becomes smaller. The reason for this behavior lies in the geometry of the problem and not in the fitting algorithms we use.

3.8.4
Segmentation of Contours into Lines, Circles, and Ellipses

So far, we have assumed that the contours to which we are fitting the geometric primitives correspond to a single primitive of the correct type, e.g., a line segment. Of course, a single contour may correspond to multiple primitives of different types. Therefore, in this section we will discuss how contours can be segmented into the different primitives.

We will start by examining how a contour can be segmented into lines. To do so, we would like to find a polygon that approximates the contour sufficiently well. Let us call the contour points $p_i = (r_i, c_i)$, for $i = 1, \ldots, n$. Approximating the contour by a polygon means that we want to find a subset p_{i_j}, for $j = 1, \ldots, m$ with $m \leq n$, of the control points of the contour that describes the contour reasonably well. Once we have found the approximating polygon, each line segment $(p_{i_j}, p_{i_{j+1}})$ of the polygon is a part of the contour that can be approximated well with a line. Hence, we can fit lines to each line segment afterwards to obtain a very accurate geometric representation of the line segments.

The question we need to answer is this: How do we define whether a polygon approximates the contour sufficiently well? A large number of different definitions have been proposed over the years. A very good evaluation of many polygonal approximation methods has been carried out by Rosin [109, 110]. In both cases, it was established that the algorithm proposed by Ramer [111], which curiously enough is one of the oldest algorithms, is the best overall method.

The Ramer algorithm performs a recursive subdivision of the contour until the resulting line segments have a maximum distance to the respective contour segments that is lower than a user-specified threshold $d_{\max}$. Figure 3.93 illustrates how the Ramer algorithm works. We start out by constructing a single line segment between the first and last contour points. If the contour is closed, we construct two segments: one from the first point to the point with index $n/2$, and the second one from $n/2$ to n. We then compute the distances of all the contour points to the line segment and find the point with the maximum distance to the line segment. If its distance is larger than the threshold we have specified, we subdivide the line segment into two segments at the point with the maximum distance. Then, this procedure is applied recursively to the new segments until no more subdivisions occur, i.e., until all segments fulfill the maximum distance criterion.

Figure 3.94 illustrates the use of the polygonal approximation in a real application. In Figure 3.94(a), a back-lit cutting tool is shown. In the application, the dimensions and angles of the cutting tool must be inspected. Since the tool consists

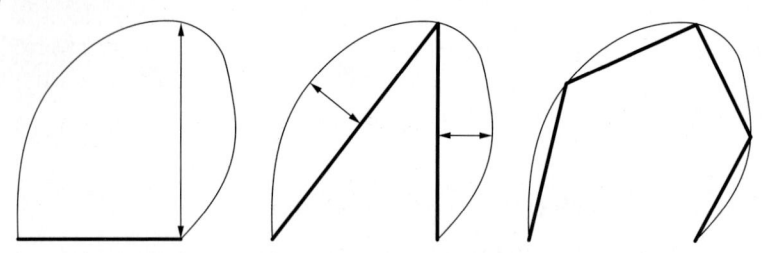

Fig. 3.93 Example of the recursive subdivision that is performed in the Ramer algorithm. The contour is displayed as a thin line, while the approximating polygon is displayed as a thick line.

of straight edges, the obvious approach is to extract edges with subpixel accuracy (Figure 3.94(b)) and to approximate them with a polygon using the Ramer algorithm. From Figure 3.94(c) we can see that the Ramer algorithm splits the edges correctly. We can also see the only slight drawback of the Ramer algorithm: it sometimes places the polygon control points into positions that are slightly offset from the true corners. In this application, this poses no problem, since to achieve maximum accuracy we must fit lines to the contour segments robustly anyway (Figure 3.94(d)). This enables us to obtain a concise and very accurate geometric description of the cutting tool. With the resulting geometric parameters, it can easily be checked whether the tool has the required dimensions.

While lines are often the only geometric primitive that occurs for the objects that should be inspected, in several cases the contour must be split into several types of primitives. For example, machined tools often consist of lines and circular arcs or lines and elliptic arcs. Therefore, we will now discuss how such a segmentation of the contours can be performed.

The approaches to segment contours into lines and circles can be classified into two broad categories. The first type of algorithm tries to identify breakpoints on the contour that correspond to semantically meaningful entities. For example, if two straight lines with different angles are next to each other, the tangent direction of the curve will contain a discontinuity. On the other hand, if two circular arcs with different radii meet smoothly, there will be a discontinuity in the curvature of the contour. Therefore, the breakpoints typically are defined as discontinuities in the contour angle, which are equivalent to maxima of the curvature, and as discontinuities in the curvature itself. The first definition covers straight lines or circular arcs that meet at a sharp angle. The second definition covers smoothly joining circles or lines and circles [112, 113]. Since the curvature depends on the second derivative of the contour, it is an unstable feature that is very prone to even small errors in the contour coordinates. Therefore, to enable these algorithms to function properly, the contour must be smoothed substantially. This, in turn, can cause the breakpoints to shift from their desired positions. Furthermore, some breakpoints may be missed. Therefore, these approaches are often

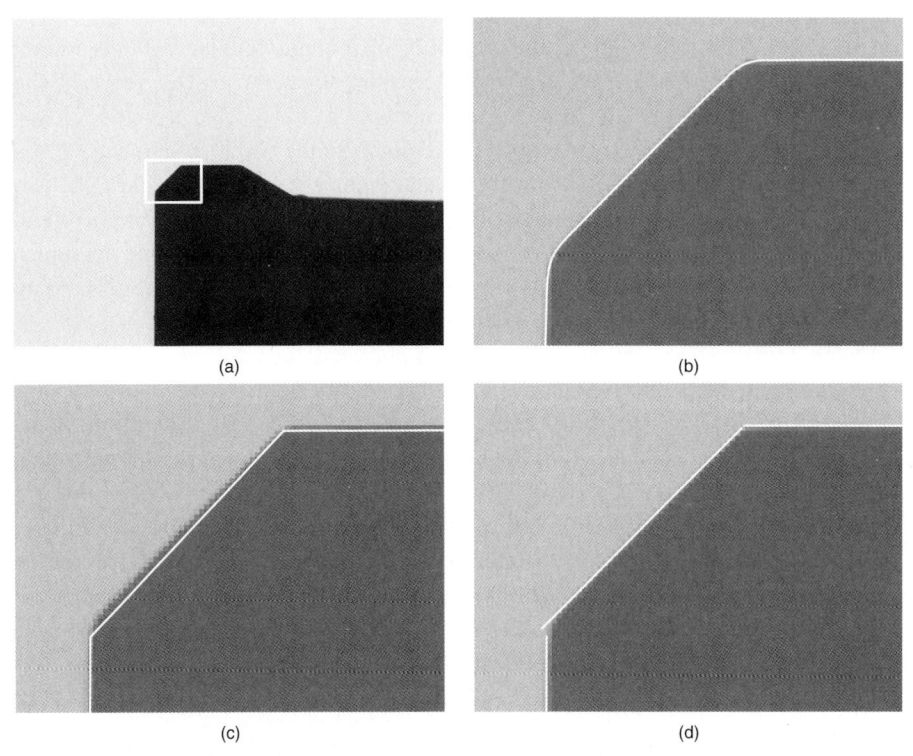

Fig. 3.94 (a) Image of a back-lit cutting tool with the part that is shown in (b)–(d) overlaid as a white rectangle. To provide a better visibility of the results, the contrast of the image has been reduced in (b)–(d). (b) Subpixel-accurate edges extracted with the Lanser filter with $\alpha = 0.7$. (c) Polygons extracted with the Ramer algorithm with $d_{\max} = 2$. (d) Lines fitted robustly to the polygon segments using the Tukey weight function.

followed by an additional splitting and merging stage and a refinement of the breakpoint positions [113, 114].

While the above algorithms work well for splitting contours into lines and circles, they are quite difficult to extend to lines and ellipses because ellipses do not have constant curvature like circles. In fact, the two points on the ellipse on the major axis have locally maximal curvature and consequently would be classified as breakpoints by the above algorithms. Therefore, if we want to have a unified approach to segmenting contours into lines and circles or ellipses, the second type of algorithm is more appropriate. They are characterized by initially performing a segmentation of the contour into lines only. This produces an over-segmentation in the areas of the contour that correspond to circles and ellipses since here many line segments are required to approximate the contour. Therefore, in a second phase the line segments are examined whether they can be merged into circles or ellipses [100, 115]. For example, the algorithm in [100] initially performs a polygonal approximation with the Ramer algorithm. Then, it checks each pair of adjacent line segments to see whether it can be better approximated by an ellipse (or, alternatively, a circle). This is done by fitting an ellipse to the part of the contour that corresponds to the two line segments. If the fitting error of the ellipse is smaller than the maximum error of the two lines, the two line segments are marked as candidates for merging. After examining all pairs of line segments, the pair with the smallest fitting error is merged. In the following iterations, the algorithm also considers pairs of line and ellipse segments. The iterative merging is continued until there are no more segments that can be merged.

Figure 3.95 illustrates the segmentation into lines and circles. Like in the previous example, the application is the inspection of cutting tools. Figure 3.95(a) displays a cutting tool that consists of two linear parts and a circular part. The result of the initial segmentation into lines with the Ramer algorithm is shown in Figure 3.95(b). Note that the circular arc is represented by four contour parts. The iterative merging stage of the algorithm successfully merges these four contour parts into a circular arc, as shown in Figure 3.95(c). Finally, the angle between the linear parts of the tool is measured by fitting lines to the corresponding contour parts, while the radius of the circular arc is determined by fitting a circle to the circular arc part. Since the camera is calibrated in this application, the fitting is actually performed in world coordinates. Hence, the radius of the arc is calculated in millimeters.

3.9
Camera Calibration

At the end of Section 3.7.4, we already discussed briefly that camera calibration is essential to obtain accurate measurements of objects. When the camera is calibrated, it is possible to correct lens distortions, which occur with different magnitudes for every lens, and to obtain the object's coordinates in metric units, e.g., in meters or millimeters. Since the previous sections have described the methods required for geometric camera calibration, we can now discuss how the calibration is performed.

To calibrate a camera, a model for the mapping of the three-dimensional (3D) points of the world to the 2D image generated by the camera, lens, and frame grabber (if used) is necessary. Two different types of lenses are relevant for machine vision tasks: regular and telecentric lenses (see Section 2.2). Regular lenses perform a perspective projection of the world into the image. We will call a camera with a regular lens a pinhole camera because of the connection between the projection performed by a regular lens and a pinhole camera that is described at the end of Section 2.2.2. Telecentric lenses perform a parallel projection of the world into the image. We will call a camera with a telecentric lens a telecentric camera.

Furthermore, two types of sensors need to be considered: line sensors and area sensors (see Section 2.3.1). For area sensors, regular as well as telecentric lenses are in common use. For line sensors, only regular lenses are commonly used. Therefore, we

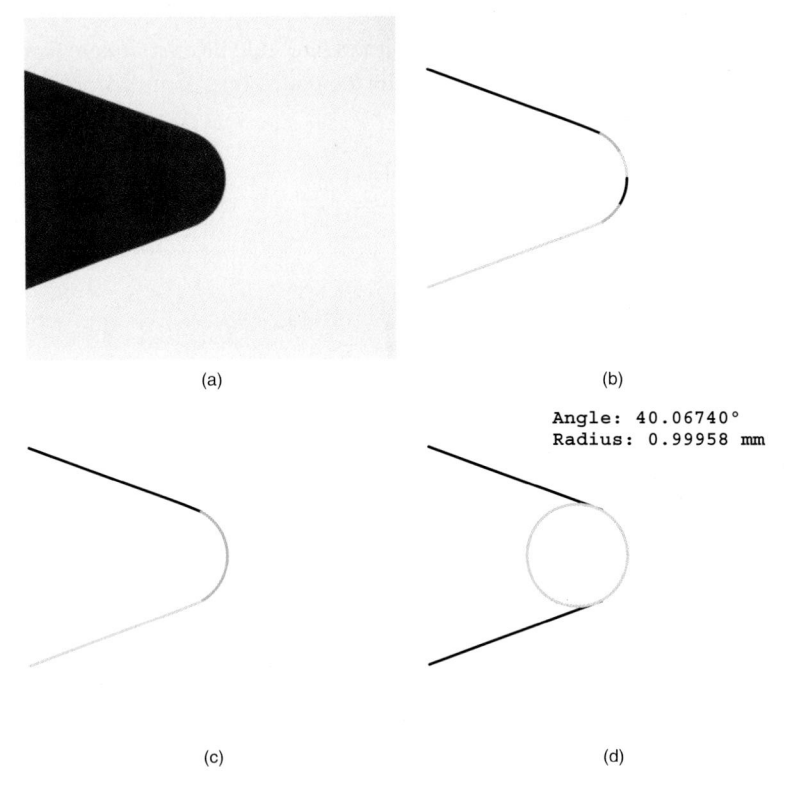

Fig. 3.95 (a) Image of a back-lit cutting tool. (b) Contour parts corresponding to the initial segmentation into lines with the Ramer algorithm. The contour parts are displayed in three different gray values. (c) Result of the merging stage of the line and circle segmentation algorithm. In this case, two lines and one circular arc are returned. (d) Geometric measurements obtained by fitting lines to the linear parts of the contour and a circle to the circular part. Because the camera was calibrated, the radius is calculated in millimeters.

will not introduce additional labels for the pinhole and telecentric cameras. Instead, it will be silently assumed that these two types of cameras use area sensors. For line sensors with regular lenses, we will simply use the term line scan camera.

For all three camera models, the mapping from 3D to 2D performed by the camera can be described by a certain number of parameters:

$$p = \pi(P_w, c_1, \ldots, c_n) \tag{3.110}$$

Here, p is the 2D image coordinate of the 3D point P_w produced by the projection π. Camera calibration is the process of determining the camera parameters $c_1, \ldots, c_n$.

3.9.1
Camera Models for Area Scan Cameras

Figure 3.96 displays the perspective projection performed by a pinhole camera. The world point P_w is projected through the projection center of the lens to the point p in the image plane. If there were no lens distortions present, p would lie on a straight line from P_w through the projection center, indicated by the dotted line. Lens distortions cause the point p to lie at a different position.

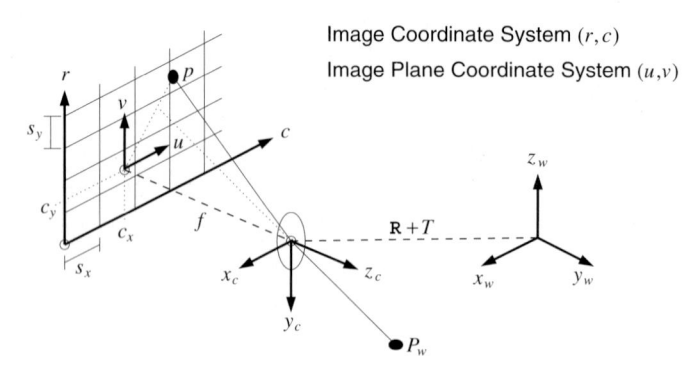

Camera Coordinate System (x_c, y_c, z_c)
World Coordinate System (x_w, y_w, z_w)

Fig. 3.96 Camera model for a pinhole camera.

The image plane is located at a distance of f behind the projection center. As explained at the end of Section 2.2.2, f is the camera constant or principal distance, and not the focal length of the lens. Nevertheless, we will use the letter f to denote the principal distance for the purposes of this section.

Although the image plane in reality lies behind the projection center of the lens, it is easier to pretend that it lies at a distance of f in front of the projection center, as shown in Figure 3.97. This causes the image coordinate system to be aligned with the pixel coordinate system (row coordinates increase downward and column coordinates to the right) and simplifies many calculations.

We are now ready to describe the mapping of objects in 3D world coordinates to the 2D image plane and the corresponding camera parameters. First, we should note that the world points P_w are given in a world coordinate system (WCS). To make the projection into the image plane possible, they need to be transformed into the camera coordinate system (CCS). The CCS is defined such that its x and y axes are parallel to the column and row axes of the image, respectively, and the z axis is perpendicular to the image plane and is oriented such that points in front of the camera have positive z coordinates. The transformation from the WCS to the CCS is a rigid transformation, i.e., a rotation followed by a translation. Therefore, the point $P_w = (x_w, y_w, z_w)^\top$ in the WCS is given by the point $P_c = (x_c, y_c, z_c)^\top$ in the CCS, where

$$P_c = R P_w + T \tag{3.111}$$

Here, $T = (t_x, t_y, t_z)^\top$ is a translation vector and $R = R(\alpha, \beta, \gamma)$ is a rotation matrix, which is determined by the three rotation angles γ (around the z axis of the CCS), β

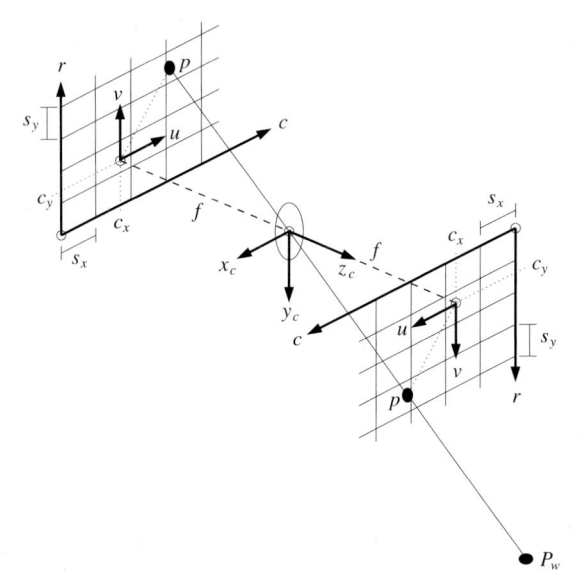

Fig. 3.97 Image plane and virtual image plane.

(around the y axis), and α (around the x axis):

$$R(\alpha,\beta,\gamma) = \begin{pmatrix} 1 & 0 & 0 \\ 0 & \cos\alpha & -\sin\alpha \\ 0 & \sin\alpha & \cos\alpha \end{pmatrix} \cdot \begin{pmatrix} \cos\beta & 0 & \sin\beta \\ 0 & 1 & 0 \\ -\sin\beta & 0 & \cos\beta \end{pmatrix} \cdot$$
$$\begin{pmatrix} \cos\gamma & -\sin\gamma & 0 \\ \sin\gamma & \cos\gamma & 0 \\ 0 & 0 & 1 \end{pmatrix} \tag{3.112}$$

The six parameters $(\alpha, \beta, \gamma, t_x, t_y, t_z)$ of R and T are called the exterior camera parameters, exterior orientation, or camera pose, because they determine the position of the camera with respect to the world.

The next step of the mapping is the projection of the 3D point P_c into the image plane coordinate system (IPCS). For the pinhole camera model, the projection is a perspective projection, which is given by

$$\begin{pmatrix} u \\ v \end{pmatrix} = \frac{f}{z_c} \begin{pmatrix} x_c \\ y_c \end{pmatrix} \tag{3.113}$$

For the telecentric camera model, the projection is a parallel projection, which is given by

$$\begin{pmatrix} u \\ v \end{pmatrix} = \begin{pmatrix} x_c \\ y_c \end{pmatrix} \tag{3.114}$$

As can be seen, there is no focal length f for telecentric cameras. Furthermore, the distance z_c of the object to the camera has no influence on the image coordinates.

After the projection to the image plane, lens distortions cause the coordinates $(u, v)^\top$ to be modified. This is a transformation that can be modeled in the image plane alone, i.e., 3D information is unnecessary. For most lenses, the distortion can be approximated sufficiently well by a radial distortion, given by [99, 100, 107]:

$$\begin{pmatrix} \tilde{u} \\ \tilde{v} \end{pmatrix} = \frac{2}{1 + \sqrt{1 - 4\kappa(u^2 + v^2)}} \begin{pmatrix} u \\ v \end{pmatrix} \quad (3.115)$$

The parameter κ models the magnitude of the radial distortions. If κ is negative, the distortion is barrel-shaped; while for positive κ it is pincushion-shaped. Figure 3.98 shows the effect of κ for an image of a calibration target that we will use below to calibrate the camera. Compare this to Figure 2.34, which shows the effect of lens distortions for circular and rectangular grids.

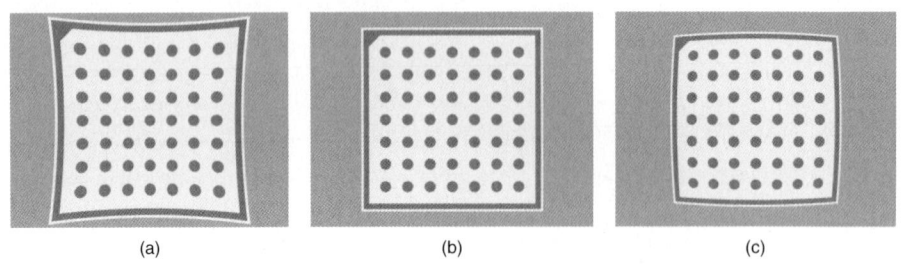

Fig. 3.98 Effects of radial distortion. (a) Pincushion distortion: $\kappa > 0$. (b) No distortion: $\kappa = 0$. (c) Barrel distortion: $\kappa < 0$.

While other models for distortions are possible [116], this model for the lens distortions describes almost all the lenses commonly encountered in machine vision applications with sufficient accuracy. In contrast to other models, it has the great advantage that the distortion correction can be calculated analytically by

$$\begin{pmatrix} u \\ v \end{pmatrix} = \frac{1}{1 + \kappa(\tilde{u}^2 + \tilde{v}^2)} \begin{pmatrix} \tilde{u} \\ \tilde{v} \end{pmatrix} \quad (3.116)$$

This will be important when we compute world coordinates from image coordinates.

Finally, the point $(u, v)^\top$ is transformed from the image plane coordinate system into the image coordinate system (ICS):

$$\begin{pmatrix} r \\ c \end{pmatrix} = \begin{pmatrix} \dfrac{\tilde{v}}{s_y} + c_y \\ \dfrac{\tilde{u}}{s_x} + c_x \end{pmatrix} \quad (3.117)$$

Here, s_x and s_y are scaling factors. For pinhole cameras, they represent the horizontal and vertical pixel pitch on the sensor (see Section 2.3.4). For telecentric cameras, they represent the size of a pixel in world coordinates (not taking into account the radial distortions). The point $(c_x, c_y)^\top$ is the principal point of the image. For pinhole cameras, this is the perpendicular projection of the projection center onto the image plane, i.e., the point in the image from which a ray through the projection center is perpendicular to the image plane. It also defines the center of the radial distortions. For telecentric cameras, no projection center exists. Therefore, the principal point is solely defined by the radial distortions.

The six parameters $(f, \kappa, s_x, s_y, c_x, c_y)$ of the pinhole camera and the five parameters $(\kappa, s_x, s_y, c_x, c_y)$ of the telecentric camera are called the interior camera parameters or interior orientation, because they determine the projection from 3D to 2D performed by the camera.

3.9.2
Camera Model for Line Scan Cameras

As described in Section 2.3.1, line scan cameras must move with respect to the object to acquire a useful image. The relative motion between the camera and the object is part of the interior orientation. By far the most frequent motion is a linear motion of the camera, i.e., the camera moves with constant velocity along a straight line relative to the object, the orientation of the camera is constant with respect to the object, and the motion is equal for all images [117]. In this case, the motion can be described by the motion vector $V = (v_x, v_y, v_z)^\top$, as shown in Figure 3.99. The vector V is best described in units of meters per scan line in the camera coordinate system. As shown in Figure 3.99, the definition of V assumes a moving camera and a fixed object. If the camera is stationary and the object is moving, e.g., on a conveyor belt, we can simply use $-V$ as the motion vector.

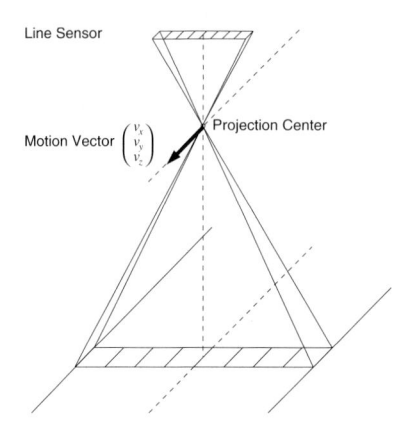

Fig. 3.99 Principle of line scan image acquisition.

The camera model for a line scan camera is displayed in Figure 3.100. The origin of the camera coordinate system (CCS) is the projection center. The z axis is identical to the optical axis and is oriented such that points in front of the camera have positive z coordinates. The y axis is perpendicular to the sensor line and to the z axis. It is oriented such that the motion vector has a positive y component, i.e., if a fixed object is assumed, the y axis points in the direction in which the camera is moving. The x axis is perpendicular to the y and z axes, so that the x, y, and z axes form a right-handed coordinate system.

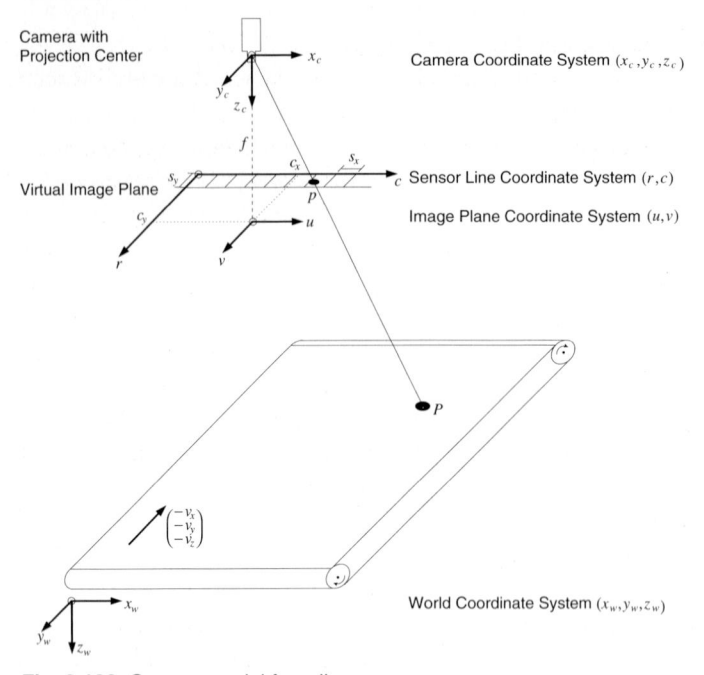

Fig. 3.100 Camera model for a line scan camera.

Similar to area scan cameras, the projection of a point given in world coordinates into the image is modeled in two steps. First, the point is transformed from the world coordinate system (WCS) into the CCS. Then, it is projected into the image.

As the camera moves over the object during the image acquisition, the camera coordinate system moves with respect to the object, i.e., each image is imaged from a different position. This means that each line has a different pose. To make things easier, we can use the fact that the motion of the camera is linear. Hence, it suffices to know the transformation from the WCS to the CCS for the first line of the image. The poses of the remaining lines can be computed from the motion vector, i.e., the motion vector V is taken into account during the projection of P_c into the image. With this, the transformation from the WCS to the CCS is identical to Eq. (3.111). Like for area scan cameras, the six parameters $(\alpha, \beta, \gamma, t_x, t_y, t_z)$ are called the exterior camera parameters or exterior orientation, because they determine the position of the camera with respect to the world.

To obtain a model for the interior geometry of a line scan camera, we can regard a line sensor as one particular line of an area sensor. Therefore, like in area scan cameras, there is an image plane coordinate system (IPCS) that lies at a distance of f (the principal distance) behind the projection center. Again, the computations can be simplified if we pretend that the image plane lies in front of the projection center, as shown in Figure 3.100. We will defer the description of the projection performed by the line scan camera until later, since it is more complicated than for area scan cameras.

Let us assume that the point P_c has been projected to the point $(u,v)^\top$ in the IPCS. Like for area scan cameras, the point is now distorted by the radial distortion Eq. (3.115), which results in a distorted point $(\tilde{u}, \tilde{v})^\top$.

Finally, like for area scan cameras, $(\tilde{u}, \tilde{v})^\top$ is transformed into the image coordinate system (ICS), resulting in the coordinates $(r,c)^\top$. Since we want to model the fact that the line sensor may not be mounted exactly behind the projection center, which often occurs in practice, we again have to introduce a principal point $(c_x, c_y)^\top$ that models how the line sensor is shifted with respect to the projection center, i.e., it describes the relative position of the principal point with respect to the line sensor. Since $(\tilde{u}, \tilde{v})^\top$ is given in metric units, e.g., meters, we need to introduce two scale factors s_x and s_y that determine how the IPCS units are converted to ICS units (i.e., pixels). Like for area scan cameras, s_x represents the horizontal pixel pitch on the sensor. As we will see below, s_y only serves as a scaling factor that enables us to specify the principal point in pixel coordinates. The value of s_y cannot be calibrated and must be set to the pixel size of the line sensor in the vertical direction.

To determine the projection of the point $P_c = (x_c, y_c, z_c)^\top$ (specified in the CCS), we first consider the case where there are no radial distortions ($\kappa = 0$), the line sensor is mounted precisely behind the projection center ($c_y = 0$), and the motion is purely in the y direction of the CCS ($V = (0, v_y, 0)^\top$). In this case, the row coordinate of the projected point p is proportional to the time it takes for the point P_c to appear directly under the sensor, i.e., to appear in the xz plane of the CCS. To determine this, we must solve $x_c - tv_y = 0$ for the "time" t (since V is specified in meters per scan line, the units of t are actually scan lines, i.e., pixels). Hence, $r = t = x_c/v_y$. Since $v_x = 0$, we also have $u = fx_c/z_c$ and $c = u/s_x + c_x$. Hence, the projection is a perspective projection in the direction of the line sensor and a parallel projection perpendicular to the line sensor (i.e., in the special motion direction $(0, v_y, 0)^\top$).

For general motion vectors (i.e., $v_x \neq 0$ or $v_z \neq 0$), non-perfectly aligned line sensors ($c_y \neq 0$), and radial distortions ($\kappa \neq 0$), the equations become significantly more complicated. As above, we need to determine the "time" t when the point P_c appears in the "plane" spanned by the projection center and the line sensor. We put "plane" in quotes since the radial distortion will cause the back-projection of the line sensor to be a curved surface in space whenever $c_y \neq 0$ and $\kappa \neq 0$. To solve this problem, we construct the optical ray through the projection center and the projected point $p = (r,c)^\top$. Let us assume that we have transformed $(r,c)^\top$ into the distorted IPCS, where we have coordinates $(\tilde{u}, \tilde{v})^\top$. Here, $\tilde{v} = s_y c_y$ is the coordinate of the principal point in metric units. Then, we undistort $(\tilde{u}, \tilde{v})^\top$ by Eq. (3.116), i.e., $(u,v)^\top = d(\tilde{u}, \tilde{v})^\top$, where $d = 1/[1 + \kappa(\tilde{u}^2 + \tilde{v}^2)]$ is the undistortion factor from Eq. (3.116). The optical ray is now given by the line equation $\lambda(u, v, f)^\top = \lambda(d\tilde{u}, d\tilde{v}, f)^\top$. The point P_c moves along the line given by $(x_c, y_c, z_c)^\top - t(v_x, v_y, v_z)^\top$ during the acquisition of the image. If p is the projection of P_c, both lines must intersect. Therefore, to determine the projection of P_c, we must solve the following nonlinear set of equations

$$\begin{aligned} \lambda d\tilde{u} &= x_c - tv_x \\ \lambda d\tilde{v} &= y_c - tv_y \\ \lambda f &= z_c - tv_z \end{aligned} \quad (3.118)$$

for λ, $\tilde{u}$, and t, where d and $\tilde{v}$ are defined above. From $\tilde{u}$ and t, the pixel coordinates can be computed by

$$\begin{pmatrix} r \\ c \end{pmatrix} = \begin{pmatrix} t \\ \dfrac{\tilde{u}}{s_x} + c_x \end{pmatrix} \quad (3.119)$$

The nine parameters $(f, \kappa, s_x, s_y, c_x, c_y, v_x, v_y, v_z)$ of the line scan camera are called the interior orientation, because they determine the projection from 3D to 2D performed by the camera.

Although the line scan camera geometry is conceptually simply a mixture of a perspective and a telecentric lens, precisely this mixture makes the line scan camera geometry much more complex than the area scan geometries. Figure 3.101 displays some of the effects that can occur. In Figure 3.101(a), the pixels are non-square because the motion is not tuned to the line frequency of the camera. In this example, either the line frequency would need to be increased or the motion speed would need to be decreased in order to obtain square pixels. Figure 3.101(b) shows the effect of a motion vector of the form $V = (v_x, v_y, 0)$ with $v_x = v_y/10$. We obtain skew pixels. In this case, the camera would need to be better aligned with the motion vector. Figures 3.101(c) and (d) show that straight lines can be projected to hyperbolic arcs [117], even if the lens has no distortions ($\kappa = 0$). This effect occurs even if $V = (0, v_y, 0)$, i.e., if the line sensor is perfectly aligned perpendicular to the motion vector. The hyperbolic arcs are more pronounced if the motion vector has non-zero v_z. Figures 3.101(e)–(h) show the effect of a lens with distortions ($\kappa \neq 0$) for the case that the sensor is located perfectly behind the projection center ($c_y = 0$) and for the case that the sensor is not perfectly aligned (here $c_y > 0$). For $c_y = 0$, the pincushion and barrel distortions only cause distortions within each row of the image. For $c_y \neq 0$, the rows are also bent.

3.9.3
Calibration Process

From the above discussion, we can see that camera calibration is the process of determining the interior and exterior camera parameters. To perform the calibration, it is necessary to know the location of a sufficiently large number of 3D points in world coordinates, and to be able to determine the correspondence between the world points

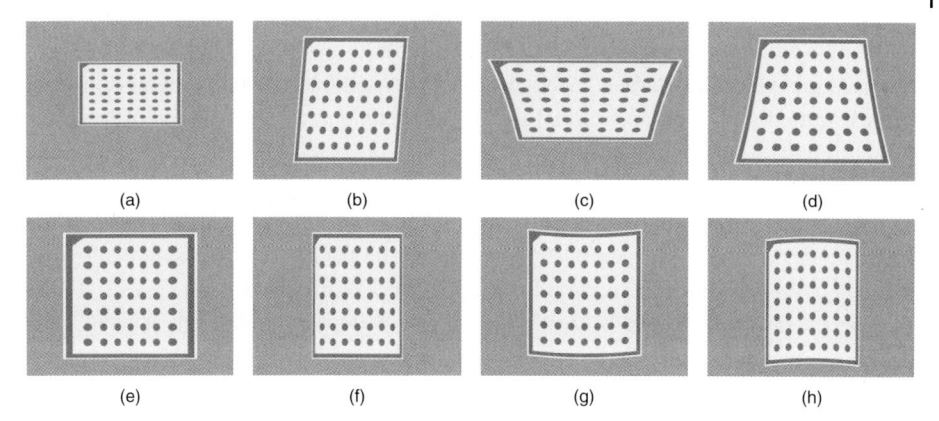

Fig. 3.101 Some effects that occur for the line scan camera geometry. (a) Non-square pixels because the motion is not tuned to the line frequency of the camera. (b) Skewed pixels because the motion is not parallel to the y axis of the CCS. (c) Straight lines can project to hyperbolic arcs, even if the line sensor is perpendicular to the motion. (d) This effect is more pronounced if the motion has a non-zero z component. Note that hyperbolic arcs occur even if the lens has no distortions ($\kappa = 0$). (e) Pincushion distortion ($\kappa > 0$) for $c_y = 0$. (f) Barrel distortion for $c_y = 0$. (g) Pincushion distortion ($\kappa > 0$) for $c_y > 0$. (h) Barrel distortion for $c_y > 0$.

and their projections in the image. To meet the first requirement, usually objects or marks that are easy to extract, e.g., circles or linear grids, must be placed into known locations. If the location of a camera must be known with respect to a given coordinate system, e.g., with respect to the building plan of, say, a factory building, then each mark location must be measured very carefully within this coordinate system. Fortunately, it is often sufficient to know the position of a reference object with respect to the camera to be able to measure the object precisely, since the absolute position of the object in world coordinates is unimportant. Therefore, a movable calibration target that has been measured accurately can be used to calibrate the camera. This has the advantage that the calibration can be performed with the camera in place, e.g., already mounted in the machine. Furthermore, the position of the camera with respect to the objects can be recalibrated if required, e.g., if the object type to be inspected changes.

The second requirement, i.e., the necessity to determine the correspondence of the known world points and their projections in the image, is in general a hard problem. Therefore, calibration targets are usually constructed in such a way that this correspondence can be determined easily. For example, a planar calibration target with $m \times n$ circular marks within a rectangular border can be used. We have already seen examples of this kind of calibration target in Figures 3.85, 3.92, 3.98, and 3.101. Planar calibration targets have several advantages. First of all, they can be handled easily. Second, they can be manufactured very accurately. Finally, they can be used for back light applications easily if a transparent medium is used as the carrier for the marks. A rectangular border around the calibration target allows the inner part of the calibration target to be found easily. A small orientation mark in one of the corners of the border enables the camera calibration algorithm to uniquely determine the orientation of the calibration target. Circular marks are used because their center point can be determined with high accuracy. Finally, the rectangular matrix layout of the rows and columns of the circles enables the camera calibration algorithm to determine the correspondence between the marks and their image points easily.

To extract the calibration target, we can make use of the fact that the border separates the inner part of the calibration target from the background. Consequently, the inner part can be found by a simple threshold operation (see Section 3.4.1). Since the correct threshold depends on the brightness of the calibration target in the image, different thresholds can be tried automatically until a region with $m \times n$ holes (corresponding to the calibration marks) has been found.

Once the region of the inner part of the calibration target has been found, the borders of the calibration marks can be extracted with subpixel-accurate edge extraction (see Section 3.7.3). Since the projections of the circular marks are ellipses, ellipses can then be fitted to the extracted edges with the algorithms described in Section 3.8.3 to obtain robustness against outliers in the edge points and to increase the accuracy. An example of the extraction of the calibration marks was already shown in Figure 3.92.

Finally, the correspondence between the calibration marks and their projections in the image can be determined easily based on the smallest enclosing quadrilateral of the extracted ellipses. Furthermore, the orientation of the marks can be determined uniquely based on the small triangular orientation mark. If it were not present, the orientation can only be determined modulo 90° for square calibration targets and modulo 180° for rectangular targets. It should be noted that the orientation of the calibration target is only important if the distances between the marks are not identical in both axes or in applications where the orientation must be determined uniquely across multiple images. One such example is stereo with cameras that are turned heavily with respect to each other around their respective optical axes.

After the correspondence between the marks and their projections has been determined, the camera can be calibrated. Let us denote the 3D positions of the centers of the marks by M_i. Since the calibration target is planar, in our case we can place the calibration target in the plane $z = 0$. However, what we describe in the following is completely general and can be used for arbitrary calibration targets. Furthermore, let us denote the projections of the centers of the marks in the image by m_i. Here, we must take into account that the projection of the center of the circle is not the center of the ellipse [118]. Finally, let us denote the camera parameters by a vector c. As described above, c consists of the interior and exterior orientation parameters of the respective camera model. For example $c = (f, \kappa, s_x, s_y, c_x, c_y, \alpha, \beta, \gamma, t_x, t_y, t_z)$ for pinhole cameras. Here, it should be noted that the exterior orientation is determined by attaching the world coordinate system to the calibration target, e.g., to the center mark, in such a way that the x and y axes of the WCS are aligned with the row and column directions of the marks on the calibration target and the z axis points in the same direction as the optical axis if the calibration target is parallel to the image plane. Then, the camera parameters can be determined by minimizing the distance of the extracted mark centers m_i and their projections $\pi(M_i, c)$:

$$d(c) = \sum_{i=1}^{k} \|m_i - \pi(M_i, c)\|^2 \to \min \qquad (3.120)$$

Here, $k = mn$ is the number of calibration marks. This is a difficult nonlinear optimization problem. Therefore, good starting values are required for the parameters. The interior orientation parameters can be determined from the specifications of the image sensor and the lens. The starting values for the exterior orientation are in general harder to obtain. For the planar calibration target described above, good starting values can be obtained based on the geometry and size of the projected circles [100, 107].

The optimization in Eq. (3.120) cannot determine all camera parameters because the physically motivated camera models we have chosen are over-parameterized. For example, for the pinhole camera, f, s_x, and s_y cannot be determined uniquely since they contain a common scale factor, as shown in Figure 3.102. For example, making the pixels twice as large and increasing the principal distance by a factor of 2 results in the same image. A similar effect happens for the line scan camera mode. The solution to this problem is to keep s_y fixed in the optimization since the image is transmitted row-by-row in the video signal (see Section 2.4). This fixes the common scale factor. On the other hand, s_x cannot be kept fixed in general since the video signal may not be sampled pixel-synchronously (see Section 2.4.1).

Even if the above problem is solved, some degeneracies remain because we use a planar calibration target. For example, as shown in Figure 3.103, if the calibration target is parallel to the image plane, f and t_z cannot be determined uniquely since they contain a common scale factor. This problem also occurs if the calibration target is not parallel to the image plane. Here, f and a combination of parameters from the exterior orientation can only be determined up to a one-parameter ambiguity. For example, if the calibration target is rotated around the x axis, f, t_z, and α cannot be determined at the same time. For telecentric cameras, it is generally impossible to determine s_x, s_y, and the rotation angles from a single image. For example, a rotation of the calibration target around the x axis can be compensated by a corresponding change in s_x. For line scan cameras, there are similar degeneracies as the ones described above plus degeneracies that include the motion vector.

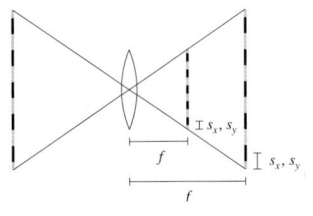

Fig. 3.102 For pinhole cameras, f, s_x, and s_y cannot be determined uniquely.

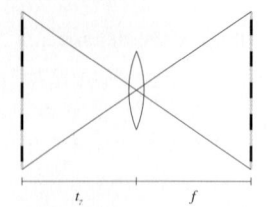

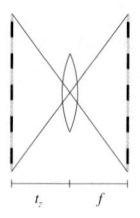

Fig. 3.103 For pinhole cameras, f and t_z cannot be determined uniquely.

To prevent the above degeneracies, the camera must be calibrated from multiple images in which the calibration target is positioned to avoid the degeneracies. For example, for pinhole cameras, the calibration targets must not be parallel to each other in all images; while for telecentric cameras, the calibration target must be rotated around all of its axes to avoid the above degeneracies. Suppose we use image l for the calibration. Then, we also have l sets of exterior orientation parameters $(\alpha_l, \beta_l, \gamma_l, t_{x,l}, t_{y,l}, t_{z,l})$ that must be determined. As above, we collect the interior orientation parameters and the l sets of exterior orientation parameters into the camera parameter vector c. Then, to calibrate the camera, we must solve the following optimization problem:

$$d(c) = \sum_{j=1}^{l} \sum_{i=1}^{k} \|m_{i,j} - \pi(M_i, c)\|^2 \to \min \quad (3.121)$$

Here, $m_{i,j}$ denotes the projection of the ith calibration mark in the jth image. If the calibration targets are placed and oriented suitably in the images, this will determine all the camera parameters uniquely. To ensure high accuracy of the camera parameters so determined, the calibration target should be placed into all four corners of the image. Since the distortion is largest in the corners, this will facilitate the determination of the radial distortion coefficient κ with the highest possible accuracy.

To conclude this section, we mention that a flexible calibration algorithm will enable one to specify a subset of the parameters that should be determined. For example, from the mechanical setup, some of the parameters of the exterior orientation may be known. One frequently encountered example is that the mechanical setup ensures with high accuracy that the calibration target is parallel to the image plane. In this case, we should set $\alpha = \beta = 0$, and the camera calibration should leave these parameters fixed. Another example is a camera with square pixels that transmits the video signal digitally. Here, $s_x = s_y$, and for pinhole cameras both parameters should be kept fixed. Finally, we will see in the next section that the calibration target determines the world plane in which measurements from a single image can be performed. In some applications, there is not enough space for the calibration target to be turned in 3D if the camera is mounted in its final position. Here, a two-step approach can be used. First, the interior orientation of the camera is determined with the camera not mounted in its final position. This ensures that the calibration target can be moved freely. (Note that this also determines the exterior orientation of the calibration target;

however, this information is discarded.) Then, the camera is mounted in its final position. Here, it must be ensured that the focus and iris diaphragm settings of the lens are not changed, since this will change the interior orientation. In the final position, a single image of the calibration target is taken and only the exterior orientation is optimized to determine the pose of the camera with respect to the measurement plane.

3.9.4
World Coordinates from Single Images

As mentioned above and at the end of Section 3.7.4, if the camera is calibrated it is possible in principle to obtain undistorted measurements in world coordinates. In general, this can only be done if two or more images of the same object are taken at the same time with cameras at different spatial positions. This is called stereo reconstruction. With this approach, discussed in Section 3.10, the reconstruction of 3D positions for corresponding points in the two images is possible because the two optical rays defined by the two optical centers of the cameras and the points in the image plane defined by the two image points can be intersected in 3D space to give the 3D position of that point. In some applications, however, it is impossible to use two cameras, e.g., because there is not enough space to mount two cameras. Nevertheless, it is possible to obtain measurements in world coordinates for objects acquired through telecentric lenses and for objects that lie in a known plane, e.g., on a conveyor belt, for pinhole and line scan cameras. Both of these problems can be solved by intersecting an optical ray (also called line of sight) with a plane. With this, it is possible to measure objects that lie in a plane, even if the plane is tilted with respect to the optical axis.

Let us first look at the problem of determining world coordinates for telecentric cameras. In this case, the parallel projection in Eq. (3.114) discards any depth information completely. Therefore, we cannot hope to discover the distance of the object from the camera, i.e., its z coordinate in the CCS. What we can recover, however, are the x and y coordinates of the object in the CCS (x_c and y_c in Eq. (3.114)), i.e., the dimensions of the object in world units. Since the z coordinate of P_c cannot be recovered, in most cases it is unnecessary to transform P_c into world coordinates by inverting Eq. (3.111). Instead, the point P_c is regarded as a point in world coordinates. To recover P_c, we can start by inverting Eq. (3.117) to transform the coordinates from the ICS into the IPCS:

$$\begin{pmatrix} \tilde{u} \\ \tilde{v} \end{pmatrix} = \begin{pmatrix} s_x(c - c_x) \\ s_y(r - c_y) \end{pmatrix} \tag{3.122}$$

Then, we can remove the radial distortions by applying Eq. (3.116) to obtain the undistorted coordinates $(u, v)^\top$ in the image plane. Finally, the coordinates of P_c are given by

$$P_c = (x_c, y_c, z_c)^\top = (u, v, 0)^\top \tag{3.123}$$

Note that the above procedure is equivalent to intersecting the optical ray given by the point $(u, v, 0)^\top$ and the direction perpendicular to the image plane, i.e., $(0, 0, 1)^\top$, with the plane $z = 0$.

The determination of world coordinates for pinhole cameras is slightly more complicated, but uses the same principle of intersecting an optical ray with a known plane. Let us look at this problem by using the application where this procedure is most useful. In many applications, the objects to be measured lie in a plane in front of the camera, e.g., a conveyor belt. Let us assume for the moment that the location and orientation of this plane (its pose) is known. We will describe below how to obtain this pose. The plane can be described by its origin and a local coordinate system, i.e., three orthogonal vectors, one of which is perpendicular to the plane. To transform the coordinates in the coordinate system of the plane (the WCS) into the CCS, a rigid transformation given by Eq. (3.111) must be used. Its six parameters $(\alpha, \beta, \gamma, t_x, t_y, t_z)$ describe the pose of the plane. If we want to measure objects in this plane, we need to determine the object coordinates in the WCS defined by the plane. Conceptually, we need to intersect the optical ray corresponding to an image point with the plane. To do so, we need to know the two points that define the optical ray. Recalling Figures 3.96 and 3.97, obviously the first point is given by the projection center, which has the coordinates $(0, 0, 0)^\top$ in the CCS. To obtain the second point, we need to transform the point $(r, c)^\top$ from the ICS into the IPCS. This transformation is given by Eqs. (3.122) and (3.116). To obtain the 3D point that corresponds to this point in the image plane, we need to take into account that the image plane lies at a distance of f in front of the optical center. Hence, the coordinates of the second point on the optical ray are given by $(u, v, f)^\top$. Therefore, we can describe the optical ray in the CCS by

$$L_c = (0, 0, 0)^\top + \lambda (u, v, f)^\top \tag{3.124}$$

To intersect this line with the plane, it is best to express the line L_c in the WCS of the plane, since in the WCS the plane is given by the equation $z = 0$. Therefore, we need to transform the two points $(0, 0, 0)^\top$ and $(u, v, f)^\top$ into the WCS. This can be done by inverting Eq. (3.111) to obtain

$$P_w = R^{-1}(P_c - T) = R^\top (P_c - T) \tag{3.125}$$

Here, $R^{-1} = R^\top$ is the inverse of the rotation matrix R in Eq. (3.111). Let us call the transformed optical center O_w, i.e., $O_w = R^\top ((0, 0, 0)^\top - T) = -R^\top T$, and the transformed point in the image plane I_w, i.e., $I_w = R^\top ((u, v, f)^\top - T)$. With this, the optical ray is given by

$$L_w = O_w + \lambda (I_w - O_w) = O_w + \lambda D_w \tag{3.126}$$

in the WCS. Here, D_w denotes the direction vector of the optical ray. With this, it is a simple matter to determine the intersection of the optical ray in Eq. (3.126) with the plane $z = 0$. The intersection point is given by

$$P_w = \begin{pmatrix} o_x - o_z d_x/d_z \\ o_y - o_z d_y/d_z \\ 0 \end{pmatrix} \quad (3.127)$$

where $O_w = (o_x, o_y, o_z)^\top$ and $D_w = (d_x, d_y, d_z)^\top$.

Up to now, we have assumed that the pose of the plane in which we want to measure objects is known. Fortunately, the camera calibration gives us this pose almost immediately since a planar calibration target is used. If the calibration target is placed onto the plane, e.g., the conveyor belt, in one of the images used for calibration, the exterior orientation of the calibration target in that image almost defines the pose of the plane we need in the above derivation. The pose would be the true pose of the plane if the calibration target were infinitely thin. To take the thickness of the calibration target into account, the WCS defined by the exterior orientation must be moved by the thickness of the calibration target in the positive z direction. This modifies the transformation from the WCS to the CCS in Eq. (3.111) as follows: the translation T simply becomes $RD + T$, where $D = d(0, 0, 1)^\top$, and d is the thickness of the calibration target. This is the pose that must be used in Eq. (3.125) to transform the optical ray into the WCS. An example of computing edge positions in world coordinates for pinhole cameras is given at the end of Section 3.7.4 (see Figure 3.86).

For line scan cameras, the procedure to obtain world coordinates in a given plane is conceptually similar to the approaches described above. First, the optical ray is constructed from Eqs. (3.119) and (3.118). Then, it is transformed into the WCS and intersected with the plane $z = 0$.

In addition to transforming image points, e.g., 1D edge positions or subpixel-precise contours, into world coordinates, sometimes it is also useful to transform the image itself into world coordinates. This creates an image that would have resulted if the camera had looked perfectly perpendicularly without distortions onto the world plane. This image rectification is useful for applications that must work on the image data itself, e.g., region processing, template matching, or optical character recognition. It can be used whenever the camera cannot be mounted perpendicular to the measurement plane. To rectify the image, we conceptually cut out a rectangular region of the world plane $z = 0$ and sample it with a specified distance, e.g., 200 µm. We then project each sample point into the image with the equations of the relevant camera model, and obtain the gray value through interpolation, e.g., bilinear interpolation. Figure 3.104 shows an example of this process. In Figure 3.104(a), the image of a caliper together with the calibration target that defines the world plane is shown. The unrectified and rectified images of the caliper are shown in Figures 3.104(b) and (c), respectively. Note that the rectification has removed the perspective and radial distortions from the image.

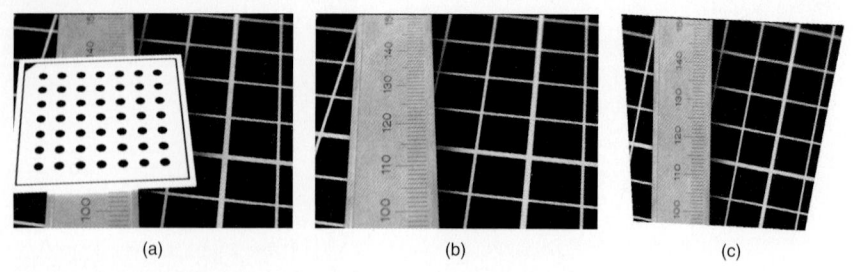

Fig. 3.104 (a) Image of a caliper with a calibration target. (b) Unrectified image of the caliper. (c) Rectified image of the caliper.

3.9.5
Accuracy of the Camera Parameters

We conclude the discussion of camera calibration by discussing two different aspects: the accuracy of the camera parameters, and the changes in the camera parameters that result from adjusting the focus and iris diaphragm settings on the lens.

As was already noted in Section 3.9.3, there are some cases where an inappropriate placement of the calibration target can result in degenerate configurations where one of the camera parameters or a combination of some of the parameters cannot be determined. These configurations must obviously be avoided if the camera parameters should be determined with high accuracy. Apart from this, the main influencing factor for the accuracy of the camera parameters is the number of images that is used to calibrate the camera. This is illustrated in Figure 3.105, where the standard deviations of the principal distance f, the radial distortion coefficient κ, and the principal point (c_x, c_y) are plotted as functions of the number of images that are used for calibration. To obtain this data, 20 images of a calibration target were taken. Then, every possible subset of l images ($l = 2, \ldots, 19$) from the 20 images was used to calibrate the camera. The standard deviations were calculated from the resulting camera parameters when l of the 20 images were used for the calibration. From Figure 3.105 it is obvious that the accuracy of the camera parameters increases significantly as the number of images l increases. This is not surprising if we consider that each image serves to constrain the parameters. If the images were independent measurements, we could expect the standard deviation to decrease proportionally to $l^{-0.5}$. In this particular example, f decreases roughly proportionally to $l^{-1.5}$, κ decreases roughly proportionally to $l^{-1.1}$, and c_x and c_y decrease roughly proportionally to $l^{-1.2}$, i.e., much faster than $l^{-0.5}$.

From Figure 3.105, it can also be seen that a comparatively large number of calibration images is required to determine the camera parameters accurately. This happens because there are non-negligible correlations between the camera parameters, which can only be resolved through multiple independent measurements. To obtain accurate camera parameters, it is important that the calibration target covers the entire field of view and that it tries to cover the range of exterior orientations as well as possible.

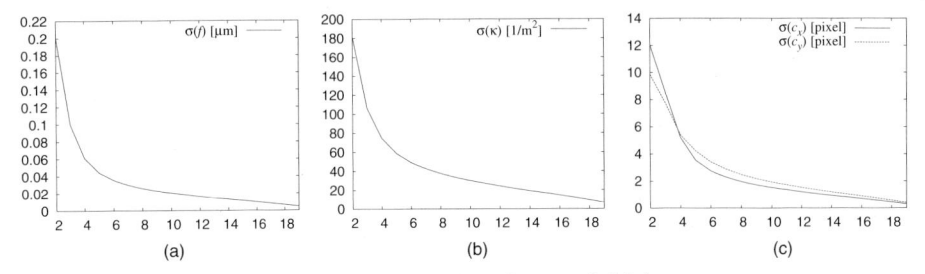

Fig. 3.105 Standard deviations of (a) the principal distance f, (b) the radial distortion coefficient κ, and (c) the principal point (c_x, c_y) as functions of the number of images that are used for calibration.

In particular, κ can be determined more accurately if the calibration target is placed into each corner of the image. Furthermore, all parameters can be determined more accurately if the calibration target covers a large depth range. This can be achieved by turning the calibration target around its x and y axes, and by placing it at different depths relative to the camera.

We now turn to a discussion of whether changes in the lens settings change the camera parameters. Figure 3.106 displays the effect of changing the focus setting of the lens. Here, a 12.5 mm lens with a 1 mm extension tube was used. The lens was set to the nearest and farthest focal settings. In the near focus setting, the camera was calibrated with a calibration target of size 1 cm × 1 cm, while for the far focus setting a calibration target of size 3 cm × 3 cm was used. This resulted in the same size of the calibration target in the focusing plane for both settings. Care was taken to use images of calibration targets with approximately the same range of depths and positions in the images. For each setting, 20 images of the calibration target were taken. To be able to evaluate statistically whether the camera parameters are different, all 20 subsets of 19 of the 20 images were used to calibrate the camera. As can be expected from the discussion in Section 2.2.2, changing the focus will change the principal distance. From Figure 3.106(a), we can see that this clearly is the case. Furthermore, the radial distortion coefficient κ also changes significantly. From Figure 3.106(b), it can also be seen that the principal point changes. While it may seem that this change is not as significant, the probability that the two means of the respective principal points are identical is less than 10^{-7}. Therefore, the principal point also changes significantly for this lens.

Finally, we examine what can happen when the iris diaphragm on the lens is changed. To test this, a similar setup as above was used. The camera was set to f-numbers $f/4$ and $f/11$. For each setting, 20 images of a 3 cm × 3 cm calibration target were taken. Care was taken to position the calibration targets in similar positions for the two settings. The lens is an 8.5 mm lens with a 1 mm extension tube. Again, all 20 subsets of 19 of the 20 images were used to calibrate the camera. Figure 3.107(a) displays the principal distance and radial distortion coefficient. Clearly, the two parameters change in a statistically significant way. Figure 3.107(b) also shows that the

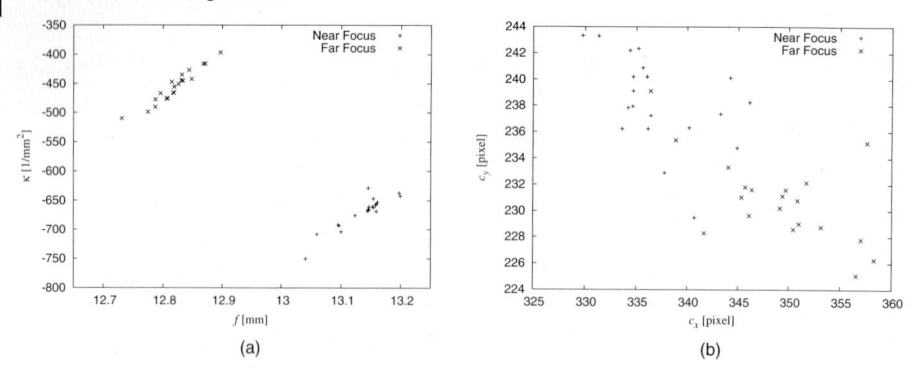

Fig. 3.106 (a) Principal distances and radial distortion coefficients and (b) principal points for a lens with two different focus settings.

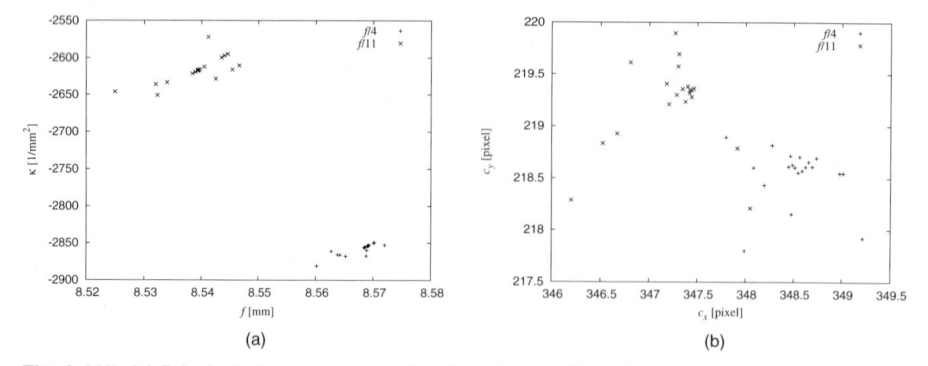

Fig. 3.107 (a) Principal distances and radial distortion coefficients and (b) principal points for a lens with two different iris diaphragm settings (f-numbers: $f/4$ and $f/11$).

principal point changes significantly. All parameters have a probability of being equal that is less than 10^{-7}. The changes in the parameters mean that there is an overall difference in the point coordinates across the image diagonal of approximately 1.5 pixels. Therefore, we can see that changing the f-number on the lens requires a recalibration, at least for some lenses.

3.10
Stereo Reconstruction

In Sections 3.9 and 3.7.4, we have seen that we can perform very accurate measurements from a single image by calibrating the camera and by determining its exterior orientation with respect to a plane in the world. We could then convert the image measurements to world coordinates within the plane by intersecting optical rays with the plane. Note, however, that these measurements are still 2D measurements within the world plane. In fact, from a single image we cannot reconstruct the 3D geometry of the scene because we can only determine the optical ray for each point in the image.

We do not know at which distance on the optical ray the point lies in the world. In the approach in Section 3.9.4, we had to assume a special geometry in the world to be able to determine the distance of a point along the optical ray. Note that this is not a true 3D reconstruction. To perform a 3D reconstruction, we must use at least two images of the same scene taken from different positions. Typically, this is done by simultaneously taking the images with two cameras. This process is called stereo reconstruction. In this section, we will examine the case of binocular stereo, i.e., we will concentrate on the two-camera case. Throughout this section, we will assume that the cameras have been calibrated, i.e, their interior orientations and relative orientation are known. While uncalibrated reconstruction is also possible [66, 67], the corresponding methods have not yet been used in industrial applications.

3.10.1
Stereo Geometry

Before we can discuss the stereo reconstruction, we must examine the geometry of two cameras, as shown in Figure 3.108. Since the cameras are assumed to be calibrated, we know their interior orientations, i.e., their principal points, focal lengths (more precisely, their principal distances, i.e., the distance between the image plane and the projection center), pixel size, and distortion coefficient. In Figure 3.108, the principal points are visualized by the points C_1 and C_2 in the first and second image, respectively. Furthermore, the projection centers are visualized by the points O_1 and O_2. The dashed line between the projection centers and principal points visualizes the principal distances. Note that, since the image planes physically lie behind the projection centers, the image is turned upside down. Consequently, the origin of the image coordinate system lies in the lower right corner, with the row axis pointing upward and the column axis pointing leftward. The camera coordinate system axes are defined such that the x axis points to the right, the y axis points downwards, and the z axis points forward from the image plane, i.e., along the viewing direction. The position and orientation of the two cameras with respect to each other are given by the relative orientation, which is a rigid 3D transformation specified by the rotation matrix R_r and the translation vector T_r. It can be interpreted either as the transformation of the camera coordinate system of the first camera into the camera coordinate system of the second camera or as a transformation that transforms point coordinates in the camera coordinate system of the second camera into point coordinates of the camera coordinate system of the first camera: $P_{c1} = R_r P_{c2} + T_r$. The translation vector T_r, which specifies the translation between the two projection centers, is also called the base. With this, we can see that a point P_w in the world is mapped to a point P_1 in the first image and to a point P_2 in the second image. If there is no distortion in the lens (which we will assume for the moment), the points P_w, O_1, O_2, P_1, and P_2 all lie in a single plane.

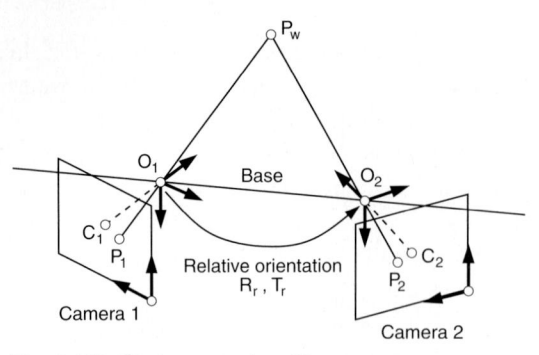

Fig. 3.108 Stereo geometry of two cameras.

To calibrate the stereo system, we can extend the method of Section 3.9.3 as follows. Let M_i denote the positions of the calibration marks. We extract their projections in both images with the methods described in Section 3.9.3. Let us denote the projection of the centers of the marks in the first set of calibration images by $m_{i,j,1}$ and in the second set by $m_{i,j,2}$. Furthermore, let us denote the camera parameters by a vector c. The camera parameters c include the interior orientation of the first and second camera, the exterior orientation of the l calibration targets in the second image, and the relative orientation of the two cameras. From the above discussion of the relative orientation, it follows that these parameters determine the mappings $\pi_1(M_i, c)$ and $\pi_2(M_i, c)$ into the first and second images completely. Hence, to calibrate the stereo system, the following optimization problem must be solved:

$$d(c) = \sum_{j=1}^{l} \sum_{i=1}^{k} \|m_{i,j,1} - \pi_1(M_i, c)\|^2 + \|m_{i,j,2} - \pi_2(M_i, c)\|^2 \to \min \quad (3.128)$$

To illustrate the relative orientation and the stereo calibration, Figure 3.109 shows an image pair taken from a sequence of 15 image pairs that are used to calibrate a binocular stereo system. The calibration returns a translation vector of $(0.1534\,\text{m}, -0.0037\,\text{m}, 0.0449\,\text{m})$ between the cameras, i.e., the second camera is 15.34 cm to the right, 0.37 cm above, and 4.49 cm in front of the first camera, expressed in the camera coordinates of the first camera. Furthermore, the calibration returns a rotation angle of $40.1139°$ around the axis $(-0.0035, 1.0000, 0.0008)$, i.e., almost around the vertical y axis of the camera coordinate system. Hence, the cameras are verging inwards, like in Figure 3.108.

To reconstruct 3D points, we must find corresponding points in the two images. "Corresponding" means that the two points P_1 and P_2 in the images belong to the same point P_w in the world. At first, it might seem that, given a point P_1 in the first image, we would have to search in the entire second image for the corresponding point P_2. Fortunately, this is not the case. In Figure 3.108 we already noted that the points P_w, O_1, O_2, P_1, and P_2 all lie in a single plane. The situation of trying to find a

Fig. 3.109 One image pair taken from a sequence of 15 image pairs that are used to calibrate a binocular stereo system. The calibration returns a translation vector (base) of $(0.1534\,\text{m}, -0.0037\,\text{m}, 0.0449\,\text{m})$ between the cameras and a rotation angle of $40.1139°$ around the axis $(-0.0035, 1.0000, 0.0008)$, i.e., almost around the y axis of the camera coordinate system. Hence, the cameras are verging inwards.

corresponding point for P_1 is shown in Figure 3.110. We can note that we know P_1, O_1, and O_2. We do not know at which distance the point P_w lies on the optical ray defined by P_1 and O_1. However, we know that P_w is coplanar to the plane spanned by P_1, O_1, and O_2 (the epipolar plane). Hence, we can see that the point P_2 can only lie on the projection of the epipolar plane into the second image. Since O_2 lies on the epipolar plane, the projection of the epipolar plane is a line called the epipolar line.

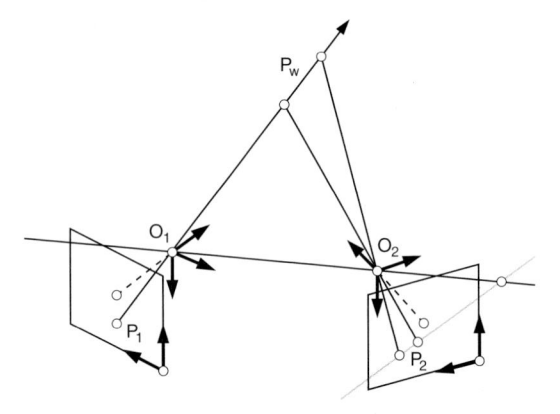

Fig. 3.110 Epipolar geometry of two cameras. Given the point P_1 in the first image, the point P_2 in the second image can only lie on the epipolar line of P_1, which is the projection of the epipolar plane spanned by P_1, O_1, and O_2 into the second image.

It is obvious that the above construction is symmetric for both images, as shown in Figure 3.111. Hence, given a point P_2 in the second image, the corresponding point can only lie on the epipolar line in the first image. Furthermore, from Figure 3.111, we can see that different points typically define different epipolar lines. We can also see that all epipolar lines of one image intersect at a single point called the epipole. The epipoles are the projections of the opposite projective centers into the respective image. Note that, since all epipolar planes contain O_1 and O_2, the epipoles lie on the line defined by the two projection centers (the base line).

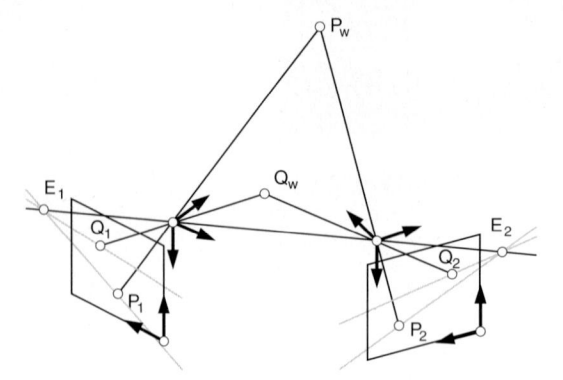

Fig. 3.111 The epipolar geometry is symmetric between the two images. Furthermore, different points typically define different epipolar lines. All epipolar lines intersect at the epipoles E_1 and E_2, which are the projections of the opposite projective centers into the respective image.

Figure 3.112 shows an example of the epipolar lines. The stereo geometry is identical to Figure 3.109. The images show a PCB. In Figure 3.112(a), four points are marked. They have been selected manually to lie at the tips of the triangles on the four small ICs, as shown in the detailed view in Figure 3.112(c). The corresponding epipolar lines in the second image are shown in Figures 3.112(b) and (d). Note that the epipolar lines pass through the tips of the triangles in the second image.

As noted above, we have so far assumed that the lenses have no distortions. In reality, this is very rarely true. In fact, by looking closely at Figure 3.112(b), we can already perceive a curvature in the epipolar lines because the camera calibration has determined the radial distortion coefficient for us. If we set the displayed image part as in Figure 3.113, we can clearly see the curvature of the epipolar lines in real images. Furthermore, we can see the epipole of the image clearly.

From the above discussion, we can see that the epipolar lines are different for different points. Furthermore, because of radial distortions, they typically are not even straight. This means that, when we try to find corresponding points, we must compute a new, complicated epipolar line for each point that we are trying to match, typically for all points in the first image. The construction of the curved epipolar lines would be much too time-consuming for real-time applications. Hence, we can ask ourselves whether the construction of the epipolar lines can be simplified for particular stereo geometries. This is indeed the case for the stereo geometry shown in Figure 3.114. Here, both image planes lie in the same plane and are vertically aligned. Furthermore, it is assumed that there are no lens distortions. Note that this implies that the two principal distances are identical, that the principal points have the same row coordinate, that the images are rotated such that the column axis is parallel to the base, and that

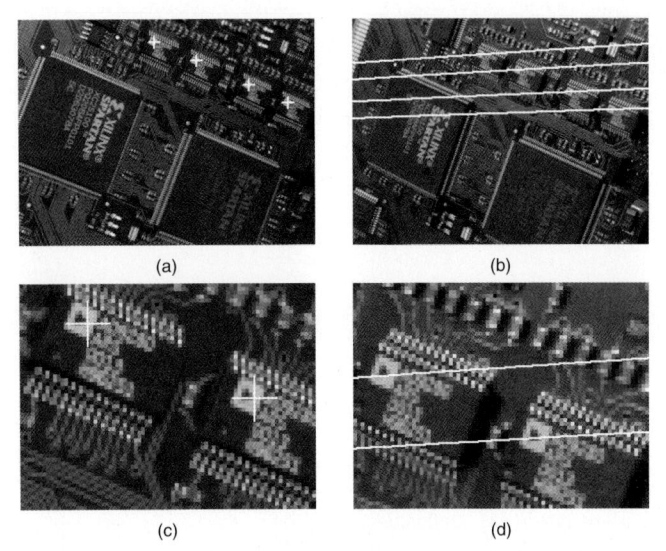

Fig. 3.112 Stereo image pair of a PCB. (a) Four points marked in the first image. (b) Corresponding epipolar lines in the second image. (c) Detail of (a). (d) Detail of (b). The four points in (a) have been selected manually at the tips of the triangles on the four small ICs. Note that the epipolar lines pass through the tips of the triangles in the second image.

the relative orientation contains only a translation in the x direction and no rotation. Since the image planes are parallel to each other, the epipoles lie infinitely far away on the base line. It is easy to see that this stereo geometry implies that the epipolar line for a point is simply the line that has the same row coordinate as the point, i.e.,

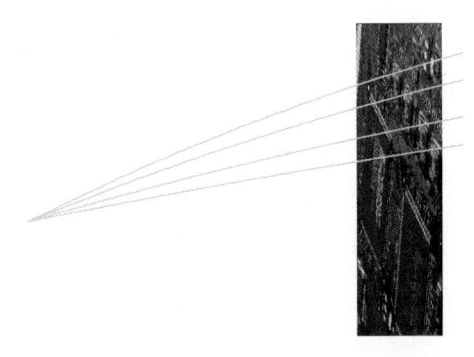

Fig. 3.113 Because of lens distortions, the epipolar lines are generally not straight. The image shows the same image as Figure 3.112(b). The zoom has been set so that the epipole is shown in addition to the image. The aspect ratio has been chosen so that the curvature of the epipolar lines is clearly visible.

the epipolar lines are horizontal and vertically aligned. Hence, they can be computed without any overhead at all. Since almost all stereo matching algorithms assume this particular geometry, we can call it the epipolar standard geometry.

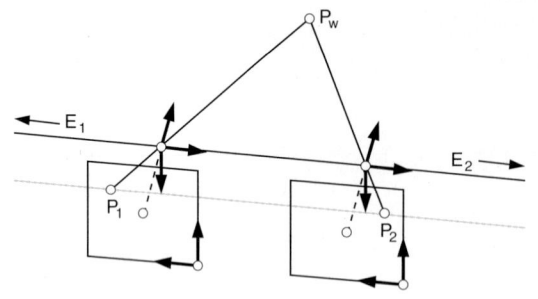

Fig. 3.114 The epipolar standard geometry is obtained if both image planes lie in the same plane and are vertically aligned. Furthermore, it is assumed that there are no lens distortions. In this geometry, the epipolar line for a point is simply the line that has the same row coordinate as the point, i.e., the epipolar lines are horizontal and vertically aligned.

While the epipolar standard geometry results in very simple epipolar lines, it is extremely difficult to align real cameras into this configuration. Furthermore, it is quite difficult and expensive to obtain distortion-free lenses. Fortunately, almost any stereo configuration can be transformed into the epipolar standard geometry, as indicated in Figure 3.115 [119]. The only exceptions are if an epipole happens to lie within one of the images. This typically does not occur in practical stereo configurations. The process of transforming the images to the epipolar standard geometry is called image rectification. To rectify the images, we need to construct two new image planes that lie in the same plane. To keep the 3D geometry identical, the projective centers must remain at the same positions in space, i.e., $O_{r1} = O_1$ and $O_{r2} = O_2$. Note, however, that we need to rotate the camera coordinate systems such that their x axes become identical to the base line. Furthermore, we need to construct two new principal points C_{r1} and C_{r2}. Their connecting vector must be parallel to the base. Furthermore, the vectors from the principal points to the projection centers must be perpendicular to the base. This leaves us two degrees of freedom. First of all, we must choose a common principal distance. Second, we can rotate the common plane in which the image planes lie around the base. These parameters can be chosen by requiring that the image distortion should be minimized [119]. The image dimensions are then typically chosen such that the original images are completely contained within the rectified images. Of course, we must also remove the lens distortions in the rectification.

To obtain the gray value for a pixel in the rectified image, we construct the optical ray for this pixel and intersect it with the original image plane. This is shown, for example, for the points P_{r1} and P_1 in Figure 3.115. Since this typically results in sub-

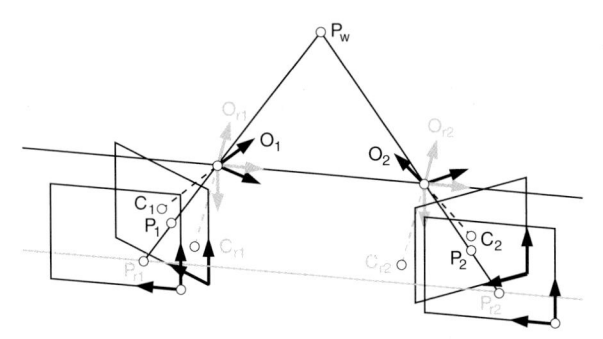

Fig. 3.115 Transformation of a stereo configuration into the epipolar standard geometry.

pixel coordinates, the gray values must be interpolated with the techniques described in Section 3.3.3.

While it may seem that image rectification is a very time-consuming process, the entire transformation can be computed once offline and stored in a table. Hence, images can be rectified very efficiently online.

Figure 3.116 shows an example of the image rectification. The input image pair is shown in Figures 3.116(a) and (b). The images have the same relative orientation as the images in Figure 3.109. The principal distances of the cameras are 13.05 mm and 13.16 mm, respectively. Both images have dimensions 320×240. Their principal points are $(155.91, 126.72)$ and $(163.67, 119.20)$, i.e, they are very close to the image center. Finally, the images have a slight barrel-shaped distortion. The rectified images are shown in Figures 3.116(c) and (d). Their relative orientation is given by the translation vector $(0.1599\,\text{m}, 0\,\text{m}, 0\,\text{m})$. As expected, the translation is solely along the x axis. Of course, the length of the translation vector is identical to Figure 3.109, since the position of the projective centers has not changed. The new principal distance of both images is 12.27 mm. The new principal points are given by $(-88.26, 121.36)$ and $(567.38, 121.36)$. As can be expected from Figure 3.115, they lie well outside the rectified images. Also as expected, the row coordinates of the principal points are identical. The rectified images have dimensions 336×242 and 367×242, respectively. Note that they exhibit a trapezoidal shape that is characteristic of the verging camera configuration. The barrel-shaped distortion has been removed from the images. Clearly, the epipolar lines are horizontal in both images.

Apart from the fact that rectifying the images results in a particularly simple structure for the epipolar lines, it also results in a very simple reconstruction of the depth, as shown in Figure 3.117. In this figure, the stereo configuration is displayed as viewed along the direction of the row axis of the images, i.e., the y axis of the camera coordinate system. Hence, the image planes are shown as the lines at the bottom of the figure. The depth of a point is quite naturally defined as its z coordinate in the camera coordinate system. By examining the similar triangles $O_1 O_2 P_w$ and $P_1 P_2 P_w$, we can see that the depth of P_w only depends on the difference of the column coordi-

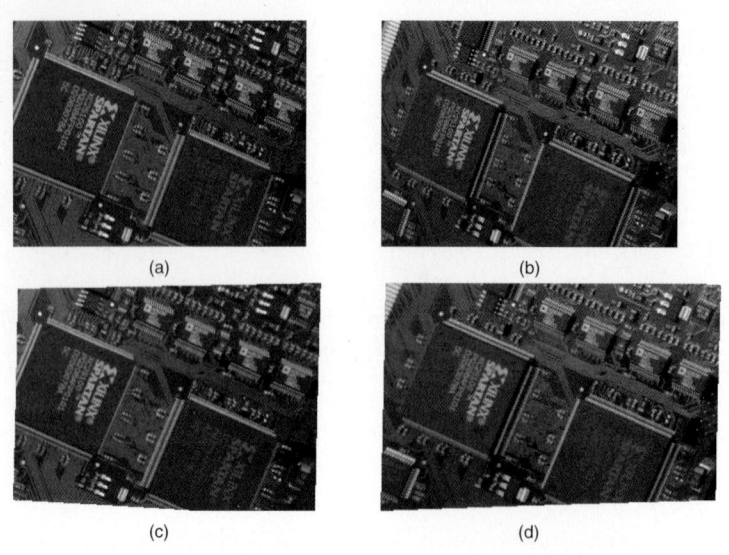

Fig. 3.116 Example of the rectification of a stereo image pair. The images in (a) and (b) have the same relative orientation as the images in Figure 3.109. The rectified images are shown in (c) and (d). Note the trapezoidal shape of the rectified images that is caused by the verging cameras. Also note that the rectified images are slightly wider than the original images.

nates of the points P_1 and P_2 as follows. From the similarity of the triangles, we have $z/b = (z+f)/(d_w + b)$. Hence, the depth is given by $z = bf/d_w$. Here, b is the length of the base, f is the principal distance, and d_w is the sum of the signed distances of the points P_1 and P_2 to the principal points C_1 and C_2. Since the coordinates of the principal points are given in pixels, but d_w is given in world units, e.g., meters, we have to convert d_w to pixel coordinates by scaling it with the size of the pixels in the x direction: $d_p = d_w/s_x$. Now, we can easily see that $d_p = (c_{c1} - c_1) + (c_2 - c_{c2})$, where c_1 and c_2 denote the column coordinates of the points P_1 and P_2, while c_{c1} and c_{c2} denote the column coordinates of the principal points. Rearranging the terms, we find

$$d_p = (c_{c1} - c_{c2}) + (c_2 - c_1)$$

Since $c_{c1} - c_{c2}$ is constant for all points and known from the calibration and rectification, we can see that the depth z depends only on the difference of the column coordinates $d = c_2 - c_1$. This difference is called the disparity. Hence, we can see that, to reconstruct the depth of a point, we must determine its disparity.

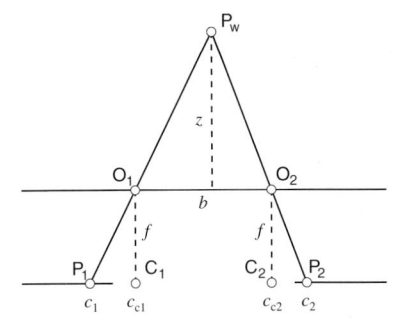

Fig. 3.117 Reconstruction of the depth z of a point only depends on the disparity $d = c_2 - c_1$ of the points, i.e., the difference of the column coordinates in the rectified images.

3.10.2 Stereo Matching

As we have seen in the previous section, the main step in the stereo reconstruction is the determination of the disparity of each point in one of the images, typically the first image. Since one calculates, or at least attempts to calculate, a disparity for each point, these algorithms are called dense reconstruction algorithms. It should be noted that there is another class of algorithms that only tries to reconstruct the depth for selected features, e.g., straight lines or points. Since these algorithms require a typically expensive feature extraction, they are seldom used in industrial applications. Therefore, we will concentrate on dense reconstruction algorithms. An overview of currently available algorithms is given in [120].

Since the goal of dense reconstruction is to find the disparity for each point in the image, the determination of the disparity can be regarded as a template matching problem. Given a rectangular window of size $(2n+1) \times (2n+1)$ around the current point in the first image, we must find the most similar window along the epipolar line in the second image. Hence, we can use the techniques that will be described in greater detail in Section 3.11 to match a point. The gray value matching methods described in Section 3.11.1 are of particular interest because they do not require a costly model generation step, which would have to be performed for each point in the first image. Therefore, the gray value matching methods typically are the fastest methods for stereo reconstruction. The simplest similarity measures are the SAD (sum of absolute gray value differences) and SSD (sum of squared gray value differences) measures described later (see Eqs. (3.133) and (3.134)). For the stereo matching problem, they are given by

$$sad(r,c,d) = \frac{1}{(2n+1)^2} \sum_{j=-n}^{n} \sum_{i=-n}^{n} |g_1(r+i, c+j) - g_2(r+i, c+j+d)| \quad (3.129)$$

and

$$ssd(r,c,d) = \frac{1}{(2n+1)^2} \sum_{j=-n}^{n} \sum_{i=-n}^{n} [g_1(r+i, c+j) - g_2(r+i, c+j+d)]^2 \quad (3.130)$$

As will be discussed in Section 3.11.1, these two similarity measures can be computed very quickly. Fast implementations for stereo matching using the SAD are given in [121, 122]. Unfortunately, these similarity measures have the disadvantage that they are not robust against illumination changes, which frequently happen in stereo reconstruction because of the different viewing angles along the optical rays. Consequently, in some applications it may be necessary to use the normalized cross-correlation (NCC) described later (see Eq. (3.135)) as the similarity measure. For the stereo matching problem, it is given by

$$ncc(r,c,d) = \frac{1}{(2n+1)^2} \sum_{i=-n}^{n} \sum_{j=-n}^{n} \frac{g_1(r+i, c+j) - m_1(r+i, c+j)}{\sqrt{s_1(r+i, c+j)^2}}$$
$$\cdot \frac{g_2(r+i, c+j+d) - m_2(r+i, c+j+d)}{\sqrt{s_2(r+i, c+j+d)^2}} \quad (3.131)$$

Here, m_i and s_i ($i = 1, 2$) denote the mean and standard deviation of the window in the first and second images. They are calculated analogously to their template matching counterparts in Eqs. (3.136) and (3.137). The advantage of the normalized cross-correlation is that it is invariant against linear illumination changes. However, is is more expensive to compute.

From the above discussion, it might appear that, to match a point, we would have to compute the similarity measure along the entire epipolar line in the second image. Fortunately, this is not the case. Since the disparity is inversely related to the depth of a point, and we typically know in which range of distances the objects we are interested in occur, we can restrict the disparity search space to a much smaller interval than the entire epipolar line. Hence, we have $d \in [d_{min}, d_{max}]$, where d_{min} and d_{max} can be computed from the minimum and maximum expected distance in the images. Therefore, the length of the disparity search space is given by $l = d_{max} - d_{min} + 1$.

After we have computed the similarity measure for the disparity search space for a point to be matched, we might be tempted to simply use the disparity with the minimum (SAD and SSD) or maximum (NCC) similarity measure as the match for the current point. However, this typically will lead to many false matches, since some windows may not have a good match in the second image. In particular, this happens if the current point is occluded because of perspective effects in the second image. Therefore, it is necessary to threshold the similarity measure, i.e., to accept matches only if their similarity measure is below (SAD and SSD) or above (NCC) a threshold. Obviously, if we perform this thresholding, some points will not have a reconstruction, and consequently the reconstruction will not be completely dense.

With the above search strategy, the matching process has a complexity of $O(whln^2)$. This is much too expensive for real-time performance. Fortunately, it can be shown that with a clever implementation the above similarity measures can be computed recursively. With this, the complexity can be made independent of the window size n, and becomes $O(whl)$. With this, real-time performance becomes possible. The interested reader is referred to [123] for details.

Once we have computed the match with an accuracy of one disparity step from the extremum (minimum or maximum) of the similarity measure, the accuracy can be refined with an approach similar to the subpixel extraction of matches that will be described in Section 3.11.3. Since the search space is 1D in the stereo matching, a parabola can be fitted through the three points around the extremum, and the extremum of the parabola can be extracted analytically. Obviously, this will also result in a more accurate reconstruction of the depth of the points.

To perform the stereo matching, we need to set one parameter: the size of the gray value windows n. It has a major influence on the result of the matching, as shown by the reconstructed depths in Figure 3.118. Here, window sizes of 3×3, 17×17, and 31×31 have been used with the NCC as the similarity measure. We can see that, if the window size is too small, many erroneous results will be found, despite the fact that a threshold of 0.4 has been used to select good matches. This happens because the matching requires a sufficiently distinctive texture within the window. If the window is too small, the texture is not distinctive enough, leading to erroneous matches. From Figure 3.118(b), we see that the erroneous matches are mostly removed by the 17×17 window. However, because there is no texture in some parts of the image, especially in the lower left corners of the two large ICs, some parts of the image cannot be reconstructed. Note also that the areas of the leads around the large ICs are broader than in Figure 3.118(a). This happens because the windows now straddle height discontinuities in a larger part of the image. Since the texture of the leads is more significant than the texture on the ICs, the matching finds the best matches at the depth of the leads. To fill the gaps in the reconstruction, we could try to increase the window size further, since this leads to more positions in which the windows have a significant texture. The result of setting the window size to 31×31

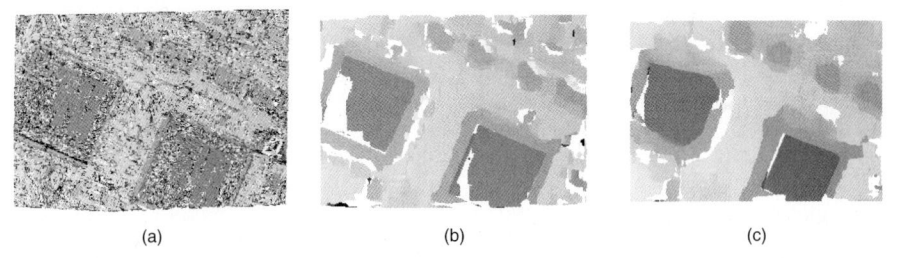

Fig. 3.118 Distance reconstructed for the rectified image pair in Figures 3.116(c) and (d) with the NCC. (a) Window size 3×3. (b) Window size 17×17. (c) Window size 31×31. White areas correspond to the points that could not be matched because the similarity was too small.

is shown in Figure 3.118(c). Note that now most of the image can be reconstructed. Unfortunately, the lead area has broadened even more, which is undesirable.

From the above example, we can see that too small window sizes lead to many erroneous matches. In contrast, larger window sizes generally lead to fewer erroneous matches and a more complete reconstruction in areas with little texture. Furthermore, larger window sizes lead to a smoothing of the result, which may sometimes be desirable. However, larger window sizes lead to worse results at height discontinuities, which effectively limits the window sizes that can be used in practice.

Despite the fact that larger window sizes generally lead to fewer erroneous matches, they typically cannot be excluded completely based on the window size alone. Therefore, additional techniques are sometimes desirable to reduce the number of erroneous matches even further. Erroneous matches occur mainly for two reasons: weak texture and occlusions. Erroneous matches caused by weak texture can sometimes be eliminated based on the matching score. However, in general it is best to exclude windows with weak texture a priori from the matching. Whether a window contains a weak texture can be decided based on the output of a texture filter. Typically, the standard deviation of the gray values within the window is used as the texture filter. It has the advantage that it is computed in the NCC anyway, while it can be computed with just a few extra operations in the SAD and SSD. Therefore, to exclude windows with weak textures, we require that the standard deviation of the gray values within the window should be large.

The second reason why erroneous matches can occur are perspective occlusions, which, for example, occur at height discontinuities. To remove these errors, we can perform a consistency check that works as follows. First, we find the match from the first to the second image as usual. We then check whether matching the window around the match in the second image results in the same disparity, i.e., finds the original point in the first image. If this is implemented naively, the runtime increases by a factor of 2. Fortunately, with a little extra bookkeeping the disparity consistency check can be performed with very few extra operations, since most of the required data have already been computed during the matching from the first to the second image.

Figure 3.119 shows the results of the different methods to increase robustness. For comparison, Figure 3.119(a) displays the result of the standard matching from the first to the second image with a window size of 17×17 using the NCC. The result of applying a texture threshold of 5 is shown in Figure 3.119(b). It mainly removes untextured areas on the two large ICs. Figure 3.119(c) shows the result of applying the disparity consistency check. Note that it mainly removes matches in the areas where occlusions occur.

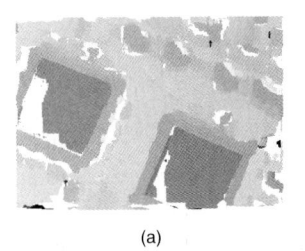

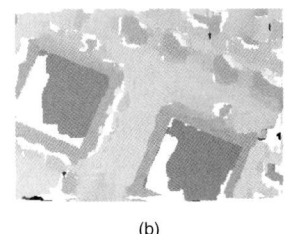

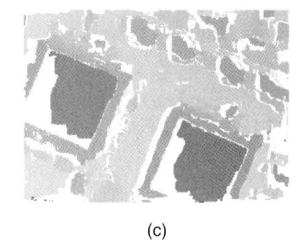

Fig. 3.119 Increasing levels of robustness of the stereo matching. (a) Standard matching from the first to the second image with a window size of 17 × 17 using the NCC. (b) Result of requiring that the standard deviations of the windows is ≥ 5. (c) Result of performing the check that matching from the second to the first image results in the same disparity.

3.11 Template Matching

In the previous sections we have discussed various techniques that can be combined to write algorithms to find objects in an image. While these techniques can in principle be used to find any kind of object, writing a robust recognition algorithm for a particular type of object can be quite cumbersome. Furthermore, if the objects to be recognized change frequently, a new algorithm must be developed for each type of object. Therefore, a method to find any kind of object that can be configured simply by showing the system a prototype of the class of objects to be found would be extremely useful.

The above goal can be achieved by template matching. Here, we describe the object to be found by a template image. Conceptually, the template is found in the image by computing the similarity between the template and the image for all relevant poses of the template. If the similarity is high, an instance of the template has been found. Note that the term similarity is used here in a very general sense. We will see below that it can be defined in various ways, e.g., based on the gray values of the template and the image, or based on the closeness of template edges to image edges.

Template matching can be used for several purposes. First of all, it can be used to perform completeness checks. Here, the goal is to detect the presence or absence of the object. Furthermore, template matching can be used for object discrimination, i.e., to distinguish between different types of objects. In most cases, however, we already know which type of object is present in the image. In these cases, template matching is used to determine the pose of the object in the image. If the orientation of the objects can be fixed mechanically, the pose is described by a translation. In most applications, however, the orientation cannot be fixed completely, if at all. Therefore, often the orientation of the object, described by rotation, must also be determined. Hence, the complete pose of the object is described by a translation and a rotation. This type of transformation is called a rigid transformation. In some applications, additionally, the size of the objects in the image can change. This can happen if the distance of the objects to the camera cannot be kept fixed, or if the real size of the objects can change. Hence, a uniform scaling must be added to the pose in these applications. This type of pose (translation, rotation, and uniform scaling) is called a similarity transformation. If even the 3D orientation of the camera with respect to the objects can change and the objects to be recognized are planar, the pose is described by a projective transformation (see Section 3.3.2). Consequently, for the purposes of this chapter, we can regard the pose of the objects as a specialization of an affine or projective transformation.

In most applications, a single object is present in the search image. Therefore, the goal of the template matching is to find this single instance. In some applications, more than one object is present in the image. If we know a priori how many objects are present, we want to find exactly this number of objects. If we do not have this knowledge, we typically must find all instances of the template in the image. In this mode, one of the goals is also to determine how many objects are present in the image.

3.11.1
Gray-Value-Based Template Matching

In this section, we will examine the simplest kind of template matching algorithms, which are based on the raw gray values in the template and the image. As mentioned above, template matching is based on computing a similarity between the template and the image. Let us formalize this notion. For the moment, we will assume that the object's pose is described by a translation. The template is specified by an image $t(r,c)$ and its corresponding ROI T. To perform the template matching, we can visualize moving the template over all positions in the image and computing a similarity measure **s** at each position. Hence, the similarity measure **s** is a function that takes the gray values in the template $t(r,c)$ and the gray values in the shifted ROI of the template at the current position in the image $f(r+u, c+v)$ and calculates a scalar value that measures the similarity based on the gray values within the respective ROI. With this approach, a similarity measure is returned for each point in the transformation space, which for translations can be regarded as an image. Hence, formally, we have

$$s(r,c) = \mathbf{s}\{t(u,v), f(r+u, c+v); (u,v) \in T\} \qquad (3.132)$$

To make this abstract notation concrete, we will discuss several possible gray-value-based similarity measures [124].

The simplest similarity measures are to sum the absolute or squared gray value differences between the template and the image (SAD and SSD). They are given by

$$sad(r,c) = \frac{1}{n} \sum_{(u,v) \in T} |t(u,v) - f(r+u, c+v)| \qquad (3.133)$$

and

$$ssd(r,c) = \frac{1}{n} \sum_{(u,v) \in T} [t(u,v) - f(r+u, c+v)]^2 \qquad (3.134)$$

In both cases, n is the number of points in the template ROI, i.e., $n = |T|$. Note that both similarity measures can be computed very efficiently with just two operations per pixel. These similarity measures have similar properties: if the template and the image are identical, they return a similarity measure of 0. If the image and template are not identical, a value greater than 0 is returned. As the dissimilarity increases, the value of the similarity measure increases. Hence, in this case the similarity measure should probably better be called a dissimilarity measure. To find instances of the template in the search image, we can threshold the similarity image $sad(r,c)$ with a certain upper threshold. This typically gives us a region that contains several adjacent pixels. To obtain a unique location for the template, we must select the local minima of the similarity image within each connected component of the thresholded region.

Figure 3.120 shows a typical application for template matching. Here, the goal is to locate the position of a fiducial mark on a PCB. The ROI used for the template is displayed in Figure 3.120(a). The similarity computed with the sum of the SAD, given by Eq. (3.133), is shown in Figure 3.120(b). For this example, the SAD was computed with the same image from which the template was generated. If the similarity is thresholded with a threshold of 20, only a region around the position of the fiducial mark is returned (Figure 3.120(c)). Within this region, the local minimum of the SAD must be computed (not shown) to obtain the position of the fiducial mark.

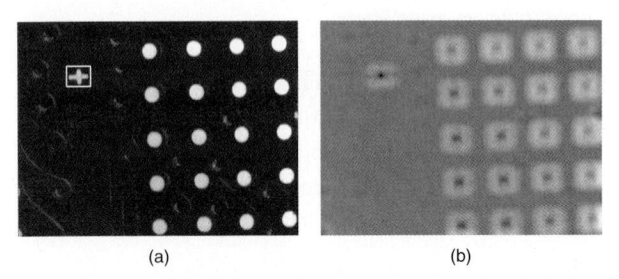

Fig. 3.120 (a) Image of a PCB with a fiducial mark, which is used as the template (indicated by the white rectangle). (b) SAD computed with the template in (a) and the image in (a). (c) Result of thresholding (b) with a threshold of 20. Only a region around the fiducial is selected.

The SAD and SSD similarity measures work very well as long as the illumination can be kept constant. However, if the illumination can change, they both return larger values, even if the same object is contained in the image, because the gray values are no longer identical. This effect is illustrated in Figure 3.121. Here, a darker and brighter image of the fiducial mark are shown. They were obtained by adjusting the illumination intensity. The SAD computed with the template of Figure 3.120(a) is displayed in Figures 3.121(b) and (e). The result of thresholding them with a threshold of 35 is shown in Figures 3.121(c) and (f). The threshold was chosen such that the true fiducial mark is extracted in both cases. Note that, because of the contrast change, many extraneous instances of the template have been found.

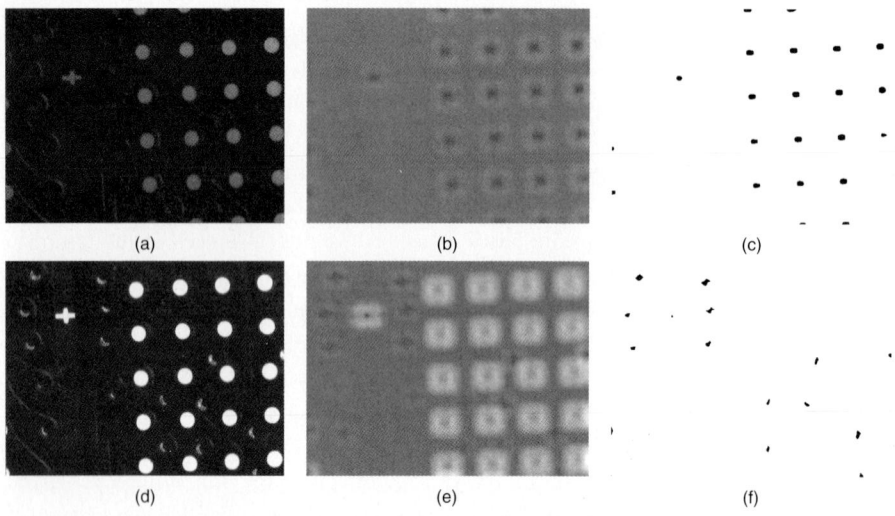

Fig. 3.121 (a) Image of a PCB with a fiducial mark with lower contrast. (b) SAD computed with the template in Figure 3.120(a) and the image in (a). (c) Result of thresholding (b) with a threshold of 35. (d) Image of a PCB with a fiducial mark with higher contrast. (e) SAD computed with the template of Figure 3.120(a) and the image in (d). (f) Result of thresholding (e) with a threshold of 35. In both cases, it is impossible to select a threshold that only returns the region of the fiducial.

As we can see from the above examples, the SAD and SSD similarity measures work well as long as the illumination can be kept constant. In applications where this cannot be ensured, a different kind of similarity measure is required. Ideally, this similarity measure should be invariant to all linear illumination changes (see Section 3.2.1). A similarity measure that achieves this is the normalized cross-correlation (NCC), given by

$$ncc(r,c) = \frac{1}{n} \sum_{(u,v) \in T} \frac{t(u,v) - m_t}{\sqrt{s_t^2}} \cdot \frac{f(r+u, c+v) - m_f(r,c)}{\sqrt{s_f^2(r,c)}} \qquad (3.135)$$

Here, m_t is the mean gray value of the template and s_t^2 is the variance of the gray values, i.e.

$$m_t = \frac{1}{n} \sum_{(u,v) \in T} t(u,v)$$

$$s_t^2 = \frac{1}{n} \sum_{(u,v) \in T} [t(u,v) - m_t]^2 \qquad (3.136)$$

Analogously, $m_f(r,c)$ and $s_f^2(r,c)$ are the mean value and variance in the image at a shifted position of the template ROI:

$$m_f(r,c) = \frac{1}{n} \sum_{(u,v)\in T} f(r+u, c+v)$$

$$s_f^2(r,c) = \frac{1}{n} \sum_{(u,v)\in T} [f(r+u, c+v) - m_f(r,c)]^2 \qquad (3.137)$$

The NCC has a very intuitive interpretation. First of all, we should note that $-1 \leq ncc(r,c) \leq 1$. Furthermore, if $ncc(r,c) = \pm 1$, the image is a linearly scaled version of the template:

$$ncc(r,c) = \pm 1 \quad \Leftrightarrow \quad f(r+u, c+v) = at(u,v) + b$$

For $ncc(r,c) = 1$ we have $a > 0$, i.e., the template and the image have the same polarity; while $ncc(r,c) = -1$ implies that $a < 0$, i.e., the polarity of the template and image are reversed. Note that this property of the NCC implies the desired invariance against linear illumination changes. The invariance is achieved by explicitly subtracting the mean gray values, which cancels additive changes, and by dividing by the standard deviation of the gray values, which cancels multiplicative changes.

While the template matches the image perfectly only if $ncc(r,c) = \pm 1$, large absolute values of the NCC generally indicate that the template closely corresponds to the image part under examination, while values close to zero indicate that the template and image do not correspond well.

Figure 3.122 displays the results of computing the NCC for the template in Figure 3.120(a) (reproduced in Figure 3.122(a)). The NCC is shown in Figure 3.122(b), while the result of thresholding the NCC with a threshold of 0.75 is shown in Figure 3.122(c). This selects only a region around the fiducial mark. In this region, the local maximum of the NCC must be computed to derive the location of the fiducial mark (not shown). The results for the darker and brighter images in Figure 3.121 are not shown because they are virtually indistinguishable from the results in Figures 3.122(b) and (c).

In the above discussion, we have assumed that the similarity measures must be evaluated completely for every translation. This is, in fact, unnecessary, since the result of calculating the similarity measure will be thresholded with a threshold t_s later on. For example, thresholding the SAD in Eq. (3.133) means that we require

$$sad(r,c) = \frac{1}{n} \sum_{i=1}^{n} |t(u_i, v_i) - f(r+u_i, c+v_i)| \leq t_s \qquad (3.138)$$

Here, we have explicitly numbered the points $(u,v) \in T$ by (u_i, v_i). We can multiply both sides by n to obtain

$$sad'(r,c) = \sum_{i=1}^{n} |t(u_i, v_i) - f(r+u_i, c+v_i)| \leq nt_s \qquad (3.139)$$

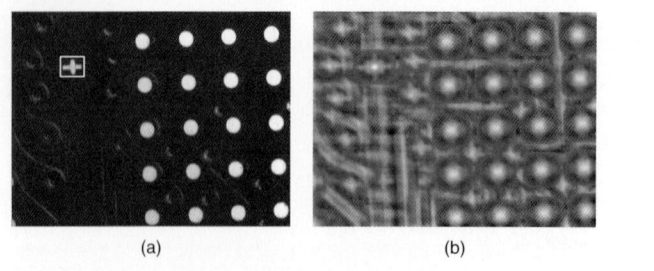

Fig. 3.122 (a) Image of a PCB with a fiducial mark, which is used as the template (indicated by the white rectangle). This is the same image as in Figure 3.120(a). (b) NCC computed with the template in (a) and the image in (a). (c) Result of thresholding (b) with a threshold of 0.75. The results for the darker and brighter images in Figure 3.121 are not shown because they are virtually indistinguishable from the results in (b) and (c).

Suppose we have already evaluated the first j terms in the sum in Eq. (3.139). Let us call this partial result $sad'_j(r,c)$. Then, we have

$$sad'(r,c) = sad'_j(r,c) + \underbrace{\sum_{i=j+1}^{n} |t(u_i,v_i) - f(r+u_i, c+v_i)|}_{\geq 0} \leq nt_s \qquad (3.140)$$

Hence, we can stop the evaluation as soon as $sad'_j(r,c) > nt_s$ because we are certain that we can no longer achieve the threshold. If we are looking for a maximum number of m instances of the template, we can even adapt the threshold t_s based on the instance with the mth best similarity found so far. For example, if we are looking for a single instance with $t_s = 20$ and we have already found a candidate with $sad(r,c) = 10$, we can set $t_s = 10$ for the remaining poses that need to be checked. Of course, we need to calculate the local minima of $sad(r,c)$ and use the corresponding similarity values to ensure that this approach works correctly if more than one instance should be found.

For the normalized cross-correlation, there is no simple criterion to stop the evaluation of the terms. Of course, we can use the fact that the mean m_t and standard deviation $\sqrt{s_t^2}$ of the template can be computed once offline because they are identical for every translation of the template. The only other optimization we can make, analogously to the SAD, is that we can adapt the threshold t_s based on the matches we have found so far [125].

The above stopping criteria enable us to stop the evaluation of the similarity measure as soon as we are certain that the threshold can no longer be reached. Hence, they prune unwanted parts of the space of allowed poses. It is interesting to note that improvements for the pruning of the search space are still actively being investigated. For example, in [125] further optimizations for the pruning of the search space when using the NCC are discussed. In [126] and [127] strategies for pruning the search space when using SAD or SSD are discussed. They rely on transforming the image into a representation in which a large portion of the SAD and SSD can be computed with very few evaluations so that the above stopping criteria can be reached as soon as possible.

3.11.2
Matching Using Image Pyramids

The evaluation of the similarity measures on the entire image is very time-consuming, even if the stopping criteria discussed above are used. If they are not used, the runtime complexity is $O(whn)$, where w and h are the width and height of the image and n is the number of points in the template. The stopping criteria typically result in a constant factor for the speed-up, but do not change the complexity. Therefore, a method to further speed up the search is necessary to be able to find the template in real time.

To derive a faster search strategy, we note that the runtime complexity of the template matching depends on the number of translations, i.e., poses, that need to be checked. This is the $O(wh)$ part of the complexity. Furthermore, it depends on the number of points in the template. This is the $O(n)$ part. Therefore, to gain a speed-up, we can try to reduce the number of poses that need to be checked as well as the number of template points. Since the templates typically are large, one way to do this would be to take into account only every ith point of the image and template in order to obtain an approximate pose of the template, which could later be refined by a search with a finer step size around the approximate pose. This strategy is identical to subsampling the image and template. Since subsampling can cause aliasing effects (see Sections 3.2.4 and 3.3.3), this is not a very good strategy because we might miss instances of the template because of the aliasing effects. We have seen in Section 3.3.3 that we must smooth the image to avoid the aliasing effects. Furthermore, typically it is better to scale the image down multiple times by a factor of 2 than only once by a factor of $i > 2$. Scaling down the image (and template) multiple times by a factor of 2 creates a data structure that is called an image pyramid. Figure 3.123 displays why the name was chosen: we can visualize the smaller versions of the image stacked on top of each other. Since their width and height are halved in each step, they form a pyramid.

When constructing the pyramid, speed is essential. Therefore, the smoothing is performed by applying a 2×2 mean filter, i.e., by averaging the gray value of each 2×2 block of pixels [128]. The smoothing could also be performed by a Gaussian filter [129]. Note, however, that, in order to avoid the introduction of unwanted shifts into the image pyramid, the Gaussian filter must have an even mask size. Therefore, the smallest mask size would be a 4×4 filter. Hence, using the Gaussian filter would incur a severe speed penalty in the construction of the image pyramid. Furthermore, the 2×2 mean filter does not have the frequency response problems that the larger

Level 4

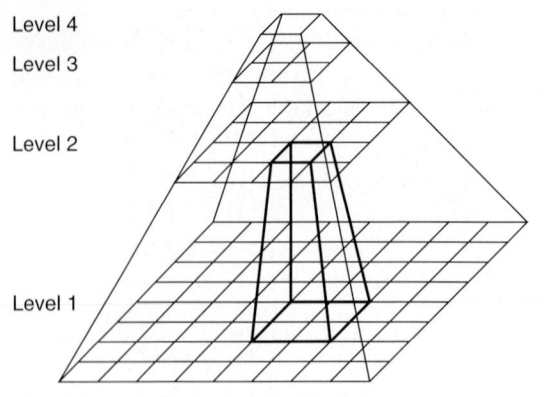

Level 3

Level 2

Level 1

Fig. 3.123 An image pyramid is constructed by successively halving the resolution of the image and combining 2 × 2 blocks of pixels in a higher resolution into a single pixel at the next lower resolution.

versions of the filter have (see Section 3.2.3). In fact, it drops off smoothly toward a zero response for the highest frequencies, like the Gaussian filter. Finally, it simulates the effects of a perfect camera with a fill factor of 100%. Therefore, the mean filter is the preferred filter for constructing image pyramids.

Figure 3.124 displays the image pyramid levels 2–5 of the image in Figure 3.120(a). We can see that on levels 1–4 the fiducial mark can still be discerned from the BGA pads. This is no longer the case on level 5. Therefore, if we want to find an approximate location of the template, we can start the search on level 4.

The above example produces the expected result: the image is progressively smoothed and subsampled. The fiducial mark we are interested in can no longer be recognized as soon as the resolution becomes too low. Sometimes, however, creating an image pyramid can produce results that are unexpected at first glance. One example of this behavior is shown in Figure 3.125. Here, the image pyramid levels 1–4 of an image of a PCB are shown. We can see that on pyramid level 4 all the tracks are suddenly merged into large components with identical gray values. This happens because of the smoothing that is performed when the pyramid is constructed. Here, the neighboring thin lines start to interact with each other once the smoothing is large enough, i.e., once we reach a pyramid level that is large enough. Hence, we can see that sometimes valuable information is destroyed by the construction of the image pyramid. If we were interested in matching, say, the corners of the tracks, we could only go as high as level 3 in the image pyramid.

Based on the image pyramids, we can define a hierarchical search strategy as follows. First, we calculate an image pyramid on the template and search image with an appropriate number of levels. How many levels can be used is mainly defined by the objects we are trying to find. On the highest pyramid level, the relevant structures of the object must still be discernible. Then, a complete matching is performed on the highest pyramid level. Here, of course, we take the appropriate stopping criterion into

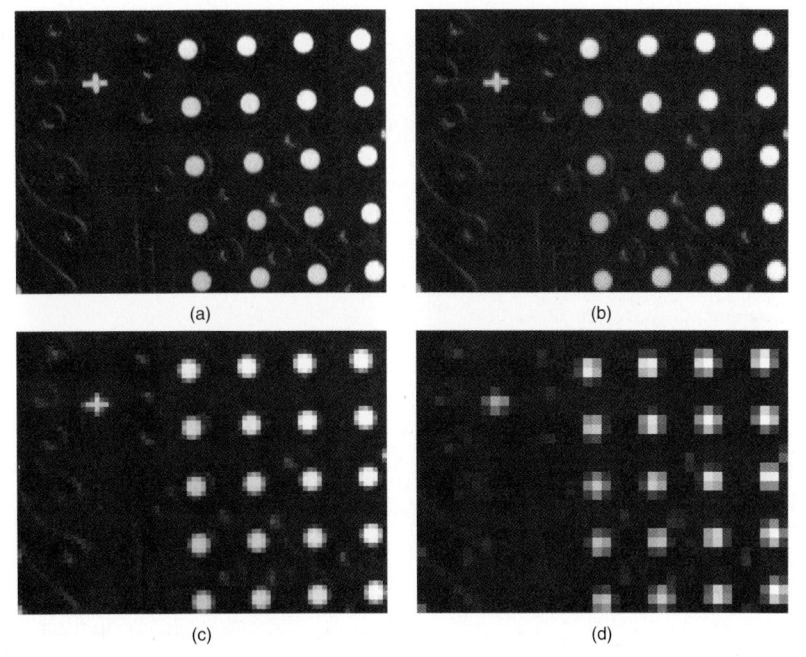

Fig. 3.124 (a)–(d) Image pyramid levels 2–5 of the image in Figure 3.120(a). Note that in level 5 the fiducial mark can no longer be discerned from the BGA pads.

account. What does this gain us? In each pyramid level, we reduce the number of image points and template points by a factor of 4. Hence, each pyramid level results in a speed-up of a factor of 16. Therefore, if we perform the complete matching, for example, on level 4, we reduce the amount of computations by a factor of 4096.

All instances of the template that have been found on the highest pyramid level are then tracked to the lowest pyramid level. This is done by projecting the match to the next lower pyramid level, i.e., by multiplying the coordinates of the found match by 2. Since there is an uncertainty in the location of the match, a search area is constructed around the match in the lower pyramid level, e.g., a 5×5 rectangle. Then, the matching is performed within this small ROI, i.e., the similarity measure is computed, thresholded, and the local extrema are extracted. This procedure is continued until the match is lost or tracked to the lowest level. Since the search spaces for the larger templates are very small, tracking the match down to the lowest level is very efficient.

While matching the template on the higher pyramid levels, we need to take the following effect into account: the gray values at the border of the object can change substantially on the highest pyramid level depending on where the object lies on the lowest pyramid level. This happens because a single pixel shift of the object translates to a subpixel shift on higher pyramid levels, which manifests itself as a change in the gray values on the higher pyramid levels. Therefore, on the higher pyramid levels

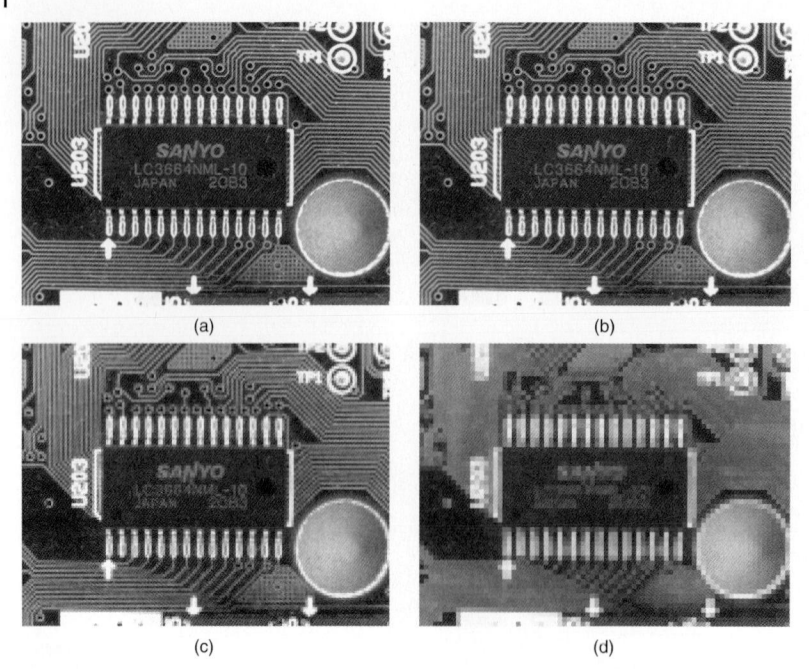

Fig. 3.125 (a)–(d) Image pyramid levels 1–4 of an image of a PCB. Note that in level 4 all the tracks are merged into large components with identical gray values because of the smoothing that is performed when the pyramid is constructed.

we need to be more lenient with the matching threshold to ensure that all potential matches are being found. Hence, for the SAD and SSD similarity measures we need to use slightly higher thresholds, and for the NCC similarity measure we need to use slightly lower thresholds on the higher pyramid levels.

The hierarchical search is visualized in Figure 3.126. The template is the fiducial mark shown in Figure 3.120(a). The template is searched in the same image from which the template was created. As discussed above, four pyramid levels are used in this case. The search starts on level 4. Here, the ROI is the entire image. The NCC and found matches on level 4 are displayed in Figure 3.126(a). As we can see, 12 potential matches are initially found. They are tracked down to level 3 (Figure 3.126(b)). The ROIs created from the matches on level 4 are shown in white. For visualization purposes, the NCC is displayed for the entire image. In reality, it is, of course, only computed within the ROIs, i.e., for a total of $12 \cdot 25 = 300$ translations. Note that on this level the true match turns out to be the only viable match. It is tracked down through levels 2 and 1 (Figures 3.126(c) and (d)). In both cases, only 25 translations need to be checked. Therefore, the match is found extremely efficiently. The zoomed part of the NCC in Figures 3.126(b)–(d) also shows that the pose of the match is progressively refined as the match is tracked down the pyramid.

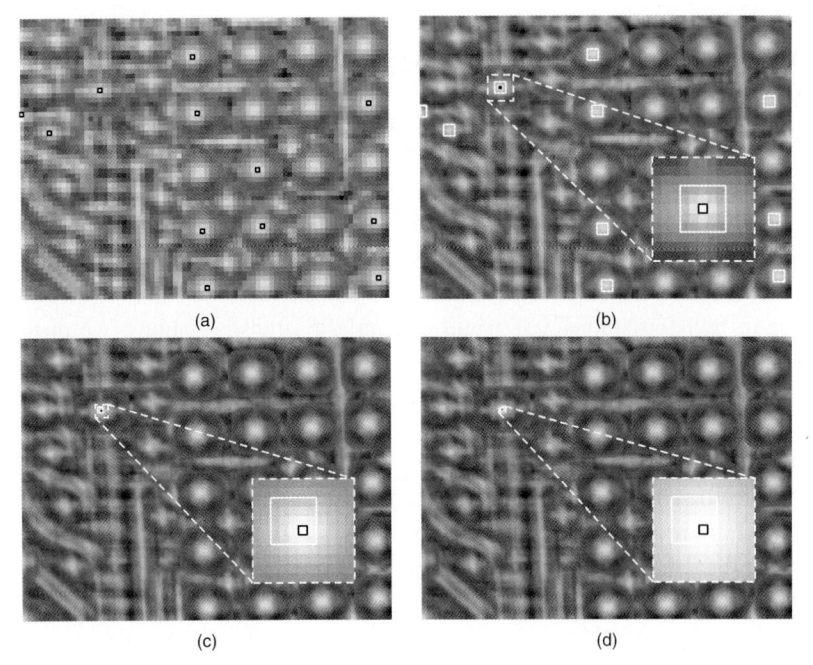

Fig. 3.126 Hierarchical template matching using image pyramids. The template is the fiducial mark shown in Figure 3.120(a). To provide a better visualization, the NCC is shown for the entire image on each pyramid level. In reality, however, it is only calculated within the appropriate ROI on each level, shown in white. The found matches are displayed in black. (a) On pyramid level 4, the matching is performed in the entire image. Here, 12 potential matches are found. (b) The matching is continued within the white ROIs on level 3. Only one viable match is found in the 12 ROIs. The similarity measure and ROI around the match are displayed zoomed in the lower right corner. (c), (d) The match is tracked through pyramid levels 2 and 1.

3.11.3
Subpixel-Accurate Gray-Value-Based Matching

So far, we have located the pose of the template with pixel precision. This has been done by extracting the local minima (SAD, SSD) or maxima (NCC) of the similarity measure. To obtain the pose of the template with higher accuracy, the local minima or maxima can be extracted with subpixel precision. This can be done in a manner that is analogous to the method we have used in edge extraction (see Section 3.7.3): we simply fit a polynomial to the similarity measure in a 3×3 neighborhood around the local minimum or maximum. Then, we extract the local minimum or maximum of the polynomial analytically. Another approach is to perform a least-squares matching of the gray values of the template and the image [130]. Since the least-squares matching of the gray values is not invariant to illumination changes, the illumination changes must be modeled explicitly, and their parameters must be determined in the least-squares fitting in order to achieve robustness to illumination changes [131].

3.11.4
Template Matching with Rotations and Scalings

Up to now, we have implicitly restricted the template matching to the case where the object must have the same orientation and scale in the template and the image, i.e., the space of possible poses was assumed to be the space of translations. The similarity measures we have discussed above can only tolerate small rotations and scalings of the object in the image. Therefore, if the object does not have the same orientation and size as the template, the object will not be found. If we want to be able to handle a larger class of transformations, e.g., rigid or similarity transformations, we must modify the matching approach. For simplicity, we will only discuss rotations, but the method can be extended to scalings and even more general classes of transformations in an analogous manner.

To find a rotated object, we can create the template in multiple orientations, i.e., we discretize the search space of rotations in a manner that is analogous to the discretization of the translations that is imposed by the pixel grid [132]. Unlike for the translations, the discretization of the orientations of the template depends on the size of the template, since the similarity measures are less tolerant to small angle changes for large templates. For example, a typical value is to use an angle step size of $1°$ for templates with a radius of 100 pixels. Larger templates must use smaller angle steps, while smaller templates can use larger angle steps. To find the template, we simply match all rotations of the template with the image. Of course, this is only done on the highest pyramid level. To make the matching in the pyramid more efficient, we can also use the fact that the templates become smaller by a factor of 2 on each pyramid level. Consequently, the angle step size can be increased by a factor of 2 for each pyramid level. Hence, if an angle step size of $1°$ is used on the lowest pyramid level, a step size of $8°$ can be used on the fourth pyramid level.

While tracking potential matches through the pyramid, we also need to construct a small search space for the angles in the next lower pyramid level, analogously to the small search space that we already use for the translations. Once we have tracked the match to the lowest pyramid level, we typically want to refine the pose to an accuracy that is higher than the resolution of the search space we have used. In particular, if rotations are used, the pose should consist of a subpixel translation and an angle that is more accurate than the angle step size we have chosen. The techniques for subpixel-precise localization of the template described above can easily be extended for this purpose.

3.11.5
Robust Template Matching

The above template matching algorithms have served for many years as the methods of choice to find objects in machine vision applications. Recently, however, there has been an increasing demand to find objects in images even if they are occluded or disturbed in other ways so that parts of the object are missing. Furthermore, the objects should be found even if there are a large number of disturbances on the object itself. These disturbances are often referred to as clutter. Finally, objects should be found if there are severe nonlinear illumination changes. The gray-value-based template matching algorithms we have discussed so far cannot handle these kinds of disturbances. Therefore, in the remainder of this section, we will discuss several approaches that have been designed to find objects in the presence of occlusion, clutter, and nonlinear illumination changes.

We have already discussed a feature that is robust to nonlinear illumination changes in Section 3.7: edges are not (or at least very little) affected by illumination changes. Therefore, they are frequently used in robust matching algorithms. The only problem when using edges is the selection of a suitable threshold to segment the edges. If the threshold is chosen too low, there will be many clutter edges in the image. If it is chosen too high, important edges of the object will be missing. This has the same effect as if parts of the object are occluded. Since the threshold can never be chosen perfectly, this is another reason why the matching must be able to handle occlusions and clutter robustly.

To match objects using edges, several strategies exist. First of all, we can use the raw edge points, possibly augmented with some features per edge point, for the matching (see Figure 3.127(b)). Another strategy is to derive geometric primitives by segmenting the edges with the algorithms discussed in Section 3.8.4, and to match these to segmented geometric primitives in the image (see Figure 3.127(c)). Finally, based on a segmentation of the edges, we can derive salient points and match them to salient points in the image (see Figure 3.127(d)). It should be noted that the salient points can also be extracted directly from the image without extracting edges first [133, 134].

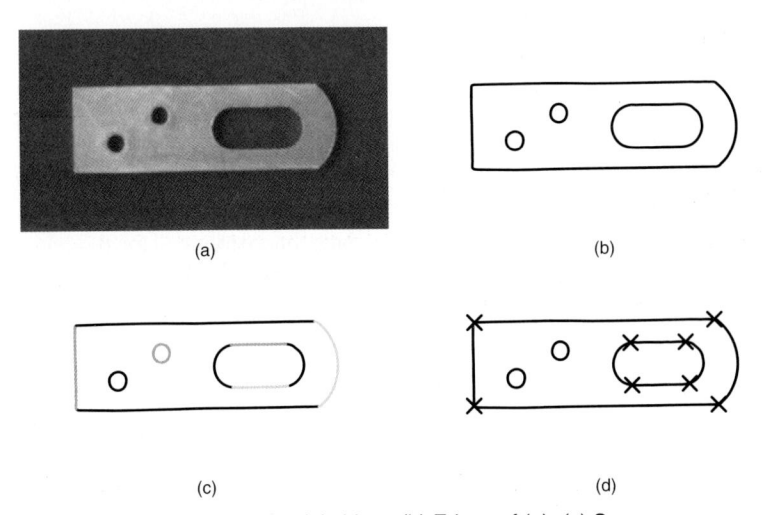

Fig. 3.127 (a) Image of a model object. (b) Edges of (a). (c) Segmentation of (b) into lines and circles. (d) Salient points derived from the segmentation in (c).

A large class of algorithms for edge matching is based on the distance of the edges in the template to the edges in the image. These algorithms typically use the raw edge points for the matching. One natural similarity measure based on this idea is to minimize the mean squared distance between the template edge points and the closest image edge points [135]. Hence, it appears that we must determine the closest image edge point for every template edge point, which would be extremely costly. Fortunately, since we are only interested in the distance to the closest edge point and not in which point is the closest point, this can be done in an efficient manner by calculating the distance transform of the background of the segmented search image [135]. A model is considered as being found if the mean distance of the template edge points to the image edge points is below a threshold. Of course, to obtain a unique location of the template, we must calculate the local minimum of this similarity measure. If we want to formalize this similarity measure, we can denote the edge points in the model by T and the distance transform of the background of the segmented search image (i.e., the complement of the segmented edge region in the search image) by $d(r,c)$. Hence, the mean squared edge distance (SED) for the case of translations is given by

$$sed(r,c) = \frac{1}{n} \sum_{(u,v) \in T} d(r+u, c+v)^2 \qquad (3.141)$$

Note that this is very similar to the SSD similarity measure in Eq. (3.134) if we set $t(u,v) = 0$ there and use the distance transform image for $f(u,v)$. Consequently, the SED matching algorithm can be implemented very easily if we already have an implementation of the SSD matching algorithm. Of course, if we use the mean distance instead of the mean squared distance, we could use an existing implementation of the SAD matching, given by Eq. (3.133), for the edge matching.

We can now ask ourselves whether the SED fulfills the above criteria for robust matching. Since it is based on edges, it is robust to arbitrary illumination changes. Furthermore, since clutter, i.e., extra edges in the search image, can only decrease the distance to the closest edge in the search image, it is robust to clutter. However, if edges are missing in the search image, the distance of the missing template edges to the closest image edges may become very large, and consequently the model may not be found. This is illustrated in Figure 3.128. Imagine what happens when the model in Figure 3.128(a) is searched in a search image in which some of the edges are missing (Figures 3.128(c) and (d)). Here, the missing edges will have a very large squared distance, which will increase the mean squared edge distance significantly. This will make it quite difficult to find the correct match.

Because of the above problems of the SED, edge matching algorithms using a different distance have been proposed. They are based on the Hausdorff distance of two point sets. Let us call the edge points in the template T and the edge points in the image E. Then, the Hausdorff distance of the two point sets is given by

$$H(T,E) = \max(h(T,E), h(E,T)) \qquad (3.142)$$

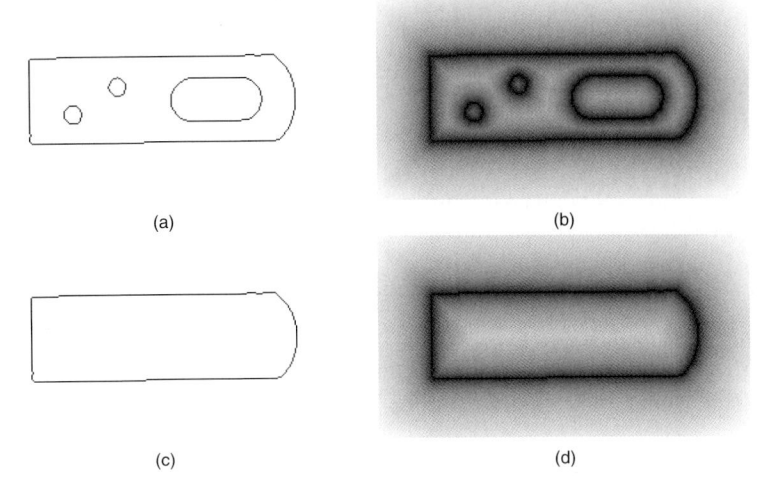

Fig. 3.128 (a) Template edges. (b) Distance transform of the background of (a). For better visualization, a square root look-up table (LUT) is used. (c) Search image with missing edges. (d) Distance transform of the background of (c). If the template in (a) is matched to a search image in which the edges are complete and which possibly contains more edges than the template, the template will be found. If the template in (a) is matched to a search image in which template edges are missing, the template may not be found because a missing edge will have a large distance to the closest existing edge.

where

$$h(T, E) = \max_{t \in T} \min_{e \in E} \|t - e\| \qquad (3.143)$$

and $h(E, T)$ is defined symmetrically. Hence, the Hausdorff distance consists of determining the maximum of two distances: the maximum distance of the template edges to the closest image edges, and the maximum distance of the image edges to the closest template edges [136]. It is immediately clear that, to achieve a low overall distance, every template edge point must be close to an image edge point and vice versa. Therefore, the Hausdorff distance is robust to neither occlusion nor clutter. With a slight modification, however, we can achieve the desired robustness. The reason for the bad performance for occlusion and clutter is that in Eq. (3.143) the maximum distance of the template edges to the image edges is calculated. If we want to achieve robustness to occlusion, instead of computing the largest distance, we can compute a distance with a different rank, e.g., the fth largest distance, where $f = 0$ denotes the largest distance. With this, the Hausdorff distance will be robust to $100 f / n \%$ occlusion, where n is the number of edge points in the template. To make the Hausdorff distance robust to clutter, we can similarly modify $h(E, T)$ to use the rth largest distance. However, normally the model covers only a small part of the search image. Consequently, typically there are many more image edge points than template edge points, and hence r would have to be chosen very large to achieve the desired robustness against clutter. Therefore, $h(E, T)$ must be modified to be calculated only within a small ROI around the template. With this, the Hausdorff distance can be made robust to $100r/m\,\%$ clutter, where m is the number of edge points in the ROI around the template [136]. Like the SED, the Hausdorff distance can be computed based on distance transforms: one for the edge region in the image and one for each pose (excluding translations) of the template edge region. Therefore, we must compute either a very large number of distance transforms offline, which requires an enormous amount of memory, or the distance transforms of the model during the search, which requires a large amount of computation.

As we can see, one of the drawbacks of the Hausdorff distance is the enormous computational load that is required for the matching. In [136], several possibilities are discussed to reduce the computational load, including pruning regions of the search space that cannot contain the template. Furthermore, a hierarchical subdivision of the search space is proposed. This is similar to the effect that is achieved with image pyramids. However, the method in [136] only subdivides the search space, but does not scale the template or image. Therefore, it is still very slow. A Hausdorff distance matching method using image pyramids is proposed in [137].

The major drawback of the Hausdorff distance, however, is that, even with very moderate amounts of occlusion, many false instances of the template will be detected in the image [138]. To reduce the false detection rate, in [138] a modification of the Hausdorff distance is proposed that takes the orientation of the edge pixels into account. Conceptually, the edge points are augmented with a third coordinate that represents the edge orientation. Then, the distance of these augmented 3D points and the corresponding augmented 3D image points is calculated as the modified Hausdorff distance. Unfortunately, this requires the calculation of a 3D distance transform, which makes the algorithm too expensive for machine vision applications. A further drawback of all approaches based on the Hausdorff distance is that it is quite difficult to obtain the pose with subpixel accuracy based on the interpolation of the similarity measure.

Another algorithm to find objects that is based on the edge pixels themselves is the generalized Hough transform proposed by Ballard in [139]. The original Hough transform [140, 141] is a method that was designed to find straight lines in segmented edges. It was later extended to detect other shapes that can be described analytically, e.g., circles or ellipses. The principle of the generalized Hough transform can be best explained by looking at a simple case. Let us try to find circles with a known radius in an edge image. Since circles are rotationally symmetric, we only need to consider translations in this case. If we want to find the circles as efficiently as possible, we can observe that, for circles that are brighter than the background, the gradient vector of the edge of the circle is perpendicular to the circle. This means that it points in the direction of the center of the circle. If the circle is darker than its background, the negative gradient vector points toward the center of the circle. Therefore, since we know the radius of the circle, we can theoretically determine the center of the circle from a single point on the circle. Unfortunately, we do not know which points lie on the circle (this is actually the task we would like to solve). However, we can detect the circle by observing that all points on the circle will have the property that, based on the gradient vector, we can construct the circle center. Therefore, we can accumulate evidence provided by all edge points in the image to determine the circle. This can be done as follows. Since we want to determine the circle center (i.e., the translation of the circle), we can set up an array that accumulates the evidence that a circle is present as a particular translation. We initialize this array with zeros. Then, we loop through all the edge points in the image and construct the potential circle center based on the edge position, the gradient direction, and the known circle radius. With this information, we increment the accumulator array at the potential circle center by one. After we have processed all the edge points, the accumulator array should contain a large amount of evidence, i.e., a large number of votes, at the locations of the circle centers. We can then threshold the accumulator array and compute the local maxima to determine the circle centers in the image.

An example of this algorithm is shown in Figure 3.129. Suppose we want to locate the circle on top of the capacitor in Figure 3.129(a) and that we know that it has a

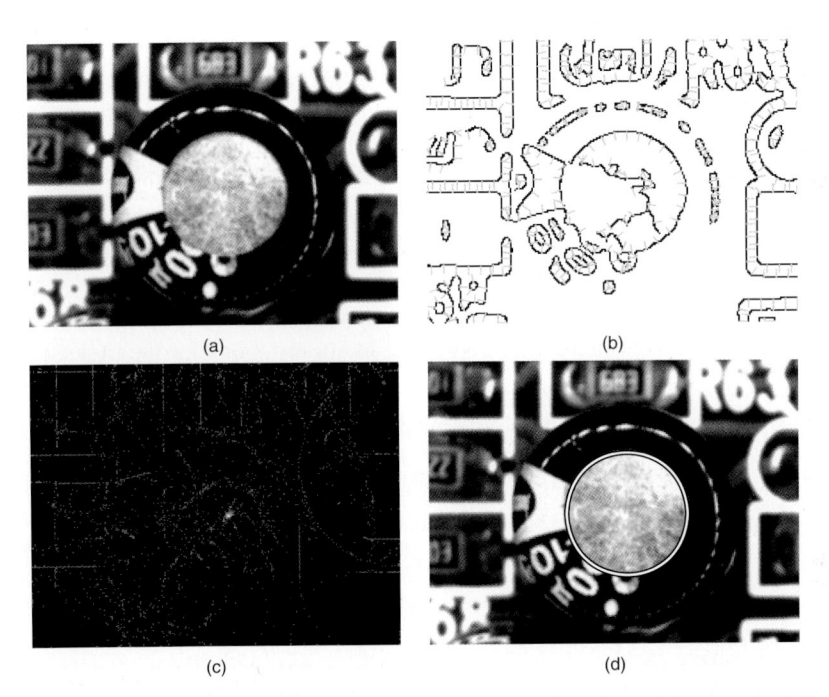

Fig. 3.129 Using the Hough transform to detect a circle. (a) Image of a PCB showing a capacitor. (b) Detected edges. For every eighth edge point, the corresponding orientation is visualized by displaying the gradient vector. (c) Hough accumulator array obtained by performing the Hough transform using the edge points and orientations. A square root LUT is used to make the less populated regions of the accumulator space more visible. If a linear LUT were used, only the peak would be visible. (d) Circle detected by thresholding (c) and computing the local maxima.

radius of 39 pixels. The edges extracted with a Canny filter with $\sigma = 2$ and hysteresis thresholds of 80 and 20 are shown in Figure 3.129(b). Furthermore, for every eighth edge point, the gradient vector is shown. Note that for the circle they all point toward the circle center. The accumulator array that is obtained with the algorithm described above is displayed in Figure 3.129(c). Note that there is only one significant peak. In fact, most of the cells in the accumulator array have received so few votes that a square root LUT had to be used to visualize that there are any votes at all in the rest of the accumulator array. If the accumulator array is thresholded and the local maxima are calculated, the circle in Figure 3.129(d) is obtained.

From the above example, we can see that we can find circles in the image extremely efficiently. If we know the polarity of the circle, i.e., whether it is brighter or darker than the background, we only need to perform a single increment of the accumulator array per edge point in the image. If we do not know the polarity of the edge, we need to perform two increments per edge point. Hence, the runtime is proportional to the number of edge points in the image and not to the size of the template, i.e., the size of the circle. Ideally, we would like to find an algorithm that is equally efficient for arbitrary objects.

What can we learn from the above example? First of all, it is clear that, for arbitrary objects, the gradient direction does not necessarily point to a reference point of the object like it did for circles. Nevertheless, the gradient direction of the edge point provides a constraint where the reference point of the object can be, even for arbitrarily shaped objects. This is shown in Table 3.2. Suppose we have singled out the reference point o of the object. For the circle, the natural choice would be its center. For an arbitrary object, we can, for example, use the center of gravity of the edge points. Now consider an edge point e_i. We can see that the gradient vector ∇f_i and the vector r_i from e_i to o always enclose the same angle, no matter how the object is translated, rotated and scaled. For simplicity, let us only consider translations for the moment.

Tab. 3.2 The principle of constructing the R-table in the generalized Hough transform (GHT). The R-table (on the right) is constructed based on the gradient angle ϕ_i of each edge point of the model object and the vector r_i from each edge point to the reference point o of the template.

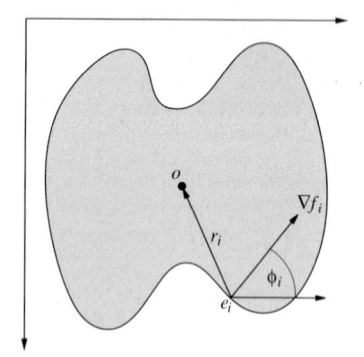

j	ϕ_j	r_i
0	0	$\{r_i \mid \phi_i = 0\}$
1	$\Delta\phi$	$\{r_i \mid \phi_i = \Delta\phi\}$
2	$2\Delta\phi$	$\{r_i \mid \phi_i = 2\Delta\phi\}$
⋮	⋮	⋮

Then, if we find an edge point in the image with a certain gradient direction or gradient angle ϕ_i, we could calculate the possible location of the template with the vector r_i and increment the accumulator array accordingly. Note that for circles the gradient vector ∇f_i has the same direction as the vector r_i. For arbitrary shapes this no longer holds. From Table 3.2, we can also see that the edge direction does not necessarily uniquely constrain the reference point, since there may be multiple points on the edges of the template that have the same orientation. For circles, this is not the case. For example, in the lower left part of the object in Table 3.2 there is a second point that has the same gradient direction as the point labeled e_i, which has a different offset vector to the reference point. Therefore, in the search we have to increment all accumulator array elements that correspond to the edge points in the template with the same edge direction. Hence, during the search we must be able to quickly determine all the offset vectors that correspond to a given edge direction in the image. This can be achieved in a preprocessing step in the template generation that is performed offline. Basically, we construct a table, called the R-table, that is indexed by the gradient angle ϕ. Each table entry contains all the offset vectors r_i of the template edges that have the gradient angle ϕ. Since the table must be discrete to enable efficient indexing, the gradient angles are discretized with a certain step size $\Delta\phi$. The concept of the R-table is also shown in Table 3.2. With the R-table, it is very simple to find the offset vectors for incrementing the accumulator array in the search: we simply calculate the gradient angle in the image and use it as an index into the R-table. After the construction of the accumulator array, we threshold the array and calculate the local maxima to find the possible locations of the object. This approach can also be extended easily to deal with rotated and scaled objects [139]. In real images, we also need to consider that there are uncertainties in the location of the edges in the image and in the edge orientations. We have already seen in Eq. (3.98) that the precision of the Canny edges depends on the signal-to-noise ratio. Using similar techniques, it can be shown that the precision of the edge angle ϕ for the Canny filter is given by $\sigma_\phi^2 = \sigma_n^2/(4\sigma^2 a^2)$. These values must be used in the online phase to determine a range of cells in the accumulator array that must be incremented to ensure that the cell corresponding to the true reference point is incremented.

The generalized Hough transform described above is already quite efficient. On average, it increments a constant number of accumulator cells. Therefore, its runtime only depends on the number of edge points in the image. However, it is still not fast enough for machine vision applications because the accumulator space that must be searched to find the objects can quickly become very large, especially if rotations and scalings of the object are allowed. Furthermore, the accumulator uses an enormous amount of memory. Consider, for example, an object that should be found in a 640×480 image with an angle range of $360°$, discretized in $1°$ steps. Let us suppose that two bytes are sufficient to store the accumulator array entries without overflow. Then, the accumulator array requires $640 \times 480 \times 360 \times 2 = 221\,184\,000$ bytes of memory, i.e., 211 MB. This is unacceptably large for most applications. Furthermore, it means that initializing this array alone will require a significant amount of processing time. For this reason, a hierarchical generalized Hough transform is proposed in [142]. It uses image pyramids to speed up the search and to reduce the size of the accumulator array by using matches found on higher pyramid levels to constrain the search on lower pyramid levels. The interested reader is referred to [142] for details of the implementation. With this hierarchical generalized Hough transform, objects can be found in real time even under severe occlusions, clutter, and almost arbitrary illumination changes.

The algorithms we have discussed so far were based on matching edge points directly. Another class of algorithms is based on matching geometric primitives, e.g., points, lines, and circles. These algorithms typically follow the hypothesize-and-test paradigm, i.e., they hypothesize a match, typically from a small number of primitives, and then test whether the hypothetical match has enough evidence in the image.

The biggest challenge that this type of algorithm must solve is the exponential complexity of the correspondence problem. Let us, for the moment, suppose that we are only using one type of geometric primitive, e.g., lines. Furthermore, let us suppose that all template primitives are visible in the image, so that potentially there is a subset of the primitives in the image that corresponds exactly to the primitives in the template. If the template consists of m primitives and there are n primitives in the image, there are $\binom{n}{m}$, i.e., $O(n^m)$, potential correspondences between the template and image primitives. If the objects in the search image can be occluded, the number of potential matches is even larger, since we must allow that multiple primitives in the image can match a single primitive in the template, because a single primitive in the template may break up into several pieces, and that some primitives in the template are not present in the search image. It is clear that, even for moderately large values of m and n, the cost of exhaustively checking all possible correspondences is prohibitive. Therefore, geometric constraints and strong heuristics must be used to perform the matching in an acceptable time.

One approach to perform the matching efficiently is called geometric hashing [143]. It was originally described for points as primitives, but can equally well be used with lines. Furthermore, the original description uses affine transformations as the set of allowable transformations. We will follow the original presentation and will note where modifications are necessary for other classes of transformations and lines as primitives. Geometric hashing is based on the observation that three points define an affine basis of the 2D plane. Thus, once we select three points e_{00}, e_{10}, and e_{01} in general positions, i.e., not collinear, we can represent every other point as a linear combination of these three points:

$$q = e_{00} + \alpha(e_{10} - e_{00}) + \beta(e_{01} - e_{00})$$

The interesting property of this representation is that it is invariant to affine transformations, i.e., (α, β) only depend on the three basis points (the basis triplet), but not on an affine transformation, i.e., they are affine invariants. With this, the values (α, β)

can be regarded as the affine coordinates of the point q. This property holds equally well for lines: three non-parallel lines can be used to define an affine basis. If we use a more restricted class of transformations, fewer points are sufficient to define a basis. For example, if we restrict the transformations to similarity transformations, two points are sufficient to define a basis. Note, however, that two lines are only sufficient to determine a rigid transformation.

The aim of geometric hashing is to reduce the amount of work that has to be performed to establish the correspondences between the template and image points. Therefore, it constructs a hash table that enables the algorithm to determine quickly the potential matches for the template. This hash table is constructed as follows. For every combination of three non-collinear points in the template, the affine coordinates (α, β) of the remaining $m - 3$ points of the template are calculated. The affine coordinates (α, β) serve as the index into the hash table. For every point, the index of the current basis triplet is stored in the hash table. If more than one template should be found, additionally the template index is stored; however, we will not consider this case further, so for our purposes only the index of the basis triplet is stored.

To find the template in the image, we randomly select three points in the image and construct the affine coordinates (α, β) of the remaining $n - 3$ points. We then use (α, β) as an index into the hash table. This returns us the index of the basis triplet. With this, we obtain a vote for the presence of a particular basis triplet in the image. If the randomly selected points do not correspond to a basis triplet of the template, the votes of all the points will not agree. However, if they correspond to a basis triplet of the template, many of the votes will agree and will indicate the index of the basis triplet. Therefore, if enough votes agree, we have a strong indication for the presence of the model. The presence of the model is then verified as described below. Since there is a certain probability that we have selected an inappropriate basis triplet in the image, the algorithm iterates until it has reached a certain probability of having found the correct match. Here, we can make use of the fact that we only need to find one correct basis triplet to find the model. Therefore, if k of the m template points are present in the image, the probability of having selected at least one correct basis triplet in t trials is approximately

$$p = 1 - \left(1 - \left(\frac{k}{n}\right)^3\right)^t \qquad (3.144)$$

If similarity transforms are used, only two points are necessary to determine the affine basis. Therefore, the inner exponent will change from 3 to 2 in this case. For example, if the ratio of visible template points to image points k/n is 0.2 and we want to find the template with a probability of 99% (i.e., $p = 0.99$), 574 trials are sufficient if affine transformations are used. For similarity transformations, 113 trials would suffice. Hence, geometric hashing can be quite efficient in finding the correct correspondences, depending on how many extra features are present in the image.

After a potential match has been obtained with the algorithm described above, it must be verified in the image. In [143], this is done by establishing point correspondences for the remaining template points based on the affine transformation given by the selected basis triplet. Based on these correspondences, an improved affine transformation is computed by a least-squares minimization over all corresponding points. This, in turn, is used to map all the edge points of the template, i.e., not only the characteristic points that were used for the geometric hashing, to the pose of the template in the image. The transformed edges are compared to the image edges. If there is sufficient overlap between the template and image edges, the match is accepted and the corresponding points and edges are removed from the segmentation. If more than one instance of the template should be found, the entire process is repeated.

The algorithm described so far works well as long as the geometric primitives can be extracted with sufficient accuracy. If there are errors in the point coordinates, an erroneous affine transformation will result from the basis triplet. Therefore, all the affine coordinates (α, β) will contain errors, and hence the hashing in the online phase will access the wrong entry in the hash table. This is probably the largest drawback of the geometric hashing algorithm in practice. To circumvent this problem, the template points must be stored in multiple adjacent entries of the hash table. Which hash table entries must be used can in theory be derived through error propagation [143]. However, in practice, the accuracy of the geometric primitives is seldom known. Therefore, estimates have to be used, which must be well on the safe side for the algorithm not to miss any matches in the online phase. This, in turn, makes the algorithm slightly less efficient because more votes will have to be evaluated during the search.

The final class of algorithms we will discuss tries to match geometric primitives themselves to the image. Most of these algorithms use only line segments as the primitives [144, 145, 146]. One of the few exceptions to this rule is the approach in [147], which uses line segments and circular arcs. Furthermore, in 3D object recognition, sometimes line segments and elliptic arcs are used [148]. As we discussed above, exhaustively enumerating all potential correspondences between the template and image primitives is prohibitively slow. Therefore, it is interesting to look at examples of different strategies that are employed to make the correspondence search tractable.

The approach in [144] segments the contours of the model object and the search image into line segments. Depending on the lighting conditions, the contours are obtained by thresholding or by edge detection. The 10 longest line segments in the template are singled out as privileged. Furthermore, the line segments in the model are ordered by adjacency as they trace the boundary of the model object. To generate a hypothesis, a privileged template line segment is matched to a line segment in the image. Since the approach is designed to handle similarity transforms, the angle, which is invariant under these transforms, to the preceding line segment in the image is compared to the angle to the preceding line segment in the template. If they are not close enough, the potential match is rejected. Furthermore, the length ratio of these two segments, which also is invariant, is used to check the validity of the hypothe-

3.11 Template Matching

sis. The algorithm generates a certain number of hypotheses in this manner. These hypotheses are then verified by trying to match additional segments. The quality of the hypotheses, including the additionally matched segments, is then evaluated based on the ratio of the lengths of the matched segments to the length of the segments in the template. The matching is stopped once a high-quality match has been found or if enough hypotheses have been evaluated. Hence, we can see that the complexity is kept manageable by using privileged segments in conjunction with their neighboring segments.

In [146], a similar method is proposed. In contrast to [144], corners (combinations of two adjacent line segments of the boundary of the template that enclose a significant angle) are matched first. To generate a matching hypothesis, two corners must be matched to the image. Geometric constraints between the corners are used to reject false matches. The algorithm then attempts to extend the hypotheses with other segments in the image. The hypotheses are evaluated based on a dissimilarity criterion. If the dissimilarity is below a threshold, the match is accepted. Hence, the complexity of this approach is reduced by matching features that have distinctive geometric characteristics first.

The approach in [145] also generates matching hypotheses and tries to verify them in the image. Here, a tree of possible correspondences is generated and evaluated in a depth-first search. This search tree is called the interpretation tree. A node in the interpretation tree encodes a correspondence between a model line segment and an image line segment. Hence, the interpretation tree would exhaustively enumerate all correspondences, which would be prohibitively expensive. Therefore, the interpretation tree must be pruned as much as possible. To do this, the algorithm uses geometric constraints between the template line segments and the image line segments. Specifically, the distances and angles between pairs of line segments in the image and in the template must be consistent. This angle is checked by using normal vectors of the line segments that take the polarity of the edges into account. This consistency check prunes a large number of branches of the interpretation tree. However, since a large number of possible matchings still remain, a heuristic is used to explore the most promising hypotheses first. This is useful because the search is terminated once an acceptable match has been found. This early search termination is criticized in [149], and various strategies to speed up the search for all instances of the template in the image are discussed. The interested reader is referred to [149] for details.

To make the principles of the geometric matching algorithms clearer, let us examine a prototypical matching procedure on an example. The template to be found is shown in Figure 3.130(a). It consists of five line segments and five circular arcs. They were segmented automatically from the image in Figure 3.127(a) using a subpixel-precise Canny filter with $\sigma = 1$ and and by splitting the edge contours into line segments and circular arcs using the method described in Section 3.8.4. The template consists of the geometric parameters of these primitives as well as the segmented contours themselves. The image in which the template should be found is shown in Figure 3.130(b).

It contains four partially occluded instances of the model along with four clutter objects. The matching starts by extracting edges in the search image and by segmenting them into line segments and circular arcs (Figure 3.130(c)). Like for the template, the geometric parameters of the image primitives are calculated. The matching now determines possible matches for all of the primitives in the template. Of these, the largest circular arc is examined first because of a heuristic that rates moderately long circular arcs as more distinctive than even long line segments. The resulting matching hypotheses are shown in Figure 3.130(d). Of course, the line segments could also have been examined first. Because in this case only rigid transformations are allowed, the matching of the circular arcs uses the radii of the circles as a matching constraint.

Since the matching should be robust to occlusions, the opening angle of the circular arcs is not used as a constraint. Because of this, the matched circles are not sufficient to determine a rigid transformation between the template and the image. Therefore, the algorithm tries to match an adjacent line segment (the long lower line segment in Figure 3.130(a)) to the image primitives while using the angle of intersection between the circle and the line as a geometric constraint. The resulting matches are shown in Figure 3.130(e). With these hypotheses, it is possible to compute a rigid transformation that transforms the template to the image. Based on this, the remaining primitives can be matched to the image based on the distances of the image primitives and the transformed template primitives. The resulting matches are shown in Figure 3.130(f). Note that, because of specular reflections, sometimes multiple parallel line segments are matched to a single line segment in the template. This could be fixed by taking the polarity of the edges into account. To obtain the rigid transformation between the template and the matches in the image as accurately as possible, a least-squares optimization of the distances between the edges in the template and the edges in the image can be used. An alternative is the minimal tolerance error zone optimization described in [147]. Note that the matching has already found the four correct instances of the template. For the algorithm, the search is not finished, however, since there might be more instances of the template in the image, especially instances for which the large circular arc is occluded more than in the leftmost instance in the image. Hence, the search is continued with other primitives as the first primitives to try. In this case, however, the search does not discover new viable matches.

After having discussed some of the approaches for robustly finding templates in an image, the question as to which of these algorithms should be used in practice naturally arises. We will say more on this topic below. From the above discussion, however, we can see that the effectiveness of a particular approach greatly depends on the shape of the template itself. Generally, the geometric matching algorithms have an advantage if the template and image contain only few, salient geometric primitives, like in the example in Figure 3.130. Here, the combinatorics of the geometric matching algorithms can work to their advantage. On the other hand, they work to their disadvantage if the template or search image contains a large number of geometric primitives.

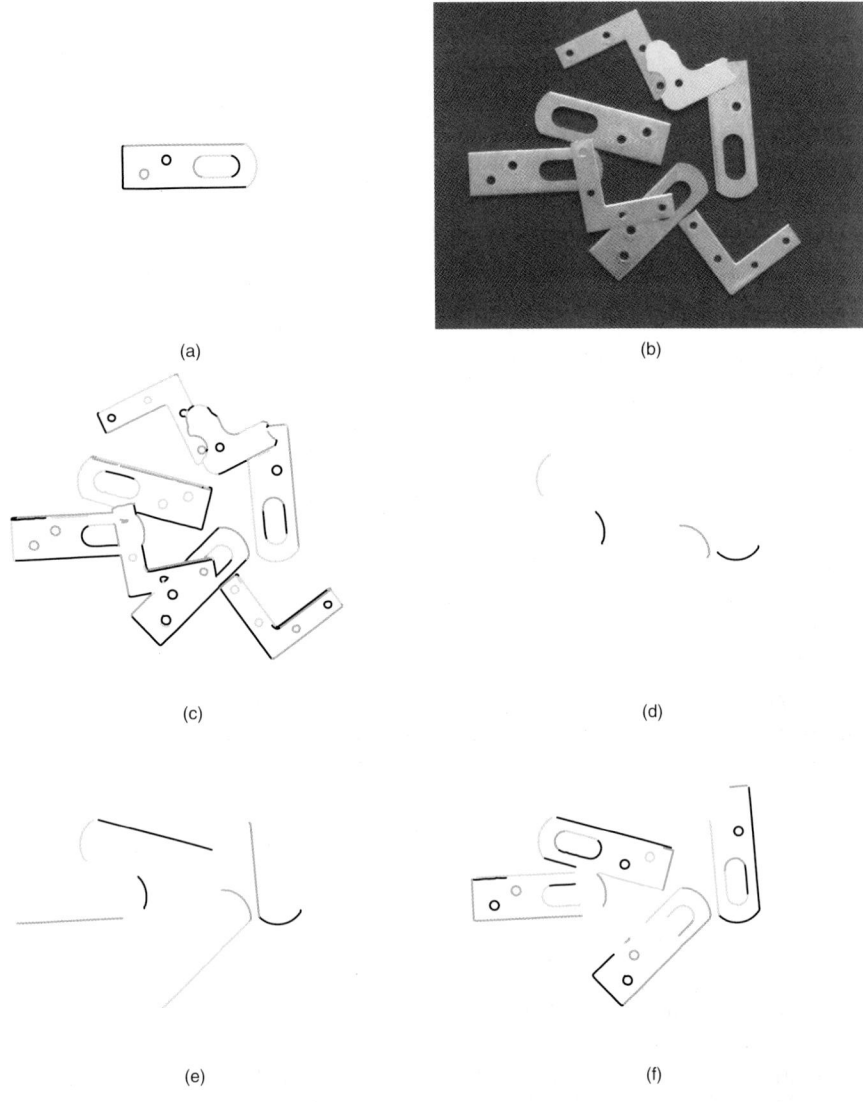

Fig. 3.130 Example of matching an object in the image using geometric primitives. (a) The template consists of five line segments and five circular arcs. The model has been generated from the image in Figure 3.127(a). (b) The search image contains four partially occluded instances of the template along with four clutter objects. (c) Edges extracted in (b) with a Canny filter with $\sigma = 1$ and split into line segments and circular arcs. (d) The matching in this case first tries to match the largest circular arc of the model and finds four hypotheses. (e) The hypotheses are extended with the lower of the long line segments in (a). These two primitives are sufficient to estimate a rigid transform that aligns the template with the features in the image. (f) The remaining primitives of the template are matched to the image. The resulting matched primitives are displayed.

A difficult model image is shown in Figure 3.131. The template contains fine structures that result in 350 geometric primitives, which are not particularly salient. Consequently, the search would have to examine an extremely large number of hypotheses that could only be dismissed after examining a large number of additional primitives. Note that the model contains 35 times as many primitives as the model in Figure 3.130, but only approximately three times as many edge points. Consequently, it could be easily found with pixel-based approaches like the generalized Hough transform.

(a) (b)

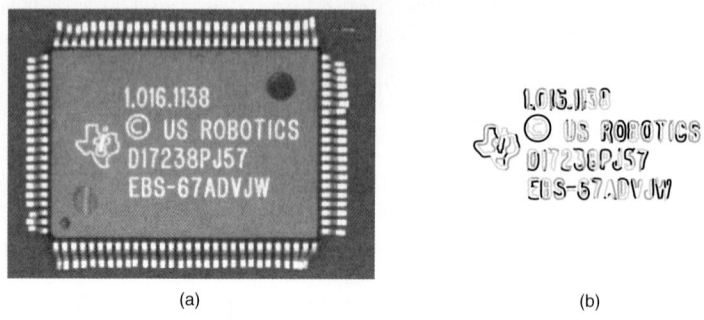

Fig. 3.131 (a) Image of a template object that is not suitable for the geometric matching algorithms. Although the segmentation of the template into line segments and circular arcs in (b) only contains approximately three times as many edge points as the template in Figure 3.130, it contains 35 times as many geometric primitives, i.e., 350.

A difficult search image is shown in Figure 3.132. Here, the goal is to find the circular fiducial mark. Since the contrast of the fiducial mark is very low, a small segmentation threshold must be used in the edge detection to find the relevant edges of the circle. This causes a very large number of edges and broken fragments that must be examined. Again, pixel-based algorithms will have little trouble with this image.

(a) (b)

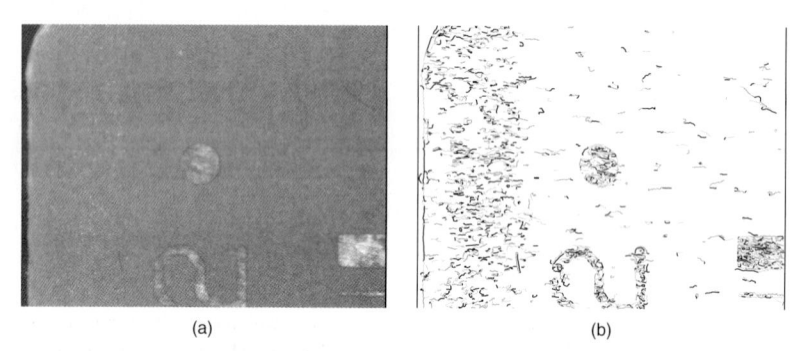

Fig. 3.132 (a) A search image that is difficult for the geometric matching algorithms. Here, because of the poor contrast of the circular fiducial, the segmentation threshold must be chosen very low so that the relevant edges of the fiducial are selected. Because of this, the segmentation in (b) contains a very large number of primitives that must be examined in the search.

From the above examples, we can see that the pixel-based algorithms have the advantage that they can represent arbitrarily shaped templates without problems. Geometric matching algorithms, on the other hand, are restricted to relatively simple shapes that can be represented with a very small number of primitives. Therefore, in the remainder of this section, we will discuss a pixel-based robust template matching algorithm called shape-based matching [150, 151, 152, 153, 154] that works very well in practice [155, 156].

One of the drawbacks of all of the algorithms that we have discussed above is that they segment the edge image. This makes the object recognition algorithm invariant only against a narrow range of illumination changes. If the image contrast is lowered, progressively fewer edge points will be segmented, which has the same effects as progressively larger occlusion. Consequently, the object may not be found for low-contrast images. To overcome this problem, a similarity measure that is robust against occlusion, clutter, and nonlinear illumination changes must be used. This similarity measure can then be used in the pyramid-based recognition strategy described in Sections 3.11.2 and 3.11.4.

To define the similarity measure, we first define the model of an object as a set of points $p_i = (r_i, c_i)^\top$ and associated direction vectors $d_i = (t_i, u_i)^\top$, with $i = 1, \ldots, n$. The direction vectors can be generated by a number of different image processing operations. However, typically edge extraction (see Section 3.7.3) is used. The model is generated from an image of the object, where an arbitrary region of interest specifies the part of the image in which the object is located. It is advantageous to specify the coordinates p_i relative to the center of gravity of the ROI of the model or to the center of gravity of the points of the model.

The image in which the model should be found can be transformed into a representation in which a direction vector $e_{r,c} = (v_{r,c}, w_{r,c})^\top$ is obtained for each image point (r, c). In the matching process, a transformed model must be compared to the image at a particular location. In the most general case considered here, the transformation is an arbitrary affine transformation (see Section 3.3.1). It is useful to separate the translation part of the affine transformation from the linear part. Therefore, a linearly transformed model is given by the points $p'_i = \mathbf{A} p_i$ and the accordingly transformed direction vectors $d'_i = (\mathbf{A}^{-1})^\top d_i$, where

$$\mathbf{A} = \begin{pmatrix} a_{11} & a_{12} \\ a_{21} & a_{22} \end{pmatrix} \qquad (3.145)$$

As discussed above, the similarity measure by which the transformed model is compared to the image must be robust to occlusions, clutter, and illumination changes. One such measure is to sum the (unnormalized) dot product of the direction vectors of the transformed model and the image over all points of the model to compute a matching score at a particular point $q = (r, c)^\top$ of the image. That is, the similarity measure of the transformed model at the point q, which corresponds to the translation part of the affine transformation, is computed as follows:

$$s = \frac{1}{n} \sum_{i=1}^{n} {d'_i}^T e_{q+p'} = \frac{1}{n} \sum_{i=1}^{n} t'_i v_{r+r'_i, c+c'_i} + u'_i w_{r+r'_i, c+c'_i} \tag{3.146}$$

If the model is generated by edge filtering and the image is preprocessed in the same manner, this similarity measure fulfills the requirements of robustness to occlusion and clutter. If parts of the object are missing in the image, there are no edges at the corresponding positions of the model in the image, i.e., the direction vectors will have a small length and hence contribute little to the sum. Likewise, if there are clutter edges in the image, there will either be no point in the model at the clutter position or it will have a small length, which means it will contribute little to the sum.

The similarity measure in Eq. (3.146) is not truly invariant against illumination changes, however, since the length of the direction vectors depends on the brightness of the image if edge detection is used to extract the direction vectors. However, if a user specifies a threshold on the similarity measure to determine whether the model is present in the image, a similarity measure with a well-defined range of values is desirable. The following similarity measure achieves this goal:

$$s = \frac{1}{n} \sum_{i=1}^{n} \frac{{d'_i}^T e_{q+p'}}{\|d'_i\| \|e_{q+p'}\|} = \frac{1}{n} \sum_{i=1}^{n} \frac{t'_i v_{r+r'_i, c+c'_i} + u'_i w_{r+r'_i, c+c'_i}}{\sqrt{{t'_i}^2 + {u'_i}^2} \sqrt{v^2_{r+r'_i, c+c'_i} + w^2_{r+r'_i, c+c'_i}}} \tag{3.147}$$

Because of the normalization of the direction vectors, this similarity measure is additionally invariant to arbitrary illumination changes, since all vectors are scaled to a length of 1. What makes this measure robust against occlusion and clutter is the fact that, if a feature is missing, either in the model or in the image, noise will lead to random direction vectors, which, on average, will contribute nothing to the sum.

The similarity measure in Eq. (3.147) will return a high score if all the direction vectors of the model and the image align, i.e., point in the same direction. If edges are used to generate the model and image vectors, this means that the model and image must have the same contrast direction for each edge. Sometimes it is desirable to be able to detect the object even if its contrast is reversed. This is achieved by:

$$s = \left| \frac{1}{n} \sum_{i=1}^{n} \frac{{d'_i}^T e_{q+p'}}{\|d'_i\| \|e_{q+p'}\|} \right| \tag{3.148}$$

In rare circumstances, it might be necessary to ignore even local contrast changes. In this case, the similarity measure can be modified as follows:

$$s = \frac{1}{n} \sum_{i=1}^{n} \frac{|{d'_i}^T e_{q+p'}|}{\|d'_i\| \|e_{q+p'}\|} \tag{3.149}$$

The normalized similarity measures in Eqs. (3.147)–(3.149) have the property that they return a number smaller than 1 as the score of a potential match. In all cases, a score of 1 indicates a perfect match between the model and the image. Furthermore, the score roughly corresponds to the portion of the model that is visible in the image. For example, if the object is 50% occluded, the score (on average) cannot exceed 0.5. This is a highly desirable property, because it gives the user the means to select an intuitive threshold for when an object should be considered as recognized.

A desirable feature of the above similarity measures in Eqs. (3.147)–(3.149) is that they do not need to be evaluated completely when object recognition is based on a user-defined threshold $s_{\min}$ for the similarity measure that a potential match must achieve. Let s_j denote the partial sum of the dot products up to the jth element of the model. For the match metric that uses the sum of the normalized dot products, this is

$$s_j = \frac{1}{n} \sum_{i=1}^{j} \frac{{d'_i}^\mathsf{T} e_{q+p'}}{\|d'_i\| \|e_{q+p'}\|} \qquad (3.150)$$

Obviously, all the remaining terms of the sum are all ≤ 1. Therefore, the partial score can never achieve the required score $s_{\min}$ if $s_j < s_{\min} - 1 + j/n$, and hence the evaluation of the sum can be discontinued after the jth element whenever this condition is fulfilled. This criterion speeds up the recognition process considerably.

As mentioned above, to recognize the model, an image pyramid is constructed for the image in which the model should be found (see Section 3.11.2). For each level of the pyramid, the same filtering operation that was used to generate the model, e.g., edge filtering, is applied to the image. This returns a direction vector for each image point. Note that the image is not segmented, i.e., thresholding or other operations are not performed. This results in true robustness to illumination changes.

As discussed in Sections 3.11.2 and 3.11.4, to identify potential matches, an exhaustive search is performed for the top level of the pyramid, i.e., all possible poses of the model are used on the top level of the image pyramid to compute the similarity measure via Eqs. (3.147), (3.148), or (3.149). A potential match must have a score larger than $s_{\min}$, and the corresponding score must be a local maximum with respect to neighboring scores. The threshold $s_{\min}$ is used to speed up the search by terminating the evaluation of the similarity measure as early as possible. Therefore, this seemingly brute-force strategy actually becomes extremely efficient.

After the potential matches have been identified, they are tracked through the resolution hierarchy until they are found at the lowest level of the image pyramid. Once the object has been recognized on the lowest level of the image pyramid, its pose is extracted with a resolution better than the discretization of the search space with the approach described in Section 3.11.3.

While the pose obtained by the extrapolation algorithm is accurate enough for most applications, in some applications an even higher accuracy is desirable. This can be achieved through a least-squares adjustment of the pose parameters. To achieve a better accuracy than the extrapolation, it is necessary to extract the model points as well as the feature points in the image with subpixel accuracy. Then, the algorithm finds the closest image point for each model point, and then minimizes the sum of the squared distances of the image points to a line defined by their corresponding model point and the corresponding tangent to the model point, i.e., the directions of the model points are taken to be correct and are assumed to describe the direction of the object's border. If an edge detector is used, the direction vectors of the model are perpendicular to the object boundary, and hence the equation of a line through a model point tangent to the object boundary is given by

$$t_i(r - r_i) + u_i(c - c_i) = 0$$

Let $q_i = (r'_i, c'_i)^\top$ denote the matched image points corresponding to the model points p_i. Then, the following function is minimized to refine the pose a:

$$d(a) = \sum_{i=1}^{n} \left[t_i(r'_i(a) - r_i) + u_i(c'_i(a) - c_i) \right]^2 \to \min \quad (3.151)$$

The potential corresponding image points in the search image are obtained without thresholding by a non-maximum suppression and are extrapolated to subpixel accuracy. By this, a segmentation of the search image is avoided, which is important to preserve the invariance against arbitrary illumination changes. For each model point, the corresponding image point in the search image is chosen as the potential image point with the smallest Euclidian distance using the pose obtained by the extrapolation to transform the model to the search image. Since the point correspondences may change by the refined pose, an even higher accuracy can be gained by iterating the correspondence search and pose refinement. Typically, after three iterations the accuracy of the pose no longer improves.

Figure 3.133 shows six examples in which the shape-based matching algorithm finds the print on the IC shown in Figure 3.131. Note that the object is found despite severe occlusions and clutter.

Extensive tests with shape-based matching have been carried out in [155, 156]. The results show that shape-based matching provides extremely high recognition rates in the presence of severe occlusion and clutter as well as in the presence of nonlinear illumination changes. Furthermore, accuracies better than 1/30 pixel and better than 1/50 degree can be achieved.

From the above discussion, we can see that the basic algorithms for implementing a robust template matching already are fairly complex. In reality, however, the complexity additionally resides in the time and effort that needs to be spent in making the algorithms very robust and fast. Additional complexity comes from the fact that, on the one hand, templates with arbitrary ROIs should be possible to exclude undesired parts from the template. On the other hand, for speed reasons, it should also be possible to specify arbitrarily shaped ROIs for the search space in the search images. Consequently, these algorithms cannot be implemented easily. Therefore, wise

Fig. 3.133 Six examples in which the shape-based matching algorithm finds an object (the print on the IC shown in Figure 3.131) despite severe occlusions and clutter.

machine vision users rely on standard software packages to provide this functionality rather than attempting to implement it themselves.

3.12
Optical Character Recognition

In quite a few applications, we face the challenge of having to read characters on the object we are inspecting. For example, traceability requirements often lead to the fact that the objects to be inspected are labeled with a serial number, and that we have to read this serial number (see, for example, Figures 3.22–3.24). In other applications, reading a serial number might be necessary to control the production flow.

Optical character recognition (OCR) is the process of reading characters in images. It consists of two tasks: segmentation of the individual characters, and the classification of the segmented characters, i.e., the assignment of a symbolic label to the segmented regions. We will examine these two tasks in this section.

3.12.1
Character Segmentation

The classification of the characters requires that we have segmented the text into individual characters, i.e., each character must correspond to exactly one region.

To segment the characters, we can use all the methods that we have discussed in Section 3.4: thresholding with fixed and automatically selected thresholds, dynamic thresholding, and the extraction of connected components.

Furthermore, we might have to use the morphological operations of Section 3.6 to connect separate parts of the same character, e.g., the dot of the character "i" to its main part (see Figure 3.43) or parts of the same character that are disconnected, e.g., because of a bad print quality. For characters on difficult surfaces, e.g., punched characters on a metal surface, gray value morphology may be necessary to segment the characters (see Figure 3.60).

Additionally, in some applications it may be necessary to perform a geometric transformation of the image to transform the characters into a standard position, typically such that the text is horizontal. This process is called image rectification. For example, the text may have to be rotated (see Figure 3.18), perspectively rectified (see Figure 3.20), or rectified with a polar transformation (see Figure 3.21).

Even though we have many segmentation strategies at our disposal, in some applications it may be difficult to segment the individual characters because the characters actually touch each other, either in reality or in the resolution at which we are looking at them in the image. Therefore, special methods to segment touching characters are sometimes required.

The simplest such strategy is to define a separate ROI for each character we are expecting in the image. This strategy can sometimes be used in industrial applications because the fonts typically have a fixed pitch (width) and we know a priori how many characters are present in the image, e.g., if we are trying to read serial numbers with a fixed length. The main problem with this approach is that the character ROIs must enclose the individual characters we are trying to separate. This is difficult if the position of the text can vary in the image. If this is the case, we first need to determine the pose of the text in the image based on another strategy, e.g., template matching to find a distinct feature in the vicinity of the text we are trying to read, and to use the pose of the text either to rectify the text to a standard position or to move the character ROIs to the appropriate position.

While defining separate ROIs for each character works well in some applications, it is not very flexible. A better method can be derived by realizing that the characters typically touch only with a small number of pixels. An example of this is shown in Figures 3.134(a) and (b). To separate these characters, we can simply count the number of pixels per column in the segmented region. This is shown in Figure 3.134(c). Since the touching part is only a narrow bridge between the characters, the number of pixels in the region of the touching part only has a very small number of pixels per column. In fact, we can simply segment the characters by splitting them vertically at the position of the minimum in Figure 3.134(c). The result is shown in Figure 3.134(d). Note that in Figure 3.134(c) the optimal splitting point is the global minimum of the number of pixels per column. However, in general this may not be the case. For example, if the strokes between the vertical bars of the letter "m" were slightly thinner the letter "m" might be split erroneously. Therefore, to make this algorithm more robust, it is typically necessary to define a search space for the splitting of the characters based on the expected width of the characters. For example, in this application the characters are approximately 20 pixels wide. Therefore, we could restrict the search space for the optimal splitting point to a range of ±4 pixels (20% of the expected width) around the expected width of the characters. This simple splitting method works very well in practice. Further approaches for segmenting characters are discussed in [157].

3.12.2
Feature Extraction

As mentioned above, the reading of the characters corresponds to the classification of the regions, i.e., the assignment of a class ω_i to a region. For the purposes of the OCR, the classes ω_i can be thought of as the interpretation of the character, i.e., the string that represents the character. For example, if an application must read serial numbers, the classes $\{\omega_1, \ldots, \omega_{10}\}$ are simply the strings $\{0, \ldots, 9\}$. If numbers and uppercase letters must be read, the classes are $\{\omega_1, \ldots, \omega_{36}\} = \{0, \ldots, 9, A, \ldots, Z\}$. Hence, classification can be thought of as a function f that maps to the set of classes $\Omega = \{\omega_i, i = 1, \ldots, m\}$. What is the input of this function? First of all, to make the above mapping well-defined and easy to handle, we require that the number n of input values to the function f is constant. The input values to f are called features. Typically they are real numbers. With this, the function f that performs the classification can be regarded as a mapping $f : \mathbb{R}^n \mapsto \Omega$.

For the OCR, the features that are used for the classification are features that we extract from the segmented characters. Any of the region features described in Sec-

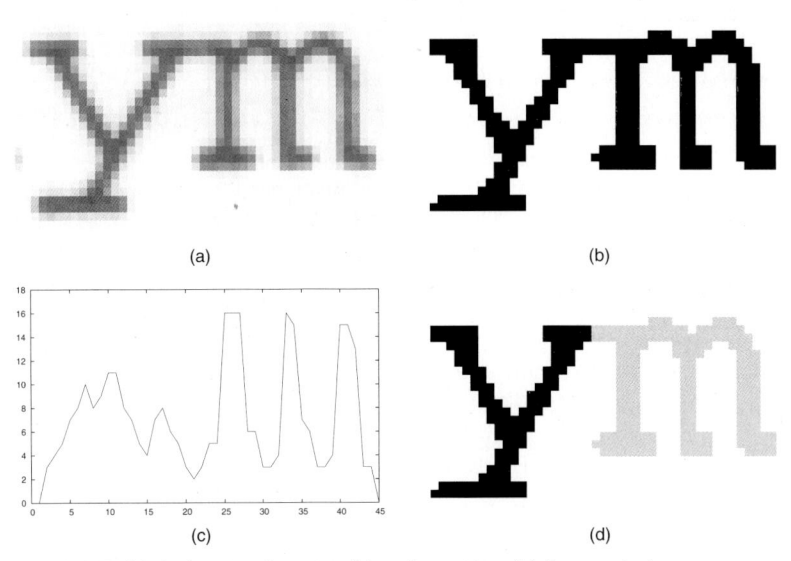

Fig. 3.134 (a) An image of two touching characters. (b) Segmented region. Note that the characters are not separated. (c) Plot of the number of pixels in each column of (b). (d) The characters have been split at the minimum of (c) at position 21.

tion 3.5.1 and the gray value features described in Section 3.5.2 can be used as features. The main requirement is that the features enable us to discern the different character classes. Figure 3.135 illustrates this point. The input image is shown in Figure 3.135(a). It contains examples of lowercase letters. Suppose that we want to classify the letters based on the region features anisometry and compactness. Figures 3.135(b) and (c) show that the letters "c" and "o" as well as "i" and "j" can be distinguished easily based on these two features. In fact, they could be distinguished solely based on their compactness. As Figures 3.135(d) and (e) show, however, these two features are not sufficient to distinguish between the classes "p" and "q" as well as "h" and "k."

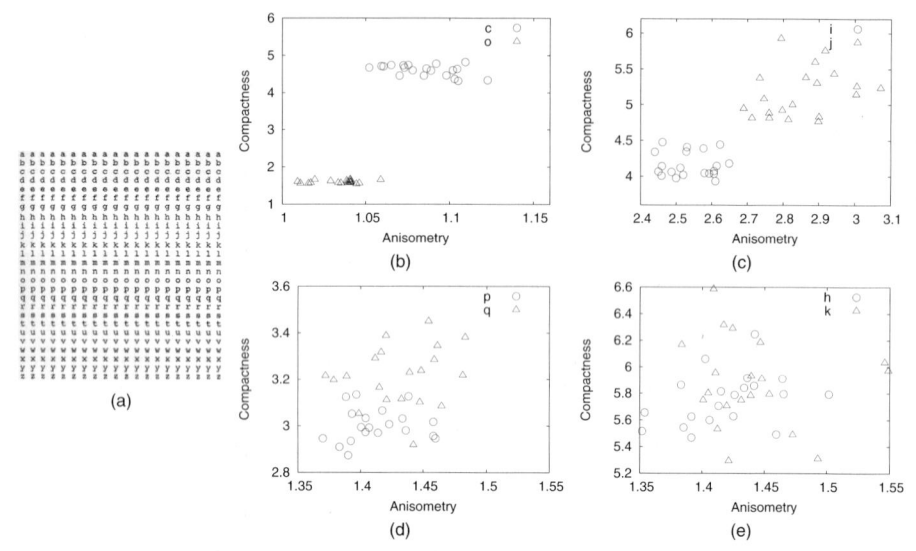

Fig. 3.135 (a) Image with lowercase letters. (b)–(e) The features anisometry and compactness plotted for the letters "c" and "o" (b), "i" and "j" (c), "p" and "q" (d), and "h" and "k" (e). Note that the letters in (b) and (c) can easily be distinguished based on the selected features, while the letters in (d) and (e) cannot be distinguished.

From the above example, we can see that the features we use for the classification must be sufficiently powerful to enable us to classify all relevant classes correctly. The region and gray value features described in Sections 3.5.1 and 3.5.2, unfortunately, are often not powerful enough to achieve this. A set of features that is sufficiently powerful to distinguish all classes of characters is the gray values of the image themselves. Using the gray values directly, however, is not possible because the classifier requires a constant number of input features. To achieve this, we can use the smallest enclosing rectangle around the segmented character, enlarge it slightly to include a suitable amount of background of the character in the features (e.g., by one pixel in each direction), and then zoom the gray values within this rectangle to a standard size, e.g., 8×10 pixels. While transforming the image, we must take care to use the

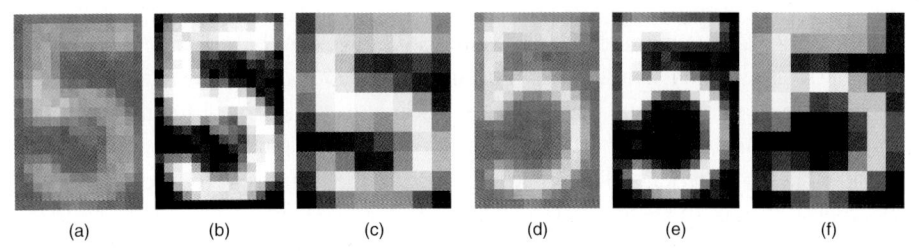

Fig. 3.136 Gray value feature extraction for the OCR. (a) Image of the letter "5" taken from the second row of characters in the image in Figure 3.23(a). (b) Robust contrast normalization of (a). (c) Result of zooming (b) to a size of 8×10 pixels. (d) Image of the letter "5" taken from the second row of characters in the image in Figure 3.23(b). (e) Robust contrast normalization of (d). (f) Result of zooming (e) to a size of 8×10 pixels.

interpolation and smoothing techniques discussed in Section 3.3.3. Note, however, that by zooming the image to a standard size based on the surrounding rectangle of the segmented character, we lose the ability to distinguish characters like "−" (minus sign) and "I" (upper case I in font without serifs). The distinction can easily be done based on a single additional feature: the ratio of the width and height of the smallest surrounding rectangle of the segmented character.

Unfortunately, the gray value features defined above are not invariant to illumination changes in the image. This makes the classification very difficult. To achieve invariance to illumination changes, two options exist. The first option is to perform a robust gray value normalization of the gray values of the character, as described in Section 3.2.1, before the character is zoomed to the standard size. The second option is to convert the segmented character into a binary image before the character is zoomed to the standard size. Since the gray values generally contain more information, the first strategy is preferable in most cases. The second strategy can be used whenever there is significant texture in the background of the segmented characters, which would make the classification more difficult.

Figure 3.136 displays two examples of the gray value feature extraction for the OCR. Figures 3.136(a) and (d) display two instances of the letter "5", taken from images with different contrast (Figures 3.23(a) and (b)). Note that the characters have different sizes (14×21 and 13×20 pixels, respectively). The result of the robust contrast normalization is shown in Figures 3.136(b) and (e). Note that both characters now have full contrast. Finally, the result of zooming the characters to a size of 8×10 pixels is shown in Figures 3.136(c) and (f). Note that this feature extraction automatically makes the OCR scale-invariant because of the zooming to a standard size.

To conclude the discussion about feature extraction, some words about the standard size are necessary. The discussion above has used the size 8×10. A large set of tests has shown that this size is a very good size to use for most industrial applications. If there are only a small number of classes to distinguish, e.g., only numbers, it may be possible to use slightly smaller sizes. For some applications involving a larger number of classes, e.g., numbers and uppercase and lowercase characters, a slightly larger size may be necessary (e.g., 10×12). On the other hand, using much larger sizes typically does not lead to better classification results because the features become progressively less robust against small segmentation errors if a large standard size is chosen. This happens because larger standard sizes imply that a segmentation error will lead to progressively larger position inaccuracies in the zoomed character as the standard size becomes larger. Therefore, it is best not to use a standard size that is much larger than the above recommendations. One exception to this rule is the recognition of an extremely large set of classes, e.g., ideographic characters like the Japanese Kanji characters. Here, much larger standard sizes are necessary to distinguish the large number of different characters.

3.12.3
Classification

As we saw in the previous section, classification can be regarded as a mapping from the feature space to the set of possible classes: thus $f : \mathbb{R}^n \mapsto \Omega$. We will now take a closer look at how the mapping can be constructed.

First of all, we can note that the feature vector x that serves as the input to the mapping can be regarded as a random variable because of the variations that the characters exhibit. In the application, we are observing this random feature vector for each character we are trying to classify. It can be shown that, to minimize the probability of erroneously classifying the feature vector, we should maximize the probability that the class ω_i occurs under the condition that we observe the feature vector x, i.e., we should maximize $P(\omega_i|x)$ over all classes ω_i, for $i = 1, \ldots, m$ [158, 159]. The probability $P(\omega_i|x)$ is also called the a posteriori probability because of the above property that it describes the probability of class ω_i given that we have observed the feature vector x. This decision rule is called the Bayes decision rule. It yields the best classifier if all errors have the same weight, which is a reasonable assumption for the OCR.

We now face the problem of how to determine the a posteriori probability. Using Bayes' theorem, $P(\omega_i|x)$ can be computed as follows:

$$P(\omega_i|x) = \frac{P(x|\omega_i)P(\omega_i)}{P(x)} \quad (3.152)$$

Hence, we can compute the a posteriori probability based on the a priori probability $P(x|\omega_i)$ that the feature vector x occurs given that the class of the feature vector is ω_i, the probability $P(\omega_i)$ that the class ω_i occurs, and the probability $P(x)$ that the feature vector x occurs. To simplify the calculations, we can note that the Bayes decision rule only needs to maximize $P(\omega_i|x)$ and that $P(x)$ is a constant if x is given. Therefore, the Bayes decision rule can be written as

$$x \in \omega_i \quad \Leftrightarrow \quad P(x|\omega_i)P(\omega_i) > P(x|\omega_j)P(\omega_j) \quad j = 1, \ldots, m, \; j \neq i \quad (3.153)$$

What does this transformation gain us? As we will see below, the probabilities $P(x|\omega_i)$ and $P(\omega_i)$ can, in principle, be determined from training samples. This enables us to evaluate $P(\omega_i|x)$, and hence to classify the feature vector x. Before we examine this point in detail, however, let us assume that the probabilities in Eq. (3.153) are known. For example, let us assume that the feature space is 1D ($n = 1$) and that there are two classes ($m = 2$). Furthermore, let us assume that $P(\omega_1) = 0.3$, $P(\omega_2) = 0.7$, and that the features of the two classes have a normal distribution $N(\mu, \sigma)$ such that $P(x|\omega_1) \sim N(-3, 1.5)$ and $P(x|\omega_2) \sim N(3, 2)$. The corresponding likelihoods $P(x|\omega_i)P(\omega_i)$ are shown in Figure 3.137. Note that features to the left of $x \approx -0.7122$ are classified as belonging to ω_1, while features to the right are classified as belonging to ω_2. Hence, there is a dividing point $x \approx -0.7122$ that separates the classes from each other.

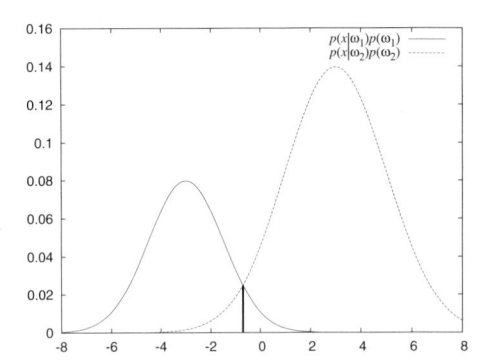

Fig. 3.137 Example of a two-class classification problem in a 1D feature space in which $P(\omega_1) = 0.3$, $P(\omega_2) = 0.7$, $P(x|\omega_1) \sim N(-3, 1.5)$, and $P(x|\omega_2) \sim N(3, 2)$. Note that features to the left of $x \approx -0.7122$ are classified as belonging to ω_1, while features to the right are classified as belonging to ω_2.

As a further example, consider a 2D feature space with three classes that have normal distributions with different means and covariances, as shown in Figure 3.138(a). Again, there are three regions in the 2D feature space in which the respective class has the highest probability, as shown in Figure 3.138(b). Note that now there are 1D curves that separate the regions in the feature space from each other.

As the above examples suggest, the Bayes decision rule partitions the feature space into mutually disjoint regions. This is obvious from the definition in Eq. (3.153): each region corresponds to the part of the feature space in which the class ω_i has the highest a posteriori probability. As also suggested by the above examples, the regions are separated by $(n - 1)$-dimensional hypersurfaces (points for $n = 1$ and curves for

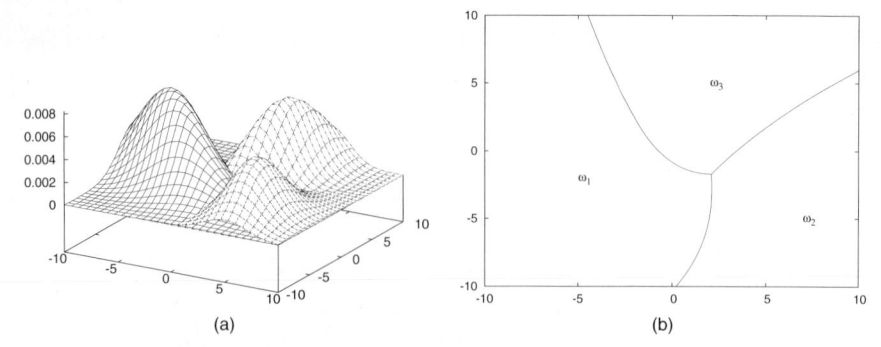

Fig. 3.138 Example of a three-class classification problem in a 2D feature space in which the three classes have normal distributions with different means and covariances. (a) The a posteriori probabilities of the occurrence of the three classes. (b) Regions in the 2D feature space in which the respective class has the highest probability.

$n = 2$, as in Figures 3.137 and 3.138). The hypersurfaces that separate the regions from each other are given by the points in which two classes are equally probable, i.e., by $P(\omega_i|x) = P(\omega_j|x)$, for $i \neq j$.

Accordingly, we can identify two different types of classifiers. The first type of classifier tries to estimate the a posteriori probabilities, typically via Bayes' theorem, from the a priori probabilities of the different classes. In contrast, the second type of classifier tries to construct the separating hypersurfaces between the classes. Below, we will examine representatives for both types of classifier.

All classifiers require a method with which the probabilities or separating hypersurfaces are determined. To do this, a training set is required. The training set is a set of sample feature vectors x_k with corresponding class labels ω_k. For the OCR, the training set is a set of character samples, from which the corresponding feature vectors can be calculated, along with the interpretation of the respective character. The training set should be representative of the data that can be expected in the application. In particular, for the OCR the characters in the training set should contain the variations that will occur later, e.g., different character sets, stroke widths, noise, etc. Since it is often difficult to obtain a training set with all variations, the image processing system must provide means to extend the training set over time with samples that are collected in the field, and should optionally also provide means to artificially add variations to the training set. Furthermore, to evaluate the classifier, in particular how well it has generalized the decision rule from the training samples, it is indispensable to have a test set that is independent from the training set. This test set is essential to determine the error rate that the classifier is likely to have in the application. Without the independent test set, no meaningful statement about the quality of the classifier can be made.

The classifiers that are based on estimating the probabilities, more precisely the probability densities, are called Bayes classifiers because they try to implement the Bayes decision rule via the probability densities. The first problem they have to solve is how to obtain the probabilities $P(\omega_i)$ of the occurrence of class ω_i. There are two basic strategies for this purpose. The first strategy is to estimate $P(\omega_i)$ from the training set. Note that, for this, the training set must be representative not only in terms of the variations of the feature vectors, but also in terms of the frequencies of the classes. Since this second requirement is often difficult to ensure, an alternative strategy for the estimation of $P(\omega_i)$ is to assume that each class is equally likely to occur, and hence to use $P(\omega_i) = 1/m$. Note that in this case the Bayes decision rule reduces to the classification according to the a priori probabilities since $P(\omega_i|x) \sim P(x|\omega_i)$ should now be maximized.

The remaining problem is how to estimate $P(x|\omega_i)$. In principle, this could be done by determining the histogram of the feature vectors of the training set in the feature space. To do so, we could subdivide each dimension of the feature space into b bins. Hence, the feature space would be divided into b^n bins in total. Each bin would count the number of occurrences of the feature vectors in the training set that lie within this bin. If the training set and b are large enough, the histogram would be a good approximation to the probability density $P(x|\omega_i)$. Unfortunately, this approach cannot be used in practice because of the so-called curse of dimensionality: the number of bins in the histogram is b^n, i.e., its size grows exponentially with the dimension of the feature space. For example, if we use the 81 features described in the previous section and subdivide each dimension into a modest number of bins, e.g., $b = 10$, the histogram would have 10^{81} bins, which is much too large to fit into any computer memory.

To obtain a classifier that can be used in practice, we can note that, in the histogram approach, the size of the bin is kept constant, while the number of samples in the bin varies. To get a different estimate for the probability of a feature vector, we can keep the number k of samples of class ω_i constant while varying the volume $v(x, \omega_i)$ of the region in space around the feature vector x that contains the k samples. Then, if there are t feature vectors in the training set, the probability of the occurrence of class ω_i is approximately given by

$$P(x|\omega_i) \approx \frac{k}{tv(x, \omega_i)}$$

Since the volume $v(x, \omega_i)$ depends on the k nearest neighbors of class ω_i, this type of density estimation is called the k nearest-neighbor density estimation. In practice, this approach is often modified as follows. Instead of determining the k nearest neighbors of a particular class and computing the volume $v(x, \omega_i)$, the k nearest neighbors in the training set of any class are determined. The feature vector x is then assigned to the class that has the largest number of samples among the k nearest neighbors. This classifier is called the k nearest-neighbor classifier (kNN classifier). For $k = 1$, we obtain the nearest-neighbor classifier (NN classifier). It can be shown that the NN classifier has an error probability that is at most twice as large as the error probability of the optimal Bayes classifier that uses the correct probability densities [158]: thus

$P_B \leq P_{NN} \leq 2P_B$. Furthermore, if P_B is small, we have $P_{NN} \approx 2P_B$ and $P_{3NN} \approx P_B + 3P_B^2$. Hence, the 3NN classifier is almost as good as the optimal Bayes classifier. Nevertheless, the kNN classifiers are seldom used in practice because they require that the entire training set is stored with the classifier (which can easily contain several hundred thousands of samples). Furthermore, the search for the k nearest neighbors is very time-consuming.

As we have seen from the above discussion, direct estimation of the probability density function is not practicable, either because of the curse of dimensionality for the histograms or because of efficiency considerations for the kNN classifier. To obtain an algorithm that can be used in practice, we can assume that $P(x|\omega_i)$ follows a certain distribution, e.g., an n-dimensional normal distribution:

$$P(x|\omega_i) = \frac{1}{(2\pi)^{n/2}|\Sigma_i|^{1/2}} \exp\left[-\tfrac{1}{2}(x-\mu_i)^\top \Sigma_i^{-1}(x-\mu_i)\right] \qquad (3.154)$$

With this, estimating the probability density function reduces to the estimation of the parameters of the probability density function. For the normal distribution, the parameters are the mean vector μ_i and the covariance matrix Σ_i of each class. Since the covariance matrix is symmetric, the normal distribution has $(n^2+3n)/2$ parameters in total. They can, for example, be estimated via the standard maximum likelihood estimators

$$\mu_i = \frac{1}{n_i}\sum_{j=1}^{n_i} x_{i,j} \qquad \Sigma_i = \frac{1}{n_i-1}\sum_{j=1}^{n_i}(x_{i,j}-\mu_i)(x_{i,j}-\mu_i)^\top \qquad (3.155)$$

Here, n_i is the number of samples for class ω_i, while $x_{i,j}$ denotes the samples for class ω_i.

While the Bayes classifier based on the normal distribution can be quite powerful, often the assumption that the classes have a normal distribution does not hold in practice. In OCR applications, this happens frequently if characters in different fonts should be recognized with the same classifier. One striking example of this is the shapes of the letters "a" and "g" in different fonts. For these letters, two basic shapes exist: a vs. a and g vs. g. It is clear that a single normal distribution is insufficient to capture these variations. In these cases, each font will typically lead to a different distribution. Hence, each class consists of a mixture of l_i different densities $P(x|\omega_i,k)$, each of which occurs with probability $P_{i,k}$:

$$P(x|\omega_i) = \sum_{k=1}^{l_i} P(x|\omega_i,k) P_{i,k} \qquad (3.156)$$

Typically, the mixture densities $P(x|\omega_i,k)$ are assumed to be normally distributed. In this case, Eq. (3.156) is called a Gaussian mixture model. If we knew to which mixture density each sample belongs, we could easily estimate the parameters of the normal distribution with the above maximum likelihood estimators. Unfortunately, in real applications we typically do not have this knowledge, i.e., we do not know k in Eq. (3.156). Hence, determining the parameters of the mixture model requires the estimation of not only the parameters of the mixture densities but also the mixture density labels k – a much harder problem, which can be solved by the expectation maximization algorithm (EM algorithm). The interested reader is referred to [158] for details. Another problem in the mixture model approach is that we need to specify how many mixture densities there are in the mixture model, i.e., we need to specify l_i in Eq. (3.156). This is quite cumbersome to do manually. Recently, algorithms that compute l_i automatically have been proposed. The interested reader is referred to [160, 161] for details.

Let us now turn our attention to classifiers that construct the separating hypersurfaces between the classes. Of all possible surfaces, the simplest ones are planes. Therefore, it is instructive to regard this special case first. Planes in the n-dimensional feature space are given by

$$w^\top x + b = 0 \tag{3.157}$$

Here, x is an n-dimensional vector that describes a point, while w is an n-dimensional vector that describes the normal vector to the plane. Note that this equation is linear. Because of this, classifiers based on separating hyperplanes are called linear classifiers.

Let us first consider the problem of classifying two classes with the plane. We can assign a feature vector to the first class ω_1 if x lies on one side of the plane, while we can assign it to the second class ω_2 if it lies on the other side of the plane. Mathematically, the test on which side of the plane a point lies is performed by looking at the sign of $w^\top x + b$. Without loss of generality, we can assign x to ω_1 if $w^\top x + b > 0$, while we assign x to ω_2 if $w^\top x + b < 0$.

For classification problems with more than two classes, we construct m separating planes (w_i, b_i) and use the following classification rule [158]:

$$x \in \omega_i \quad \Leftrightarrow \quad w_i^\top x + b_i > w_j^\top x + b_j, \quad j = 1, \ldots, m, \ j \neq i \tag{3.158}$$

Note that, in this case, the separating planes do not have the same meaning as in the two-class case, where the plane actually separates the data. The interpretation of Eq. (3.158) is that the plane is chosen such that the feature vectors of the correct class have the largest positive distance of all feature vectors from the plane.

Linear classifiers can also be regarded as neural networks, as shown in Figure 3.139 for the two-class and n-class cases. The neural network has processing units (neurons) that are visualized by circles. They first compute the linear combination of the feature vector x and the weights w: $w^\top x + b$. Then, a nonlinear activation function f is applied. For the two-class case, the activation function simply is $\text{sgn}(w^\top x + b)$, i.e., the side of the hyperplane on which the feature vector lies. Hence, the output is mapped to its essence: -1 or $+1$. Note that this type of activation function essentially thresholds the input value. For the n-class case, the activation function f is typically chosen such that input values < 0 are mapped to 0, while input values ≥ 0 are mapped to 1. The goal in this approach is that a single processing unit returns the value 1, while all other units return the value 0. The index of the unit that returns 1 indicates the class of the feature vector. Note that the plane in Eq. (3.158) needs to be modified for this activation function to work since the plane is chosen such that the feature vectors have the largest distance from the plane. Therefore, $w_j^T x + b_j$ is not necessarily < 0 for all values that do not belong to the class. Nevertheless, the two definitions are equivalent. Note that, because the neural network has one layer of processing units, this type of neural network is also called a single-layer perceptron.

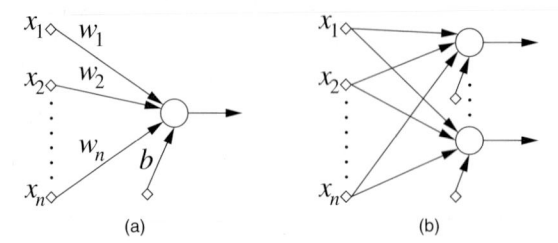

Fig. 3.139 The architecture of a linear classifier expressed as a neural network (single-layer perceptron). (a) A two-class neural network. (b) An n-class neural network. In both cases, the neural network has a single layer of processing units that are visualized by circles. They first compute the linear combination of the feature vector and the weights. After this, a nonlinear activation function is computed, which maps the output to -1 or $+1$ (two-class neural network) or 0 or 1 (n-class neural network).

While linear classifiers are simple and easy to understand, they have very limited classification capabilities. By construction, the classes must be linearly separable, i.e., separable by a hyperplane, for the classifier to produce the correct output. Unfortunately, this is rarely the case in practice. In fact, linear classifiers are unable to represent a simple function like the *xor* function, as illustrated in Figure 3.140, because there is no line that can separate the two classes. Furthermore, for n-class linear classifiers, there is often no separating hyperplane for each class against all the other classes although each pair of classes can be separated by a hyperplane. This happens, for example, if the samples of one class lie completely within the convex hull of all the other classes.

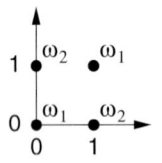

Fig. 3.140 A linear classifier is not able to represent the *xor* function because the two classes, corresponding to the two outputs of the *xor* function, cannot be separated by a single line.

To get a classifier that is able to construct more general separating hypersurfaces, one approach is simply to add more layers, to the neural network, as shown in Figure 3.141. Each layer first computes the linear combination of the feature vector or the results from the previous layer

$$a_j^{(l)} = \sum_{i=1}^{n_l} w_{ji}^{(l)} x_i^{(l-1)} + b_j^{(l)} \qquad (3.159)$$

Here, $x_i^{(0)}$ is simply the feature vector, while $x_i^{(l)}$, with $l \geq 1$, is the result vector of layer l. The coefficients $w_{ji}^{(l)}$ and $b_j^{(l)}$ are the weights of layer l. Then, the results are passed through a nonlinear activation function

$$x_j^{(l)} = f(a_j^{(l)}) \qquad (3.160)$$

Let us assume for the moment that the activation function in each processing unit is the threshold function that is also used in the single-layer perceptron, i.e., the function that maps input values < 0 to 0, while mapping input values ≥ 0 to 1. Then, it can be seen that the first layer of processing units maps the feature space to the corners of the hypercube $\{0, 1\}^p$, where p is the number of processing units in the first layer. Hence, the feature space is subdivided by hyperplanes into half-spaces [158]. The second layer of processing units separates the points on the hypercube by hyperplanes. This corresponds to intersections of half-spaces, i.e., convex polyhedra. Hence, the second layer is capable of constructing the boundaries of convex polyhedra as the separating hypersurfaces [158]. This is still not general enough, however, since the separating hypersurfaces might need to be more complex than this. If a third layer is added, the network can compute unions of the convex polyhedra [158]. Hence, three layers are sufficient to approximate any separating hypersurface arbitrarily closely if the threshold function is used as the activation function.

In practice, the above threshold function is rarely used because it has a discontinuity at $x = 0$, which is very detrimental for the determination of the network weights by

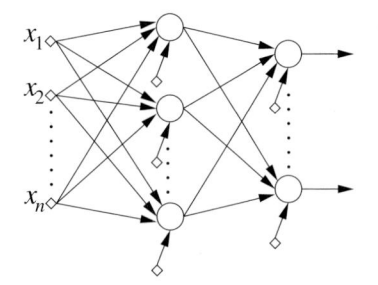

Fig. 3.141 The architecture of a multilayer perceptron. The neural network has multiple layers of processing units that are visualized by circles. They compute the linear combination of the results of the previous layer and the network weights, and then pass the results through a nonlinear activation function.

numerical optimization. Instead, often a sigmoid activation function is used, e.g., the logistic function (see Figure 3.142(a))

$$f(x) = \frac{1}{1 + e^x} \qquad (3.161)$$

Similar to the hard threshold function, it maps its input to a value between 0 and 1. However, it is continuous and differentiable, which is a requirement for most numerical optimization algorithms. Another choice for the activation functions is to use the hyperbolic tangent function (see Figure 3.142(b))

$$f(x) = \tanh(x) = \frac{e^x - e^{-x}}{e^x + e^{-x}} \qquad (3.162)$$

in all layers except the output layer, in which the softmax activation function is used [162] (see Figure 3.142(c))

$$f(x) = \frac{e^{x_i}}{\sum_{j=1}^{n} e^{x_j}} \qquad (3.163)$$

The hyperbolic tangent function behaves similarly to the logistic function. The major difference is that it maps its input to values between -1 and $+1$. There is experimental evidence that the hyperbolic tangent function leads to a faster training of the network [162]. In the output layer, the softmax function maps the input to the range $[0, 1]$, as desired. Furthermore, it ensures that the output values sum to 1, and hence have the same properties as a probability density [162]. With any of these choices of activation function, it can be shown that two layers are sufficient to approximate any separating hypersurface and, in fact, any output function with values in $[0, 1]$, arbitrarily closely [162]. The only requirement for this is that there is a sufficient number of processing units in the first layer (the "hidden layer").

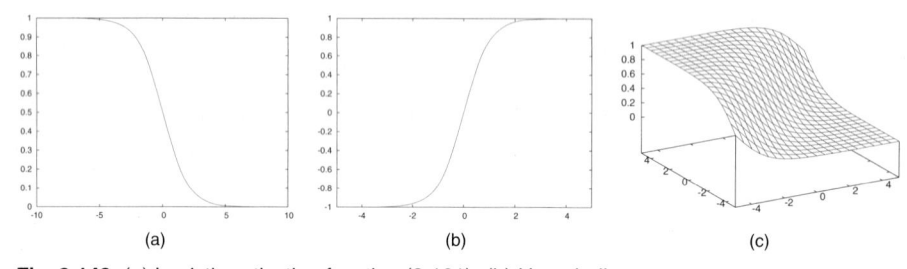

Fig. 3.142 (a) Logistic activation function (3.161). (b) Hyperbolic tangent activation function (3.162). (c) Softmax activation function (3.163) for two classes.

After having discussed the architecture of the multilayer perceptron, we can now examine how the network is trained. Training the network means that the weights $w_{ji}^{(l)}$ and $b_j^{(l)}$, with $l = 1, 2$, of the network must be determined. Let us denote the number of input features by n_i, the number of hidden units (first layer units) by n_h, and the number of output units (second layer units) by n_o. Note that n_i is the dimensionality of the feature vector, while n_o is the number of classes in the classifier. Hence, the only free parameter is the number n_h of units in the hidden layer. Note that there are $(n_i + 1)n_h + (n_h + 1)n_o$ weights in total. For example, if $n_i = 81$, $n_h = 40$, and $n_o = 10$, there are 3690 weights that must be determined. It is clear that this is a very complex problem and that we can only hope to determine the weights uniquely if the number of training samples is of the same order of magnitude as the number of weights.

As described above, the training of the network is performed based on a training set, which consists of sample feature vectors x_k with corresponding class labels ω_k, for $k = 1, \ldots, m$. The sample feature vectors can be used as they are. The class labels, however, must be transformed into a representation that can be used in an optimization procedure. As described above, ideally we would like to have the multilayer perceptron return a 1 in the output unit that corresponds to the class of the sample. Hence, a suitable representation of the classes is a target vector $y_k \in \{0, 1\}^{n_o}$, chosen such that there is a 1 at the index that corresponds to the class of the sample and a 0 in all other positions. With this, we can train the network by minimizing, for example, the squared error of the outputs of the network on all the training samples [162]. In the notation of Eq. (3.160), we would like to minimize

$$\varepsilon = \sum_{k=1}^{m} \sum_{j=1}^{n_o} (x_j^{(2)} - y_{k,j})^2 \tag{3.164}$$

Here, $y_{k,j}$ is the jth element of the target vector y_k. Note that $x_j^{(2)}$ implicitly depends on all the weights $w_{ji}^{(l)}$ and $b_j^{(l)}$ of the network. Hence, minimization of Eq. (3.164) determines the optimum weights. To minimize Eq. (3.164), for a long time the backpropagation algorithm was used, which successively inputs each training sample into the network, determines the output error, and derives a correction term for the weights from the error. It can be shown that this procedure corresponds to the steepest descent minimization algorithm, which is well known to converge extremely slowly [64]. Currently, the minimization of Eq. (3.164) is typically being performed by sophisticated numerical minimization algorithms, such as the conjugate gradient algorithm [64, 162] or the scaled conjugate gradient algorithm [162].

Another approach to obtain a classifier that is able to construct arbitrary separating hypersurfaces is to transform the feature vector into a space of higher dimension, in which the features are linearly separable, and to use a linear classifier in the higher-dimensional space. Classifiers of this type have been known for a long time as generalized linear classifiers [158]. One instance of this approach is the polynomial classifier, which transforms the feature vector by a polynomial of degree $\leq d$. For example, for $d = 2$ the transformation is

$$\Phi(x_1, \ldots, x_n) = (x_1, \ldots, x_n, x_1^2, \ldots, x_1 x_n, \ldots, x_n x_1, \ldots, x_n^2) \tag{3.165}$$

The problem with this approach is again the curse of dimensionality: the dimension of the feature space grows exponentially with the degree d of the polynomial. In fact, there are $\binom{d+n-1}{d}$ monomials of degree $= d$ alone. Hence, the dimension of the transformed feature space is

$$n' = \sum_{i=1}^{d} \binom{i+n-1}{i} = \binom{d+n}{d} - 1 \qquad (3.166)$$

For example, if $n = 81$ and $d = 5$, the dimension is 34 826 301. Even for $d = 2$ the dimension already is 3402. Hence, transforming the features into the larger feature space seems to be infeasible, at least from an efficiency point of view. Fortunately, however, there is an elegant way to perform the classification with generalized linear classifiers that avoids the curse of dimensionality. This is achieved by support vector machine (SVM) classifiers [163, 164].

Before we can take a look at how support vector machines avoid the curse of dimensionality, we have to take a closer look at how the optimal separating hyperplane can be constructed. Let us consider the two-class case. As described in Eq. (3.157) for linear classifiers, the separating hyperplane is given by $w^\top x + b = 0$. As noted above, the classification is performed based on the sign of $w^\top x + b$. Hence, the classification function is

$$f(x) = \text{sgn}(w^\top x + b) \qquad (3.167)$$

Let the training samples be denoted by x_i and their corresponding class labels by $y_i = \pm 1$. Then, a feature is classified correctly if $y_i(w^\top x + b) > 0$. However, this restriction is not sufficient to determine the hyperplane uniquely. This can be achieved by requiring that the margin between the two classes should be as large as possible. The margin is defined as the closest distance of any training sample to the separating hyperplane.

Let us look at a small example of the optimal separating hyperplane, shown in Figure 3.143. Note that, if we want to maximize the margin (shown as a dotted line), there will be samples from both classes that attain the minimum distance to the separating hyperplane defined by the margin. These samples "support" the two hyperplanes that have the margin as the distance to the separating hyperplane (shown as the dashed lines). Hence, the samples are called the support vectors.

In fact, the optimal separating hyperplane is defined entirely by the support vectors, i.e., a subset of the training samples: $w = \sum_{i=1}^{m} \alpha_i y_i x_i$, for $\alpha_i \geq 0$, where $\alpha_i > 0$ if and only if the training sample is a support vector [163]. With this, the classification function can be written as

$$f(x) = \text{sgn}(w^\top x + b) = \text{sgn}\left(\sum_{i=1}^{m} \alpha_i y_i x_i^\top x + b\right) \qquad (3.168)$$

Hence, to determine the optimal hyperplane, the coefficients α_i of the support vectors must be determined. This can be achieved by solving the following quadratic

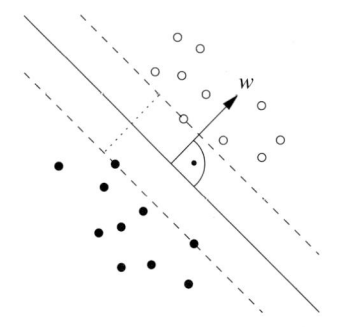

Fig. 3.143 The optimal separating hyperplane between two classes. The samples of the two classes are represented by the filled and unfilled circles. The hyperplane is visualized by the solid line. The margin is visualized by the dotted line between the two dashed lines that visualize the hyperplanes in which samples are on the margin, i.e., attain the minimum distance between the classes. The samples on the margin define the separating hyperplane. Since they "support" the margin hyperplanes, they are called support vectors.

programming problem [163]: maximize

$$\sum_{i=1}^{m} \alpha_i - \frac{1}{2} \sum_{i=1}^{m} \sum_{j=1}^{m} \alpha_i \alpha_j y_i y_j x_i^\top x_j \tag{3.169}$$

subject to

$$\alpha_i \geq 0, \quad i = 1, \ldots, m \quad \text{and} \quad \sum_{i=1}^{m} \alpha_i y_i = 0 \tag{3.170}$$

Note that in both the classification function (3.168) and the optimization function (3.169), the feature vectors x, x_i, and x_j only are present in the dot product.

We now turn our attention back to the case in which the feature vector x is first transformed into a higher-dimensional space by a function $\Phi(x)$, e.g., by the polynomial function in Eq. (3.165). Then, the only change in the above discussion is that we substitute the feature vectors x, x_i, and x_j by their transformations $\Phi(x)$, $\Phi(x_i)$, and $\Phi(x_j)$. Hence, the dot products are simply computed in the higher-dimensional space. The dot products become functions of two input feature vectors: $\Phi(x)^\top \Phi(x')$. These dot products of transformed feature vectors are called kernels in the SVM literature and are denoted by $k(x, x') = \Phi(x)^\top \Phi(x')$. Hence, the decision function becomes a function of the kernel $k(x, x')$:

$$f(x) = \text{sgn}\left(\sum_{i=1}^{m} \alpha_i y_i k(x_i, x) + b\right) \tag{3.171}$$

The same happens with the optimization function Eq. (3.169).

So far, it seems that the kernel does not gain us anything because we still have to transform the data into a feature space of a prohibitively large dimension. The ingenious trick of the SVM classification is that, for a large class of kernels, the kernel can be evaluated efficiently without explicitly transforming the features into the higher-dimensional space, thus making the evaluation of the classification function (3.171) feasible. For example, if we transform the features by a polynomial of degree d, it can easily be shown that

$$k(x, x') = (x^\top x')^d \tag{3.172}$$

Hence, the kernel can be evaluated solely based on the input features without going to the higher-dimensional space. This kernel is called a homogeneous polynomial kernel. As another example, the transformation by a polynomial of degree $\leq d$ can simply be evaluated as

$$k(x, x') = (x^\top x' + 1)^d \tag{3.173}$$

This kernel is called an inhomogeneous polynomial kernel. Further examples of possible kernels include the Gaussian radial basis function kernel

$$k(x, x') = \exp\left(-\frac{\|x - x'\|^2}{2\sigma^2}\right) \tag{3.174}$$

and the sigmoid kernel

$$k(x, x') = \tanh(\kappa x^\top x' + \vartheta) \tag{3.175}$$

Note that this is the same function that is also used in the hidden layer of the multilayer perceptron. With any of the above four kernels, support vector machines can approximate any separating hypersurface arbitrarily closely.

Note that the above training algorithm that determines the support vectors still assumes that the classes can be separated by a hyperplane in the higher-dimensional transformed feature space. This may not always be achievable. Fortunately, the training algorithm can be extended to handle overlapping classes. The reader is referred to [163] for details.

By its nature, the SVM classification can only handle two-class problems. To extend the SVM to multiclass problems, two basic approaches are possible. The first strategy is to perform a pairwise classification of the feature vector against all pairs of classes and to use the class that obtains the most votes, i.e., is selected most often as the result of the pairwise classification. Note that this implies that $l(l-1)/2$ classifications have to be performed if there are l classes. The second strategy is to perform l classifications of one class against the union of the rest of the classes. From an efficiency point of view, the second strategy may be preferable since it depends linearly on the number of classes. Note, however, that in the second strategy, typically there will be a larger number of support vectors than in the pairwise classification. Since the runtime depends linearly on the number of support vectors, the number of support vectors must grow less than quadratically for the second strategy to be faster.

After having discussed the different types of classifiers, the natural question to ask is which of the classifiers should be used in practice. First of all, it must be noted that the quality of the classification results of all the classifiers depends to a large degree on the size and quality of the training set. Therefore, to construct a good classifier, the training set should be as large and representative as possible.

If the main criterion for comparing classifiers is the classification accuracy, i.e., the error rate on an independent test set, classifiers that construct the separating hypersurfaces, i.e., the multilayer perceptron or the support vector machine, should be preferred. Of these two, the support vector machine is portrayed to have a slight advantage [163]. This advantage, however, is achieved by building certain invariances into the classifier that do not generalize to other classification tasks apart from OCR and a particular set of gray value features. The invariances built into the classifier in [163] are translations of the character by one pixel, rotations, and variations of the stroke width of the character. With the features in Section 3.12.2, the translation invariance is automatically achieved. The remaining invariances could also be achieved by extending the training set by systematically modified training samples. Therefore, neither the multilayer perceptron nor the support vector machine have a definite advantage in terms of classification accuracy.

Another criterion is the training speed. Here, support vector machines have a large advantage over the multilayer perceptron because the training times are significantly shorter. Therefore, if the speed in the training phase is important, support vector machines currently should be preferred. Unfortunately, the picture is reversed in the online phase when unknown features must be classified. As noted above, the classification time for the support vector machines depends linearly on the number of support vectors, which usually is a substantial fraction of the training samples (typically, between 10% and 40%). Hence, if the training set consists of several hundred thousands of samples, tens of thousands of support vectors must be evaluated with the kernel in the online phase. In contrast, the classification time of the multilayer perceptron only depends on the topology of the net, i.e., the number of processing units per layer. Therefore, if the speed in the online classification phase is important, the multilayer perceptron currently should be preferred.

A final criterion is whether the classifier provides a simple means to decide whether the feature vector to be classified should be rejected because it does not belong to any of the trained classes. For example, if only digits have been trained, but the segmented character is the letter "M", it is important in some applications to be able to reject the character as being no digit. Another example is an erroneous segmentation, i.e., a segmentation of an image part that corresponds to no character at all. Here, classifiers that construct the separating hypersurfaces provide no means to tell whether the feature vector is close to a class in some sense since the only criterion is on which side of the separating hypersurface the feature lies. For the support vector machines, this behavior is obvious from the architecture of the classifier. For the multilayer perceptron, this behavior is not immediately obvious since we have noted above that the softmax activation function (3.163) has the same properties as a probability density. Hence, one could assume that the output reflects the likelihood of each class, and can thus be used to threshold unlikely classes. In practice, however, the likelihood is very close to 1 or 0 everywhere, except in areas in which the classes overlap in the training samples, as shown by the example in Figure 3.144. Note that all classes have a very high likelihood for feature vectors that are far away from the training samples of each class.

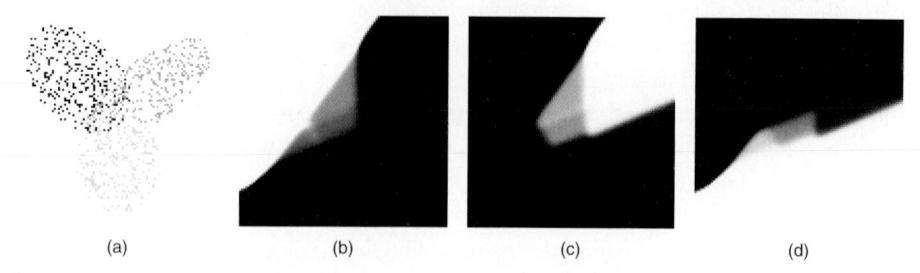

(a) (b) (c) (d)

Fig. 3.144 (a) Samples in a 2D feature space for three classes, which are visualized by three gray levels. (b) Likelihood for class 1 determined in a square region of the feature space with a multilayer perceptron with five hidden units. (c) Likelihood for class 2. (d) Likelihood for class 3. Note that all classes have a very high likelihood for feature vectors that are far away from the training samples of each class. For example, class 3 has a very high likelihood in the lower corners of the displayed portion of the feature space. This high likelihood continues to infinity because of the architecture of the multilayer perceptron.

As can be seen from the above discussion, the behaviors of the multilayer perceptron and the support vector machines are identical with respect to samples that lie far away from the training samples: they provide no means to evaluate the closeness of a sample to the training set. In applications where it is important that feature vectors can be rejected as not belonging to any class, there are two options. First, we can train an explicit rejection class. The problem with this approach is how to obtain or construct the samples for the rejection class. Second, we can use a classifier that provides a measure of closeness to the training samples, such as the Gaussian mixture model classifier. However, in this case we have to be prepared to accept slightly higher error rates for feature vectors that are close to the training samples.

We conclude this section with an example that uses the ICs that we have already used in Section 3.4. In this application, the goal is to read the characters in the last two lines of the print in the ICs. Figures 3.145(a) and (b) show images of two sample ICs. The segmentation is performed with a threshold that is selected automatically based on the gray value histogram (see Figure 3.23). After this, the last two lines of characters are selected based on the smallest surrounding rectangle of the characters. For this, the character with the largest row coordinate is determined and characters lying within an interval above this character are selected. Furthermore, irrelevant characters like the "–" are suppressed based on the height of the characters. The characters are classified with a multilayer perceptron that has been trained with several tens of thousands of samples of characters on electronic components, which do not include the characters

3.12 Optical Character Recognition | 261

(a)

(b)

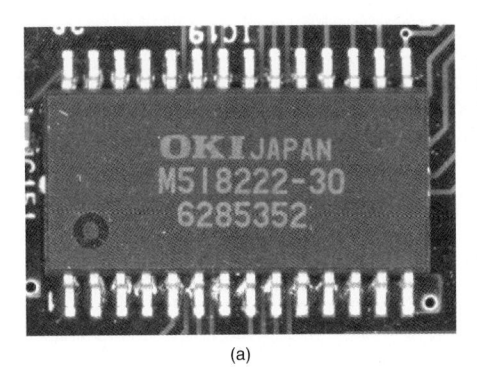

M5I8222 30
6285352
**M5 I8222 30
6285352**

**CXD 117 5AM
6 4 8 E 10 E**

(c)

(d)

Fig. 3.145 (a), (b) Images of prints on ICs. (c), (d) Result of the segmentation of the characters (light gray) and the OCR (black). The images are segmented with a threshold that is selected automatically based on the gray value histogram (see Figure 3.23). Furthermore, only the last two lines of characters are selected. Additionally, irrelevant characters like the "–" are suppressed based on the height of the characters.

on the ICs in Figure 3.145. The result of the segmentation and classification is shown in Figures 3.145(c) and (d). Note that all characters have been read correctly.

4
Machine Vision Applications

To emphasize the engineering aspects of machine vision, this chapter contains a wealth of examples and exercises that show how the machine vision algorithms discussed in Chapter 3 can be combined in non-trivial ways to solve typical machine vision applications. The examples are based on the machine vision software HALCON. As described in the preface, you can download all the applications presented in this chapter to get a hands-on experience with machine vision. Throughout this chapter, we will mention the location where you can find the example programs. We use "..." to indicate the directory into which you have installed the downloaded applications.

4.1
Wafer Dicing

Semiconductor wafers generally contain multiple dies that are arranged in a rectangular grid. To obtain the single dies, the wafer must be diced at the gaps between the dies. Because these gaps typically are very narrow ($< 100 \, \mu m$), they must be located with a very high accuracy in order not to damage the dies during the cutting process.

In this application, we locate the center lines of the gaps on a wafer (see Figure 4.1). In the first step, we determine the width and height of the rectangular dies. The exact position of the dies, and hence the position of the gaps, can then be extracted in the second step.

The algorithms used in this application are:

- Fourier transform (Section 3.2.4)
- Correlation (Section 3.2.4)
- Robust template matching (Section 3.11.5)

The corresponding example program is:
.../machine_vision_book/wafer_dicing/wafer_dicing.dev

We assume that all dies have a rectangular shape and have the same size in the image, i.e., the camera must be perpendicular to the wafer. If this camera setup cannot be realized or if the lens produces heavy radial distortions, the camera must be calibrated

Machine Vision Algorithms and Applications. Carsten Steger, Markus Ulrich, and Christian Wiedemann
Copyright © 2008 WILEY-VCH Verlag GmbH & Co. KGaA, Weinheim
ISBN: 978-3-527-40734-7

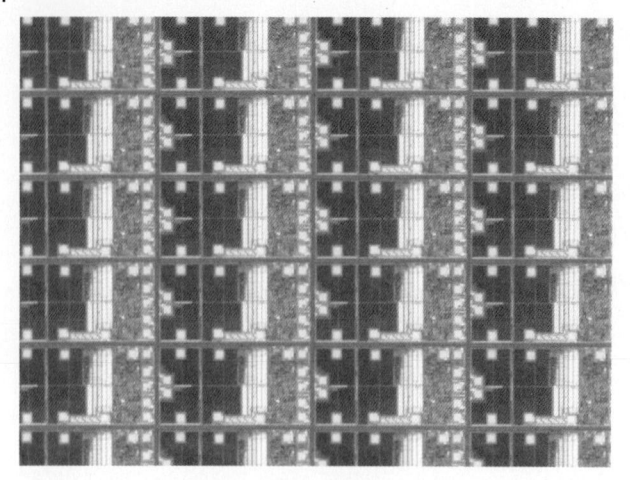

Fig. 4.1 Image of a wafer. Note that the dies are arranged in a rectangular grid, which is horizontally aligned.

and the images must be rectified (see Section 3.9). Furthermore, in wafer dicing applications, in general the wafer is horizontally aligned in the image. Consequently, we do not need to worry about the orientation of the dies in the image, which reduces the complexity of solution.

In the first step, the width and the height of the rectangular dies is determined by using the so-called autocorrelation. The autocorrelation of an image g is the correlation of the image with itself ($g \star g$), and hence can be used to find repeating patterns in the image, which in our case is the rectangular structure of the die. As described in Section 3.2.4, the correlation can be computed in the frequency domain by a simple multiplication. This can be performed by the following operations:

```
rft_generic (WaferDies, ImageFFT, 'to_freq', 'none', 'complex', Width)
correlation_fft (ImageFFT, ImageFFT, CorrelationFFT)
rft_generic (CorrelationFFT, Correlation, 'from_freq', 'n',
             'real', Width)
```

First, the input image is transformed into the frequency domain with the fast Fourier transform. Because we know that the image is real-valued, we do not have to compute the complete Fourier transform but only one half by using the real-valued Fourier transform (see Section 3.2.4). Then, the correlation is computed by multiplying the Fourier-transformed image with its complex conjugate. Finally, the resulting correlation, which is represented in the frequency domain, is transformed back into the spatial domain by using the inverse Fourier transform.

Figure 4.2(a) shows the result of the autocorrelation. The gray value of the pixel (r, c) in the autocorrelation image corresponds to the correlation value that is obtained when shifting the image by (r, c) and correlating it with the unshifted original image. Consequently, the pixel in the upper left corner (origin) of the image has a high gray

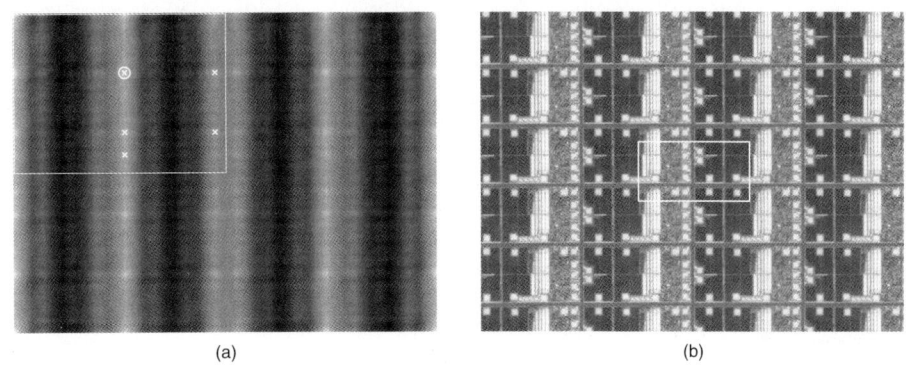

Fig. 4.2 (a) Autocorrelation of the wafer image. Higher gray values indicate higher correlation values. The region of interest for the maximum extraction is visualized by the white rectangle. The local maxima are displayed as white crosses. The white circle indicates the local maximum that represents the die size. (b) Rectangle representing the extracted die size. Note that for visualization purposes the rectangle is arbitrarily positioned in the center of the image.

value because it represents the correlation of the (unshifted) image with itself. Furthermore, if periodic rectangular structures of width and height (w, h) are present in the image, we can expect a high correlation value also at the position (w, h) in the autocorrelation image. Thus, by extracting the local maximum in the correlation image that is closest to the upper left corner, we directly obtain the size of the periodic structure from the position of the maximum. The local maxima can be extracted with subpixel precision by using the following lines of code:

```
gen_rectangle1 (Rectangle, 1, 1, Height/2, Width/2)
reduce_domain (Correlation, Rectangle, CorrelationReduced)
local_max_sub_pix (CorrelationReduced, 'gauss', 2, 5000000, Row, Col)
```

Note that, with the first two operators, the region of interest is restricted to the inner part of the correlation image, excluding a one-pixel-wide border. This prevents local maxima from being extracted at the image border. On the one hand, local maxima at the image border can only be extracted with a limited subpixel precision because not all gray values in the neighborhood can be used for subpixel interpolation. On the other hand, we would have to identify the two corresponding local maxima at the image border to determine the width and the height of the dies instead of identifying only one maximum in the diagonal direction. Additionally, the region of interest can be restricted to the upper left quarter of the autocorrelation image. This is possible because the autocorrelation is an even function with respect to both coordinate axes when assuming that the wafer is horizontally aligned in the image. The region of interest is visualized in Figure 4.2(a) by the white rectangle.

The resulting maxima are shown in Figure 4.2(a) as white crosses. From the resulting maxima, the one that is closest to the origin can easily be selected by computing the distances of all maxima to the origin. It is indicated by the white circle in Figure 4.2(a). In our example, the size of the dies is $(w, h) = (159.55, 83.86)$ pixels. For visualization purposes, a rectangle of the corresponding size is displayed in Figure 4.2(b).

The position of the dies can be determined by using HALCON's shape-based matching (see Section 3.11.5). Since we have calculated the width and height of the dies, we can create an artificial template image that contains a representation of four adjacent dies:

```
LineWidth := 7
RefRow := round(0.5*Height)
RefCol := round(0.5*Width)
for Row := -0.5 to 0.5 by 1
  for Col := -0.5 to 0.5 by 1
    gen_rectangle2_contour_xld (Rectangle,
                                RefRow + Row*DieHeight,
                                RefCol + Col*DieWidth, 0,
                                0.5*DieWidth - 0.5*LineWidth,
                                0.5*DieHeight - 0.5*LineWidth)
    paint_xld (Rectangle, Template, Template, 0)
  endfor
endfor
```

To obtain a correct representation, we additionally have to know the line width `LineWidth` between the dies. The line width is assumed to be constant for our application because the distance between the camera and the wafer does not change. Otherwise, we would have to introduce a scaling factor that can be used to compute the line width based on the determined die size. First, four black rectangles of the appropriate size are painted into an image that previously has been initialized with a gray value of 128. The center of the four rectangles is arbitrarily set to the image center.

```
LineWidthFraction := 0.6
gen_rectangle2_contour_xld (Rectangle, RefRow, RefCol, 0,
                            0.5*LineWidthFraction*LineWidth, DieHeight)
paint_xld (Rectangle, Template, Template, 0)
gen_rectangle2_contour_xld (Rectangle, RefRow, RefCol, 0, DieWidth,
                            0.5*LineWidthFraction*LineWidth)
paint_xld (Rectangle, Template, Template, 0)
gen_rectangle2 (ROI, RefRow, RefCol, 0, DieWidth+5, DieHeight+5)
reduce_domain (Template, ROI, TemplateReduced)
```

After this, two additional black rectangles that reflect the non-uniform gray value of the lines are added, each of which covers 60% of the line width. Note that `paint_xld` paints the contour onto the background using anti-aliasing. Consequently, the gray value of a pixel depends on the fraction by which the pixel is covered by the rectangle. For example, if only half of the pixel is covered, the gray value is set to the mean gray value of the background and the rectangle. The last rectangle ROI represents the region of interest that is used to create the shape model that can be used for the matching. The resulting template image together with its region of interest is shown in Figure 4.3(a). Finally, the shape model is created by using the following operation:

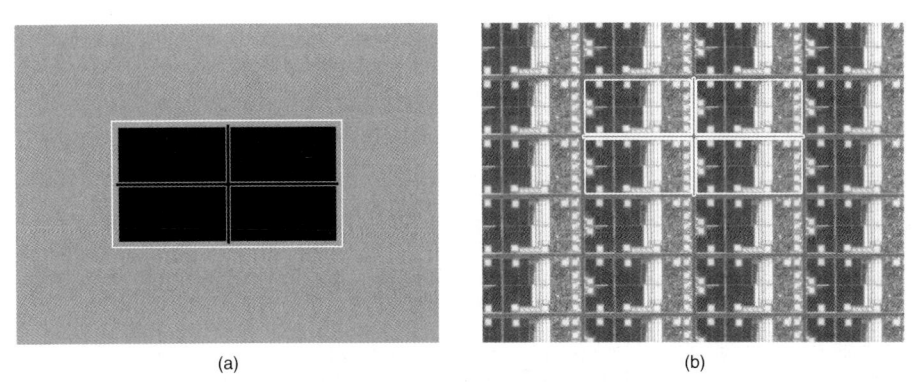

```
create_shape_model (TemplateReduced, 'auto', 0, 0, 'auto', 'auto',
                   'ignore_local_polarity', 'auto', 5, ModelID)
```

Fig. 4.3 (a) Template that is used for matching. The ROI from which the shape model is created is displayed in white. (b) Best match that is found by the shape-based matching.

The model can be used to find the position of the best match of the four adjacent dies in the original image that was shown in Figure 4.1:

```
NumMatches := 1
find_shape_model (WaferDies, ModelID, 0, 0, MinScore, NumMatches,
                  0.5, 'least_squares', 0, Greediness, MatchRow,
                  MatchColumn, MatchAngle, MatchScore)
```

The obtained match is visualized in Figure 4.3(b). Note that, although more than one instance of our model is present in the image, we are only interested in the best-fitting match, and hence the parameter NumMatches is set to 1. Also note that the coordinates MatchRow and MatchColumn are obtained with subpixel precision.

In the last step, the cutting lines can be computed based on the position of the found match, e.g., by using the following lines of code:

```
NumRowMax := ceil(Height/DieHeight)
NumColMax := ceil(Width/DieWidth)
for RowIndex := -NumRowMax to NumRowMax by 1
  RowCurrent := MatchRow + RowIndex*DieHeight
```

```
    gen_contour_polygon_xld (CuttingLine, [RowCurrent,RowCurrent],
                             [0,Width-1])
    dev_display (CuttingLine)
endfor
for ColIndex := -NumColMax to NumColMax by 1
    ColCurrent := MatchColumn + ColIndex*DieWidth
    gen_contour_polygon_xld (CuttingLine, [0,Height-1],
                             [ColCurrent,ColCurrent])
    dev_display (CuttingLine)
endfor
```

First, the maximum number of cutting lines that might be present in the image in the horizontal and vertical directions are estimated based on the size of the image and of the dies. Then, starting from the obtained match coordinates `MatchRow` and `MatchCol`, parallel horizontal and vertical lines are computed using a step width of `DieHeight` and `DieWidth`, respectively. The result is shown in Figure 4.4.

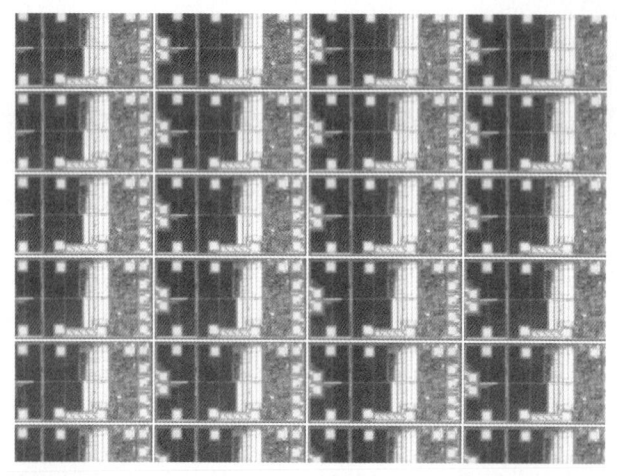

Fig. 4.4 Cutting lines that can be used to slice the wafer.

To conclude this example, it should be noted that the proposed algorithm can also be used for wafers for which the dies have a different size or even a completely different appearance, as in this example. One restriction is that the gap between the dies must have a known size. Furthermore, the gap must have a similar appearance in the image as in this example. Otherwise, a new appropriate template image would have to be created.

Exercises

1. In the above application we used the matching to find only the best-fitting match. Alternatively, the matching could be used to find all instances of the model in

the image. Modify the program such that the cutting lines are computed based on all found matches.

2. The above program assumes that the wafer is horizontally aligned in the image. However, for some applications this assumption is not valid. Modify the program so that it can also be used for images in which the wafer appears slightly rotated, e.g., up to $\pm 20°$. Tip 1: By extracting two maxima in the correlation image, the size as well as the orientation of the dies can be computed. For this, it is easiest to rearrange the correlation image such that the origin is in the center of the image before extracting the maxima (for example, by using the operators `crop_part` and `tile_images_offset`). Tip 2: The model that is used for the matching must be generated in the corresponding orientation.

4.2
Reading of Serial Numbers

In this application we read the serial number that is printed on a compact disk (CD) by using optical character recognition (OCR). The serial number is printed along a circle that is concentric with the CD's center. First, we rectify the image using a polar transformation to transform the characters into a standard position. Then, we segment the individual characters in the rectified image. In the last step, the characters are read by using a neural network classifier.

The algorithms used in this application are:

- Image smoothing (Section 3.2.3)
- Thresholding (Section 3.4.1)
- Circle fitting (Section 3.8.2)
- Polar transformation (Section 3.3.4)
- Extraction of connected components (Section 3.4.2)
- Region morphology (Section 3.6.1)
- Region features (Section 3.5.1)
- Optical character recognition (Section 3.12)

The corresponding example program is:
.../machine_vision_book/reading_of_serial_numbers/
reading_of_serial_numbers.dev

Figure 4.5 shows the image of the center part of the CD. The image was acquired using diffuse bright-field front light illumination (see Section 2.1.5). The characters of the serial number, which is printed in the outermost annulus, appear dark on a bright background.

Fig. 4.5 Image of the center part of a CD. The serial number is printed in the outermost annulus. For visualization purposes, it is highlighted with a white border.

The first task is to determine the center of the CD and to compute the radius of the outer border of the outermost annulus. Based on this information, we will later rectify the image using a polar transformation. Because the outer border is darker than its local neighborhood, it can be segmented with a dynamic thresholding operation. To estimate the gray value of the local background, a 51 × 51 mean filter is applied. The filter size is chosen relatively large in order to suppress noise in the segmentation better. Because no neighboring objects can occur outside the annulus, we do not have to worry about the maximum filter size (see Section 3.4.1).

```
mean_image (Image, ImageMean, 51, 51)
dyn_threshold (Image, ImageMean, RegionDynThresh, 15, 'dark')
fill_up (RegionDynThresh, RegionFillUp)
gen_contour_region_xld (RegionFillUp, ContourBorder, 'border')
fit_circle_contour_xld (ContourBorder, 'ahuber', -1, 0, 0, 3, 2,
                        CenterRow, CenterColumn, Radius,
                        StartPhi, EndPhi, PointOrder)
```

All image structures that are darker than the estimated background by 15 gray values are segmented. The result of the mean filter and the segmented region are shown in Figure 4.6(a). Note that the border of the outer annulus is completely contained in the result. To find the outer border, we fill up the holes in the segmentation result. Then, the center of the CD and the radius of the outer border can be obtained

by fitting a circle to the outer border of the border pixels, which can be obtained with
gen_contour_region_xld. The circle that best fits the resulting contour points is
shown in Figure 4.6(b).

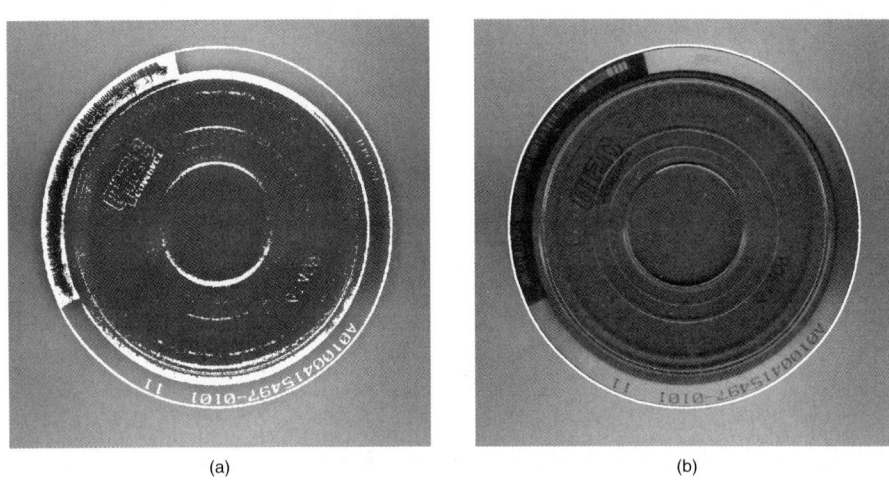

Fig. 4.6 (a) Mean image with the segmentation result of the dynamic thresholding operation overlaid in white. (b) Original image with the circle fitted to the border of the segmentation result overlaid in white. Holes in the segmentation result have previously been filled up before the circle is fitted.

Based on the circle center and the radius of the outer border, we can compute the parameters of the polar transformation that rectifies the annulus containing the characters:

```
AnnulusInner := 0.90
AnnulusOuter := 0.99
```

The inner and outer radii of the annulus to be rectified are assumed to be constant fractions (0.90 and 0.99, respectively) of the radius of the outer border. With this, we can achieve scale invariance in our application. In Figure 4.7, a zoomed part of the original image is shown. The fitted circle (solid line) and the inner and outer borders of the annulus (dashed lines) are overlaid in white. Note that the outer border of the annulus is chosen in such a way that the black line is excluded from the polar transformation, which simplifies the subsequent segmentation of the serial number.

The width and height of the transformed image are chosen such that no information is lost, i.e., the image resolution is kept constant:

```
WidthPolar  := 2*Pi*Radius*AnnulusOuter
HeightPolar := Radius*(AnnulusOuter-AnnulusInner)
```

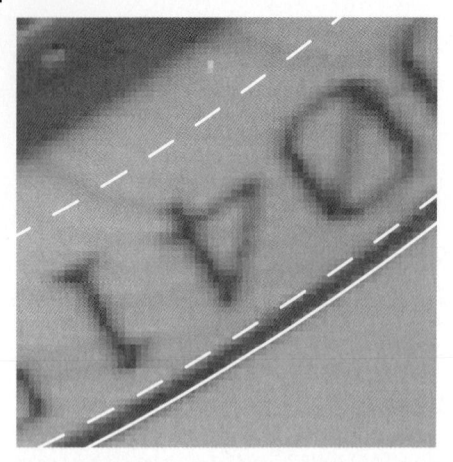

Fig. 4.7 Zoomed part of the original image. The fitted circle (solid line) and the inner and outer borders of the annulus (dashed lines) that restrict the polar transformation are overlaid in white.

Finally, the annulus arc that is to be transformed can be specified by its radius and angle interval:

```
RadiusStart := Radius*AnnulusOuter
RadiusEnd   := Radius*AnnulusInner
AngleStart  := 2*Pi - 2*Pi/WidthPolar
AngleEnd    := 0
```

Note that the start radius corresponds to the outer border of the annulus, while the end radius corresponds to its inner border. With this it is ensured that the characters appear upright in the transformed image. Also note that the end angle is smaller than the start angle. This ensures that the characters can be read from left to right, since the polar transformation is performed in the mathematically positive orientation (counterclockwise). Finally, to avoid the first column of the transformed image being identical to the last column, we exclude the last discrete angle from the transformation. Now, we are ready to perform the polar transformation:

```
polar_trans_image_ext (Image, PolarTransImage,
                       CenterRow, CenterColumn,
                       AngleStart, AngleEnd,
                       RadiusStart, RadiusEnd,
                       WidthPolar, HeightPolar, 'bilinear')
```

The center point of the polar transformation is set to the center of the CD. The transformed image is shown in Figure 4.8. Note that the characters of the serial number have been rectified successfully.

In the next step, the OCR can be applied. As described in Section 3.12, the OCR consists of the two tasks, segmentation and classification. Consequently, we first seg-

4.2 Reading of Serial Numbers

Fig. 4.8 Annulus that contains the serial number transformed to polar coordinates. Note that for visualization purposes the transformed image is split into three parts, which from top to bottom correspond to the angle intervals $[0, 2\pi/3[$, $[2\pi/3, 4\pi/3[$, and $[4\pi/3, 2\pi[$.

ment the single characters in the transformed image. For a robust segmentation, we must compute the approximate size of the characters in the image:

```
CharWidthFraction := 0.01
CharWidth := WidthPolar * CharWidthFraction
CharHeight := CharWidth
```

To maintain scale invariance, we compute the width of the characters as a constant fraction of the full circle, i.e., the width of the polar transformation. In this application, the width is approximately 1% of the full circle. Furthermore, the height of the characters is approximately the same as their width.

Because the characters are darker than their local neighborhood, we can use the dynamic thresholding again for the segmentation. In this case, the background is estimated with a mean filter that is twice the size of the characters:

```
mean_image (PolarTransImage, ImageMean, 2*CharWidth, 2*CharHeight)
dyn_threshold (PolarTransImage, ImageMean, RegionThreshold, 10, 'dark')
```

Fig. 4.9 Result of the dynamic thresholding operation for the three image parts of Figure 4.8. Note that the three parts are framed for a better visualization.

The result of the thresholding operation is shown in Figure 4.9. Note that, in addition to the characters, several unwanted image regions are segmented. The latter can be eliminated by checking the values of appropriate region features of the connected components:

```
connection (RegionThreshold, ConnectedRegions)
select_shape (ConnectedRegions, RegionChar,
              ['height','width','row'], 'and',
              [CharHeight*0.1, CharWidth*0.3, HeightPolar*0.25],
              [CharHeight*1.1, CharWidth*1.1, HeightPolar*0.75])
sort_region (RegionChar, RegionCharSort, 'character', 'true', 'row')
```

In this application, three region features are sufficient to single out the characters: the region height, the region width, and the row coordinate of the center of gravity of the region. The minimum value for the height must be set to a relatively small value because we want to segment the hyphen in the serial number. Furthermore, the minimum width must be set to a small value to include characters like "I" or "1" in the segmentation. The respective maximum values can be set to the character size enlarged by a small tolerance factor. Finally, the center of gravity must be around the center row of the transformed image. Note that the latter restriction would not be reasonable if punctuation marks like "." or "," had to be read. To obtain the characters in the correct order, the connected components are sorted from left to right. The resulting character regions are displayed in Figure 4.10.

Fig. 4.10 Segmented characters for the three image parts of Figure 4.8. The characters are displayed in three different gray values. Note that the three parts are framed for a better visualization.

In the last step, the segmented characters are classified, i.e., a symbolic label is assigned to each segmented region. HALCON provides several pretrained fonts that can be used for the OCR. In this application, we use the pretrained font Industrial, which is similar to the font of the serial number:

```
read_ocr_class_mlp ('Industrial', OCRHandle)
do_ocr_multi_class_mlp (RegionCharSort, PolarTransImage, OCRHandle,
                        Class, Confidence)
SNString := sum(Class)
```

First, the pretrained OCR classifier is read from file. The classifier is based on a multilayer perceptron, which is a special form of a neural network (see Section 3.12.3). In addition to the gray values (scaled to a fixed range), several region features were used to train the classifier, e.g., aspect ratio, anisometry, compactness, convexity, and region moments (see Section 3.5.1). Consequently, the same features are used to classify the segmented characters. The result of the classification is a tuple of class labels that are returned in Class, one for each input region. For convenience reasons, the class labels are converted into one connected string.

For visualization purposes, the character regions can be transformed back from polar coordinates into the original image:

```
polar_trans_region_inv (RegionCharSort, XYTransRegion,
                        CenterRow, CenterColumn,
                        AngleStart, AngleEnd,
                        RadiusStart, RadiusEnd,
                        WidthPolar, HeightPolar,
                        Width, Height, 'nearest_neighbor')
```

In Figure 4.11, the original image, the transformed character region, and the result of the OCR are shown.

Fig. 4.11 Result of the OCR overlaid as white text in the upper left image corner. The segmented character regions, which have been transformed back into the original image, are overlaid in white.

Exercises

1. The described program assumes that the serial number does not cross the 0° border in the polar transformation. Extend the program so that this restriction can be discarded. Tip: One possibility is to determine the orientation of the CD

in the image. Another possibility is to generate an image that holds two copies of the polar transformation in a row.

2. In many applications, the segmentation of the characters is more difficult than in the described example. Some of the problems that might occur are discussed in Section 3.12. For such cases, HALCON provides the two operators `segment_characters` and `select_characters` that ease the segmentation considerably. Modify the above program by replacing the segmentation of the characters with these two operators.

4.3
Inspection of Saw Blades

During the production process of saw blades, it is important to inspect the single saw teeth to ensure that the shape of each tooth is within predefined limits. In such applications, high inspection speed as well as high inspection accuracy are the primary concerns.

Fig. 4.12 Image of a back-lit saw blade that is used for teeth inspection.

Because in this application the essential information can be derived from the contour of the saw blades and the depth of the saw blade is small, we can take the images by using diffuse back lighting (see Section 2.1.5). Figure 4.12 shows an image of a saw blade that was taken with this kind of illumination. This simplifies the segmentation process significantly. First, the contour of the saw blade is extracted in the image. Then, the individual teeth are obtained by splitting the contour appropriately. Finally, for each tooth, the included angle between its two sides, i.e., the tooth face and the tooth back, is computed. This angle can then easily be compared to a reference value.

The algorithms used in this application are:
- Subpixel-precise thresholding (Section 3.4.3)
- Segmentation of contours into lines and circles (Section 3.8.4)
- Contour features (Section 3.5.3)
- Fitting lines (Section 3.8.1)

The corresponding example program is:
.../machine_vision_book/saw_blade_inspection/saw_blade_inspection.dev

Because we want to extract the included angles of the tooth sides with a high accuracy, it is advisable to use subpixel-precise algorithms in this application. First, we want to obtain the subpixel-precise contour of the saw blade in the image. In Chapter 3 we have seen that two possible ways to extract subpixel-precise contours are subpixel thresholding (see Section 3.4.3) and subpixel edge extraction (see Section 3.7.3). Unfortunately, the threshold that must be specified when using the subpixel-precise thresholding influences the position of the contour. Nevertheless, in our case applying the subpixel-precise thresholding is preferable because it is considerably faster than the computationally expensive edge extraction. Because of the back-lit saw blades, the background appears white while the saw blade appears black in the image. Therefore, it is not difficult to find an appropriate threshold, for example by simply using a mean gray value, e.g., 128. Moreover, the dependence of the contour position on the threshold value can be neglected in this application, as we will see later. Hence, the contour, which is visualized in Figure 4.13, is obtained by the following operator:

```
threshold_sub_pix (Image, Border, 128)
```

In the next step, we get rid of all contour parts that are not part of the tooth sides:

```
segment_contours_xld (Border, ContoursSplit, 'lines_circles', 5, 6, 4)
select_shape_xld (ContoursSplit, SelectedContours,
                  'contlength', 'and' 30, 200)
```

First, the contour is split into line segments and circular arcs by using the algorithm described in Section 3.8.4. On the one hand, this enables us to separate the linear contour parts of the tooth sides from the circularly shaped gullets between adjacent teeth. On the other hand, because polygons in the Ramer algorithm are approximated by line segments, we can also separate the face and the back of each tooth from each other. The resulting contour parts are shown in Figure 4.14(a). In the second step, the length of each such obtained contour part is calculated as described in Section 3.5.3. Because

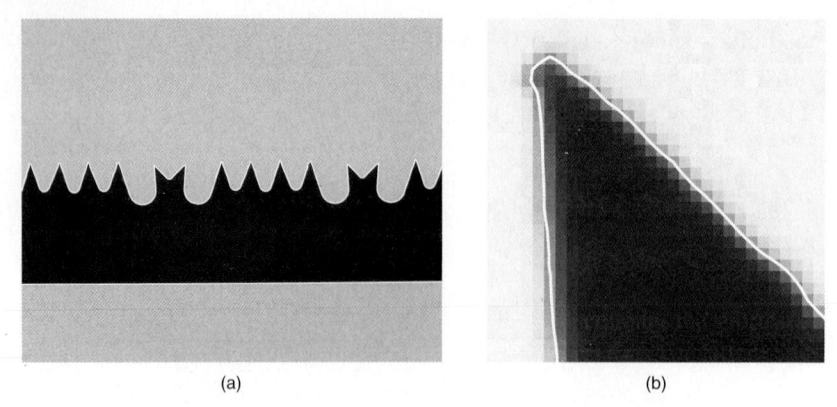

Fig. 4.13 (a) Contours that result from the subpixel-precise thresholding. To provide a better visibility of the extracted contours, the contrast of the image has been reduced. (b) Detail of (a) (original contrast).

Fig. 4.14 (a) Contour parts that are obtained by splitting the original contour into circular arcs and line segments. The contour parts are displayed in three different gray values. (b) Remaining tooth sides obtained after eliminating contour parts that are too short or too long and after eliminating circular arcs.

we approximately know the size of the tooth sides, we can discard all contour parts that are shorter than 30 and longer than 200 pixels by using `select_contours_xld`.

Finally, we can exclude all circular arcs, which represent the gullets, from further processing because we are only interested in the line segments, which represent the tooth sides:

```
Number := |SelectedContours|
ToothSides := []
for Index2 := 1 to Number by 1
    SelectedContour := SelectedContours[Index2]
    get_contour_global_attrib_xld (SelectedContour, 'cont_approx', Attrib)
```

```
    if (Attrib = -1)
      ToothSides := [ToothSides,SelectedContour]
    endif
  endfor
  sort_contours_xld (ToothSides, ToothSidesSorted, 'upper_left',
                     'true', 'column')
```

This can be achieved by querying the attribute 'cont_approx' for each contour part, which was set in segment_contours_xld. For line segments, the attribute returned in Attrib is -1, while for circular arcs it is 1. The line segments correspond to the remaining tooth sides and are collected in the array ToothSides. They are shown in Figure 4.14(b). Finally, the tooth sides are sorted with respect to the column coordinate of the upper left corner of their surrounding rectangles. The sorting is performed in ascending order. Consequently, the resulting ToothSidesSorted are sorted from left to right in the image, which later enables us to easily group the tooth sides into pairs, and hence to obtain the face and the back of each saw tooth.

In the next step, the orientation of the teeth's sides is computed. The straightforward way to do this would be to take the first and the last contour point of the contour part, which represents a line segment, and compute the orientation of the line running through both points. Unfortunately, this method would not be robust against outliers. As can be seen from Figure 4.15(a), which shows a zoomed part of a saw tooth, the tooth tip is not necessarily a perfect peak. Obviously, the last point of the tooth side lies far from the ideal tooth back, which would falsify the computation of the orientation. Figure 4.15(b) shows a line that has the same start and end points as the contour of the tooth back shown in Figure 4.15(a). A better way to compute the orientation is to use all contour points of the line segment instead of only the end points and to identify outliers. This can be achieved by robustly fitting a line through the contour points as described in Section 3.8.1. We use the Tukey weight function with a clipping factor of $2\sigma_\delta$ with five iterations:

```
  fit_line_contour_xld (ToothSidesSorted, 'tukey', -1, 0, 5, 2,
                        Rows1, Columns1, Rows2, Columns2, Nr, Nc, Dist)
```

The orientation of each fitted line can be computed with the following operation:

```
  line_orientation (Rows1, Columns1, Rows2, Columns2, Orientations)
```

At this point we can see why the threshold dependence of the contour position, which is introduced by the subpixel-precise thresholding, can be neglected. When changing the value for the threshold, the contour points of the tooth side would be shifted in a direction perpendicular to the tooth side. Because this shift would be approximately constant for all contour points, the direction of the fitted line would not change.

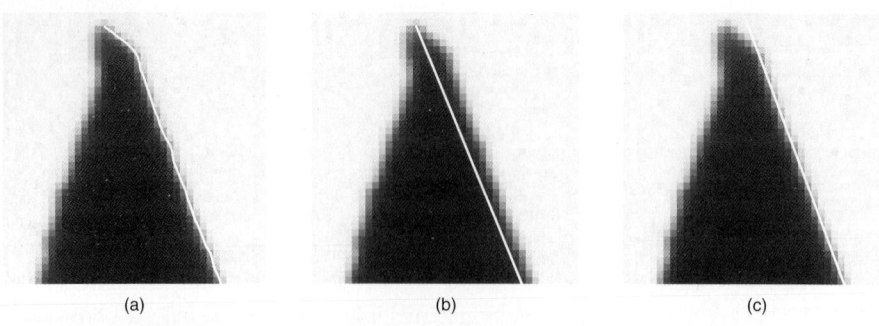

Fig. 4.15 (a) Zoomed part of a saw tooth. The contour part that represents the tooth back is displayed in white. (b) Line with the same start and end points as the contour part in (a). (c) Line robustly fitted to the contour part in (a).

In the last step, for each tooth the included angle between its face and its back can be computed with the following operations:

```
for Index2 := Start to |Orientations|-1 by 2
  Angle := abs(Orientations[Index2-1] - Orientations[Index2])
  if (Angle > PI_2)
    Angle := PI - Angle
  endif
endfor
```

Because we have sorted the tooth sides from left to right, we can simply step through the sorted array by using a step width of 2 and select two consecutive tooth sides each time. This ensures that both tooth sides belong to the same tooth, of which the absolute angle difference can be computed. Finally, the obtained `Angle` is reduced to the interval $[0, \pi/2]$. In Figure 4.16, three examples of the computed tooth angle are displayed by showing the fitted lines from which the angle is computed. For visualization purposes, the two lines are extended up to their intersection point.

Exercise

1. In the above program, only the angles of the tooth sides are computed. Often, it is also important to ensure that the teeth are of a perfect triangular shape, i.e., they do not show any indentations that exceed a certain size tolerance. Extend the program so that these irregularities of the tooth shape can be identified as well.

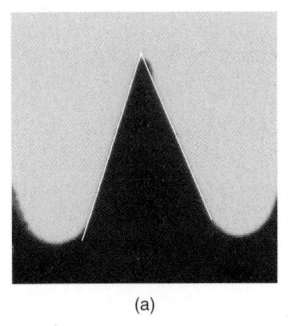

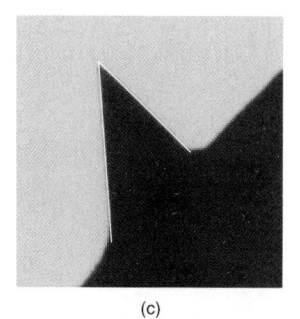

Fig. 4.16 Three example teeth with the fitted lines from which the included angle is computed: (a) 41.35°; (b) 38.08°; (c) 41.64°. Note that the contrast of the images has been reduced for visualization purposes.

4.4
Print Inspection

Print inspection is used to guarantee the quality of prints on arbitrary objects. Depending on the type of material and the printing technology, different image acquisition setups are used. For transparent objects, diffuse bright-field back light illumination may be used; while for other objects, e.g., for embossed characters or braille print, a directed dark-field front light illumination is necessary.

In this application, we inspect the textual information as well as the wiring diagram on the surface of a relay. We check for smears and splashes as well as for missing or misaligned parts of the print.

The algorithms used in this application are:

- 2D edge extraction (Section 3.7.3)
- Gray value morphology (Section 3.6.2)
- Robust template matching (Section 3.11.5)
- Image transformation (Section 3.3.3)
- Thresholding (Section 3.4.1)

The corresponding example program is:
.../machine_vision_book/print_inspection/print_inspection.dev

The image of a relay shown in Figure 4.17(a) was acquired with diffuse bright-field front light illumination (see Section 2.1.5). The printed textual information and the wiring diagram appear dark on the bright surface of the relay. Because we only want to check the print and not the boundary of the relay or the position of the print on the relay, a predefined, manually generated, region of interest is used to reduce the domain of the image.

```
reduce_domain (Image, ROI, ImageReduced)
```

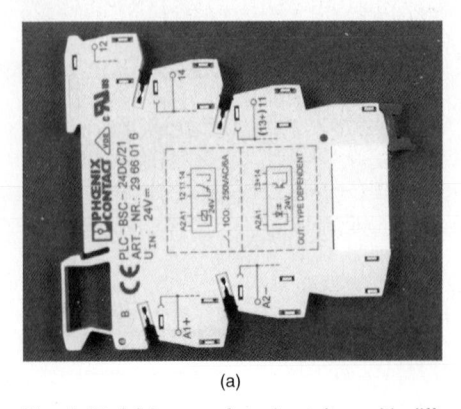

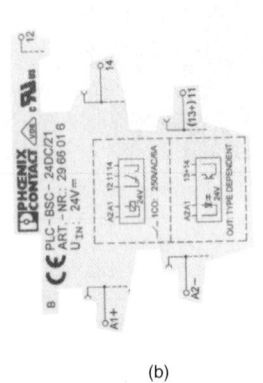

Fig. 4.17 (a) Image of a relay taken with diffuse front light bright-field illumination. (b) Reference image for which only the region of interest is shown.

The reduced image (see Figure 4.17(b)) is used as the reference image. Because only small size and position tolerances are allowed, the comparison of images of correctly printed relays with the reference image will result in larger differences only near gray value edges of the reference image. For this reason, the allowed variations can be approximated by computing the edge amplitude from the reference image:

```
Sigma := 0.5
edges_image (ImageReduced, ImaAmp, ImaDir, 'canny',
             Sigma, 'none', 20, 40)
gray_dilation_rect (ImaAmp, VariationImage, 3, 3)
```

The reference image contains edges that lie close to each other. To minimize the influence of neighboring edges on the calculated edge amplitude, a small filter mask is used. For the Canny edge filter, $\sigma = 0.5$ corresponds to a mask size of 3×3 pixels. To broaden the area of allowed differences between the test image and the reference image, a gray value dilation is applied to the amplitude image. The resulting variation image is shown in Figure 4.18.

With the reference image and the variation image, we can create the variation model directly:

```
AbsThreshold := 15
VarThreshold := 1
create_variation_model (Width, Height, 'byte', 'direct', VarModelID)
prepare_direct_variation_model (ImageReduced, VariationImage,
                                VarModelID, AbsThreshold, VarThreshold)
```

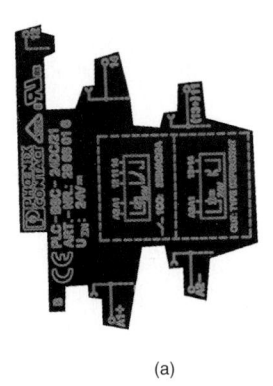

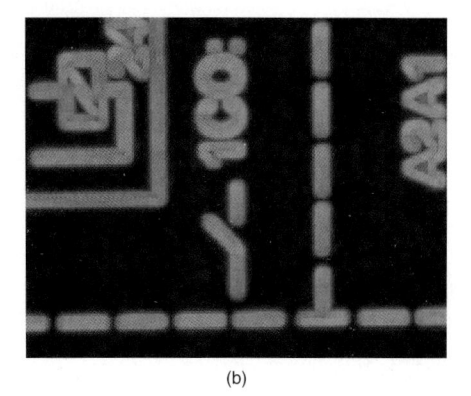

Fig. 4.18 (a) Variation image derived from the reference image shown in Figure 4.17(b). (b) Detail of the variation image.

For the comparison of the test image with the reference image, it is essential that the two images are perfectly aligned. To determine the transformation between the two images, robust template matching is used:

```
reduce_domain (Image, MatchingROI, Template)
create_shape_model (Template, 5, -rad(5), rad(10),
                    'auto', 'auto', 'use_polarity',
                    'auto', 'auto', ShapeModelID)
area_center (MatchingROI, ModelArea, ModelRow, ModelColumn)
```

To speed up the matching, we use only selected parts of the print as template. These parts surround the area to be inspected to ensure that test images can be aligned with high accuracy. The shape model is created such that the test images may be rotated by ±5 degrees with respect to the reference image.

With the preparations above described, we are ready to check the test images. For each test image, we have to determine the transformation parameters to align it with the reference image:

```
find_shape_model (Image, ShapeModelID, -rad(5), rad(10),
                  0.5, 1, 0.5, 'least_squares', 0, 0.9,
                  Row, Column, Angle, Score)
vector_angle_to_rigid (Row, Column, Angle,
                  ModelRow, ModelColumn, 0, HomMat2D)
affine_trans_image (Image, ImageAligned, HomMat2D,
                  'constant', 'false')
```

First, we determine the position and orientation of the template in the test image. Using these values, we can calculate the transformation parameters of a rigid transformation that aligns the test image with the reference image. Then, we transform the test image using the determined transformation parameters. Now, we compare the transformed test image with the reference image:

```
reduce_domain (ImageAligned, ROI, ImageAlignedReduced)
compare_variation_model (ImageAlignedReduced, RegionDiff, VarModelID)
```

The resulting region may contain some clutter, and the parts of the regions that indicate defects may be disconnected. To separate the defect regions from the clutter, we assume that the latter is equally distributed while defect regions that belong to the same defect lie close to each other. Therefore, we can close small gaps in the defect regions by performing a dilation:

```
MinComponentSize := 5
dilation_circle (RegionDiff, RegionDilation, 3.5)
connection (RegionDilation, ConnectedRegions)
intersection (ConnectedRegions, RegionDiff, RegionIntersection)
select_shape (RegionIntersection, SelectedRegions,
              'area', 'and', MinComponentSize, ModelArea)
```

After the dilation, previously interrupted defect regions are merged. From the merged regions, the connected components are computed. To obtain the original shape of the defects, the connected components are intersected with the original non-dilated region. Note that the connectivity of the components is preserved during the intersection computation. Thus, with the dilation we can simply extend the neighborhood definition that is used in the computation of the connected components. Finally, we select all components with an area exceeding a predefined minimum size.

Results are shown in Figure 4.19(a) and (b). The detected errors are indicated by black ellipses along with the expected edges from the reference image, which are outlined in white.

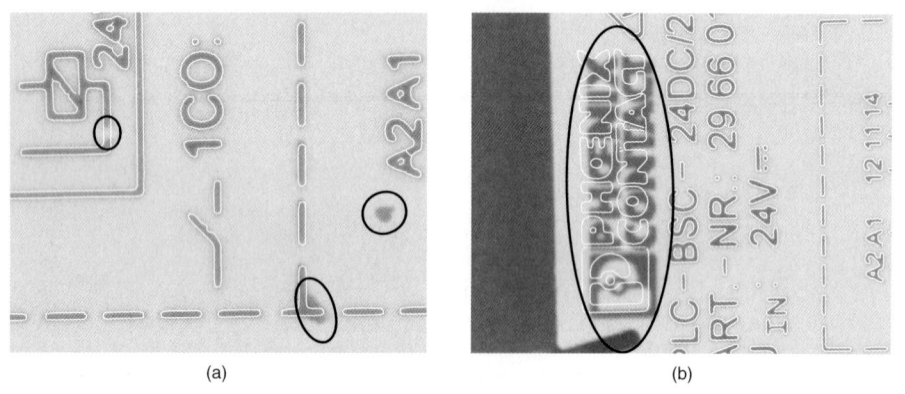

Fig. 4.19 Results of the print inspection. Detected errors are indicated by black ellipses along with the expected edges from the reference image, which are outlined in white. Note that the contrast of the images has been reduced for visualization purposes. (a) Missing parts of the print, smears, and splashes. (b) Misaligned parts of the print.

Exercises

1. In the above application, the variation image has been created by computing the edge amplitude image of the reference image. Another possibility for the creation of the variation image from one reference image is to create artificial variations of the model. Modify the example program such that the variation image is created from multiple artificially created reference images.

2. In the example program described above, the application of the variation model approach is only possible if a non-varying illumination of the object can be assured. Extend the program such that it works even for illumination conditions that vary over time.

4.5
Inspection of Ball Grid Arrays (BGA)

A ball grid array (BGA) is a chip package having solder balls on the underside for mounting on a circuit board. To ensure a proper connection to the circuit board, it is important that all individual balls are at the correct position, have the correct size, and are not damaged in any way.

In this application, we inspect BGAs. First, we check the size and shape of the balls. Then, we test the BGA for missing or extraneous balls as well as for wrongly positioned balls.

The algorithms used in this application are:

- Thresholding (Section 3.4.1)
- Extraction of connected components (Section 3.4.2)
- Region features (Section 3.5.1)
- Subpixel-precise thresholding (Section 3.4.3)
- Contour features (Section 3.5.3)
- Geometric transformations (Section 3.3)

The corresponding example program is:
.../machine_vision_book/bga_inspection/bga_inspection.dev

For BGA inspection, the images are typically acquired with a directed dark-field front light illumination (see Section 2.1.5). With this kind of illumination, the balls appear as doughnut-like structures, while the surroundings of the balls are dark (see Figure 4.20). To ensure that all correct balls appear in the same size and that they form a rectangular grid in the image, the image plane of the camera must be parallel to the BGA. If this camera setup cannot be realized for the image acquisition or if the lenses

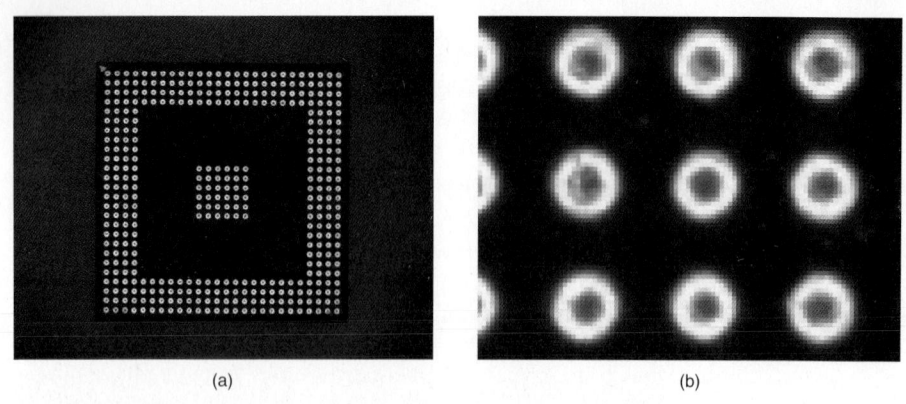

Fig. 4.20 Image of a BGA. The image has been acquired with directed dark-field front light illumination. The balls appear as doughnut-like structures, while the surroundings of the balls appear dark. (a) The whole image. (b) Upper left part of the BGA.

produce heavy radial distortions, the camera must be calibrated and the images must be rectified (see Section 3.9).

First, we will determine wrongly sized and damaged balls. For this, we have to segment the balls in the image. Then we calculate features for the segmented regions, which describe their size and shape. If the feature values of a region are within a given range, the respective ball is assumed to be correct. Otherwise, it will be marked as wrongly sized or damaged.

There are different ways to perform this check. One possibility is to carry out a pixel-precise segmentation and to derive the features either from the segmented regions (see Section 3.5.1) or from the gray values that lie inside the segmented regions (see Section 3.5.2). Another possibility is to use a subpixel-precise segmentation (see Section 3.4.3) and to derive the features from the contours (see Section 3.5.3). In this application, we use pixel-precise segmentation with region-based features as well as subpixel-precise segmentation.

The pixel-precise segmentation of the balls can be performed as follows:

```
threshold (Image, Region, BrighterThan, 255)
connection (Region, ConnectedRegions)
fill_up (ConnectedRegions, RegionFillUp)
select_shape (RegionFillUp, Balls, ['area','circularity'], 'and',
              [0.8*MinArea,0.75], [1.2*MaxArea,1.0])
```

First, the region that is brighter than a given threshold is determined. Then, the connected components of this region are calculated and all holes in these connected components are filled up. Finally, the components that have an area in a given range and that have a more or less circular shape are selected. To be able to detect wrongly sized balls, we select regions even if they are slightly smaller or larger than expected.

A part of a test image that contains wrongly sized and shaped balls along with the extracted balls is displayed in Figure 4.21(a).

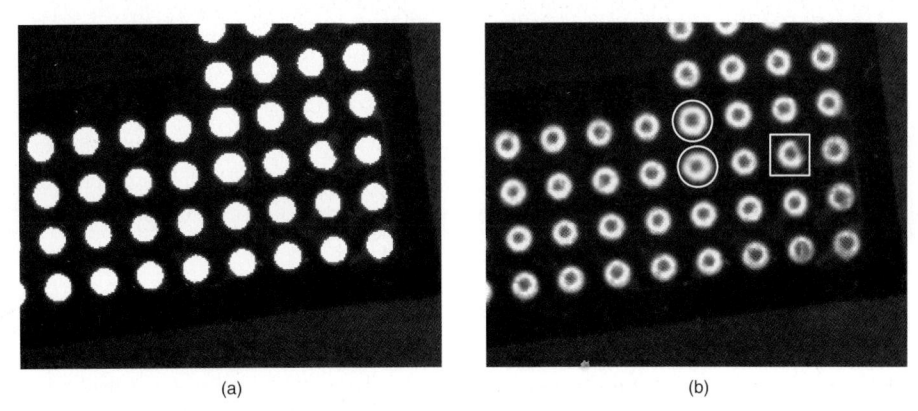

Fig. 4.21 Cutout of an image that contains wrongly sized and shaped balls. (a) Segmentation result. (b) Wrongly sized balls (indicated by white circles) and wrongly shaped balls (indicated by white squares).

Now we select the regions that have an incorrect size or shape:

```
select_shape (Balls, WrongAreaBalls, ['area','area'],
              'or', [0,MaxArea], [MinArea,10000])
select_shape (Balls, WrongAnisometryBalls, 'anisometry',
              'and', MaxAnisometry, 100)
```

The first operator selects all regions that have an area less than MinArea pixels or larger than MaxArea pixels. These regions represent the wrongly sized balls. The second operator selects the regions that have an anisometry larger than a given value, e.g., MaxAnisometry = 1.2. The detected defective balls are shown in Figure 4.21(b). Wrongly sized balls are indicated by white circles around them and wrongly shaped balls by white squares.

The segmentation and feature extraction can also be done with subpixel-precise segmentation methods and contour-based features. First, we use the balls extracted above to define a region of interest for the subpixel-precise extraction of the boundaries of the balls. In the image that is reduced to this region of interest, we perform the following operations:

```
threshold_sub_pix (ImageReduced, Boundary, BrighterThan)
select_shape_xld (Boundary, Balls, 'area', 'or',
                  0.8*MinArea, 1.2*MaxArea)
```

With subpixel-precise thresholding, the boundaries of the balls are extracted. To eliminate clutter, only the boundaries that enclose a suitable area are selected.

As above, we can now select the boundaries that have an incorrect size or shape:

```
select_shape_xld (Balls, WrongAreaBalls, ['area','area'],
                 'or', [0,MaxArea], [MinArea,10000])
select_shape_xld (Balls, WrongAnisometryBalls, 'anisometry',
                 'or', MaxAnisometry, 100)
```

The above two methods – pixel-precise and subpixel-precise segmentation and feature extraction – yield similar results if the regions to be extracted are large enough. In this case, the pixel-precise approximation of the regions is sufficient. However, this does not hold for small regions, where the subpixel-precise approach provides better results.

Up to now, we have not detected falsely positioned and missing balls. To be able to detect such balls, we need a model of a correct BGA. This model can be stored in matrix representation. In this matrix, a value of -1 indicates that there is no ball at the respective position of the BGA. If there is a ball, the respective entry of the matrix contains its (non-negative) index. The index points to two other arrays that hold the exact reference coordinates of the ball given, for example, in millimeters.

For BGAs with concentric squares of balls, we can define the BGA layout in a tuple that contains a 1 if the respective square contains balls and 0 if not. The first entry of this tuple belongs to the innermost square, the last entry to the outermost square of balls. Furthermore, we need to define the distance between neighboring balls.

```
BgaLayout      := [1,1,1,0,0,0,0,0,0,1,1,1,1]
BallDistRowRef := 0.05 * 25.4
BallDistColRef := 0.05 * 25.4
```

Using this information, we can create the BGA model. We store the matrix in linearized form.

```
BallsPerRow   := 2*|BgaLayout|
BallsPerCol   := 2*|BgaLayout|
BallMatrixRef := gen_tuple_const(BallsPerRow*BallsPerCol,-1)
BallsRowsRef  := []
BallsColsRef  := []
CenterRow     := (BallsPerRow-1)*0.5
CenterCol     := (BallsPerCol-1)*0.5
I := 0
for R := 0 to BallsPerRow-1 by 1
  for C := 0 to BallsPerCol-1 by 1
    Dist := max(int(fabs([R-CenterRow,C-CenterCol])))
    if (BgaLayout[Dist])
      BallMatrixRef[R*BallsPerCol+C] := I
      BallsRowsRef := [BallsRowsRef,R*BallDistRowRef]
      BallsColsRef := [BallsColsRef,C*BallDistColRef]
      I := I+1
    endif
  endfor
endfor
```

4.5 Inspection of Ball Grid Arrays (BGA)

First, the numbers of balls per row and column are determined from the given BGA layout tuple. Then, the whole BGA model matrix is initialized with -1, and empty fields for the x and y coordinates are created. For any position on the BGA, the index of the respective square in the BGA layout tuple is given by the chessboard distance (see Section 3.6.1) between the ball position and the center of the BGA. If a ball exists at the current position, its index I stored in the linearized model matrix and its coordinates are appended to the fields of row and column coordinates.

We can display the BGA matrix easily with the following lines of code (see Figure 4.22).

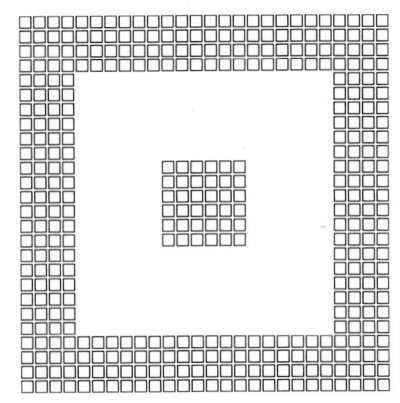

```
Scale := RectangleSize / (0.8*min([BallDistanceRow,BallDistanceCol]))
gen_rectangle2_contour_xld (Matrix,
                            RectangleSize/2.0+BallsRows*Scale,
                            RectangleSize/2.0+BallsCols*Scale,
                            gen_tuple_const(|BallsRows|,0),
                            gen_tuple_const(|BallsRows|,
                                            RectangleSize/2.0),
                            gen_tuple_const(|BallsRows|,
                                            RectangleSize/2.0))
dev_display (Matrix)
```

Fig. 4.22 The model for the BGA shown in Figure 4.20. The rectangles indicate the positions where balls must lie.

To detect erroneously placed and missing balls, we must create a similar matrix for the image to be investigated. This BGA matrix contains the actual presence and exact position of the balls.

To build this matrix, we have to relate the position of the balls in the image to the matrix positions, i.e., we need to know the relative position of the balls in the BGA. For this, we must know the size of the BGA in the current image:

```
area_center_xld (BallsSubPix, Area, BallsRows, BallsCols, PointOrder)
gen_region_points (RegionBallCenters, BallsRows, BallsCols)
smallest_rectangle2 (RegionBallCenters, RowBGARect, ColumnBGARect,
                    PhiBGARect, Length1BGARect, Length2BGARect)
```

We create a region that contains all extracted ball positions and determine the smallest surrounding rectangle of this region. With this, we can determine the transformation of the ball positions into indices of the BGA matrix. Note that this only works if no extraneous balls lie outside the BGA.

```
hom_mat2d_identity (HomMat2DIdentity)
hom_mat2d_rotate (HomMat2DIdentity, -PhiBGARect,
                 RowBGARect, ColumnBGARect, HomMat2DRotate)
hom_mat2d_translate (HomMat2DRotate,
                    -RowBGARect+Length2BGARect,
                    -ColumnBGARect+Length1BGARect,
                    HomMat2DTranslate)
BallDistCol := 2*Length1BGARect/(BallsPerCol-1)
BallDistRow := 2*Length2BGARect/(BallsPerRow-1)
hom_mat2d_scale (HomMat2DTranslate, 1/BallDistRow, 1/BallDistCol,
                0, 0, HomMat2DScale)
```

The first part of this transformation is a rotation around the center of the BGA to align the BGA with the rows and columns of the model matrix. Then, we translate the rotated ball positions such that the upper left ball lies at the origin of the BGA model. Finally, the distance between the balls in the image is calculated in the row and column directions. The inverse distances are used to scale the ball coordinates such that the distance between them in the row and column directions will be one.

Now, we transform the ball positions with the transformation matrix derived above and round them to get for each extracted ball its index in the BGA model matrix.

```
affine_trans_point_2d (HomMat2DScale, BallsRows, BallsCols,
                      RowNormalized, ColNormalized)
BallRowIndex := round(RowNormalized)
BallColIndex := round(ColNormalized)
```

Finally, we set the indices of the balls in the BGA matrix:

```
BallMatrix := gen_tuple_const(BallsPerRow*BallsPerCol,-1)
for I := 0 to NumBalls-1 by 1
  BallMatrix[BallRowIndex[I]*BallsPerCol+BallColIndex[I]] := I
endfor
```

With this representation of the extracted balls, it is easy to detect missing or additional balls. Missing balls have a non-negative index in the BGA model matrix and a negative index in the BGA matrix of the current image, while extraneous balls have a negative index in the BGA model matrix and a non-negative index in the BGA matrix derived from the image.

```
for I := 0 to BallsPerRow*BallsPerCol-1 by 1
  if (BallMatrixRef[I] >= 0 and BallMatrix[I] < 0)
*     Missing ball.
  endif
  if (BallMatrixRef[I] < 0 and BallMatrix[I] >= 0)
*     Extraneous ball.
  endif
endfor
```

The missing and extraneous balls are displayed in Figure 4.23(a). Missing balls are indicated by white diamonds while extraneous balls are indicated by white crosses.

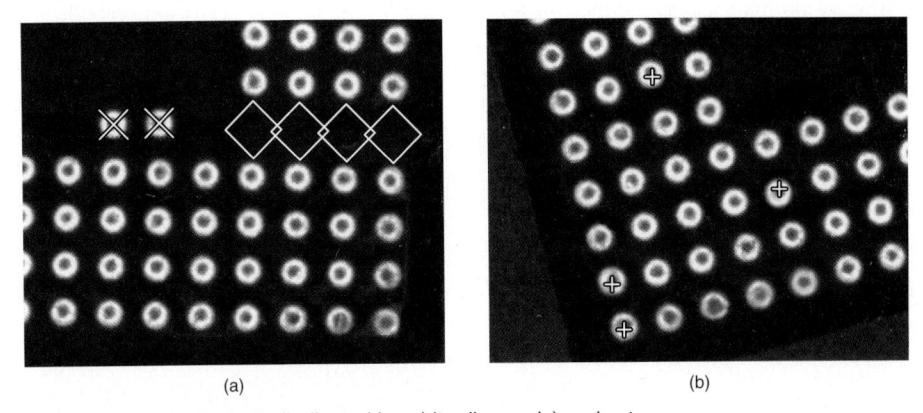

Fig. 4.23 (a) Missing balls (indicated by white diamonds) and extraneous balls (indicated by white crosses). (b) Wrongly placed balls (indicated by white plus signs at the expected position).

Finally, we check the position of the balls. For this, we must transform the pixel coordinates of the extracted balls into the world coordinate system in which the reference coordinates are given.

```
BallsMatchedRowsImage := []
BallsMatchedColsImage := []
IndexImage := []
BallsMatchedRowsRef := []
BallsMatchedColsRef := []
IndexRef := []
K := 0
for I := 0 to BallsPerRow*BallsPerCol-1 by 1
  if (BallMatrixRef[I] >= 0 and BallMatrix[I] >= 0)
    BallsMatchedRowsImage := [BallsMatchedRowsImage,
                              BallsRows[BallMatrix[I]]]
    BallsMatchedColsImage := [BallsMatchedColsImage,
                              BallsCols[BallMatrix[I]]]
```

```
        IndexImage := [IndexImage,BallMatrix[I]]
        BallsMatchedRowsRef := [BallsMatchedRowsRef,
                                BallsRowsRef[BallMatrixRef[I]]]
        BallsMatchedColsRef := [BallsMatchedColsRef,
                                BallsColsRef[BallMatrixRef[I]]]
        IndexRef := [IndexRef,BallMatrixRef[I]]
        K := K+1
    endif
endfor
```

To be able to calculate the transformation from the pixel coordinate system to the world coordinate system, we must establish a set of corresponding points in the two coordinate systems. Because we already know the mapping from the extracted ball positions to the positions of the balls in the BGA model, this can be done easily by using all points that are defined in both matrices. Using this set of corresponding points, we can determine the transformation and transform the extracted ball positions into the reference coordinate system.

```
vector_to_similarity (BallsMatchedRowsImage, BallsMatchedColsImage,
                      BallsMatchedRowsRef, BallsMatchedColsRef,
                      HomMat2D)
affine_trans_point_2d (HomMat2D,
                       BallsMatchedRowsImage, BallsMatchedColsImage,
                       BallsMatchedRowsWorld, BallsMatchedColsWorld)
```

Note that, if the camera was calibrated, the ball positions could be transformed into the world coordinate system. Then, a rigid transformation could be used instead of the similarity transformation.

The deviation of the ball positions from their reference positions can be determined simply by calculating the distance between the reference coordinates and the transformed pixel coordinates extracted from the image.

```
distance_pp (BallsMatchedRowsRef, BallsMatchedColsRef,
             BallsMatchedRowsWorld, BallsMatchedColsWorld,
             Distances)
for I := 0 to |Distances|-1 by 1
    if (Distances[I] > MaxDistance)
*       Wrongly placed ball.
    endif
endfor
```

Each ball for which the distance is larger than a given tolerance can be marked as wrongly placed. Figure 4.23(b) shows the wrongly placed balls indicated by white plus signs at the correct position.

Exercises

1. In the above application, we calculated the features in two ways: from pixel-precise regions, and from subpixel-precise contours. A third possibility is to derive the features from the gray values that lie inside the segmented regions (see Section 3.5.2). Modify the above program so that the position of the balls is determined in this way. Compare the results with the results achieved above. What conditions must be fulfilled so that the gray value features can be applied successfully to determine the center of segmented objects.

2. The creation of the BGA model described above is restricted to BGAs with balls located in concentric squares. To be more flexible, write a procedure that creates the BGA model from an image of a correct BGA and the reference distances between the balls in the row and column directions.

4.6
Surface Inspection

A typical machine vision application is to inspect the surface of an object. Often, it is essential to identify defects like scratches or ridges.

In this application, we inspect the surface of doorknobs. Other industry sectors in which surface inspection is important are, for example, optics, the automobile industry, and the metal-working industry. An example image of a doorknob with a typical scratch is shown in Figure 4.24(a). First, we have to select an appropriate illumination to highlight scratches in the surface. Then we can segment the doorknob in the image. After that, we create a region of interest that contains the planar surface of the doorknob. In the last step, we search for scratches in the surface within the region of interest.

The algorithms used in this application are:

- Image smoothing (Section 3.2.3)
- Thresholding (Section 3.4.1)
- Region features (Section 3.5.1)
- Region morphology (Section 3.6.1)
- Extraction of connected components (Section 3.4.2)
- Affine transformations (Section 3.3.1)

The corresponding example program is:
.../machine_vision_book/surface_inspection/surface_inspection.dev

For surface inspection, typically the images are acquired with a directed dark-field front light illumination (see Section 2.1.5). Figure 4.24(b) shows an image of the

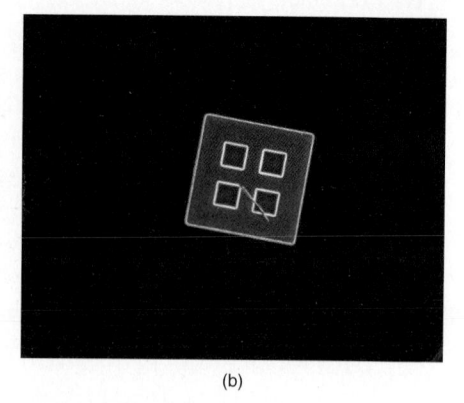

Fig. 4.24 (a) Image of a doorknob using directed bright-field front light illumination. (b) Image of the same doorknob using directed dark-field front light illumination using an LED ring light. Note that the scratch in the surface is clearly visible.

doorknob of Figure 4.24(a) acquired with a directed dark-field front light illumination by using an LED ring light. The edges of the doorknobs appear bright, while its planar surface appears dark. This simplifies the segmentation process significantly. Note that the illumination makes the scratch in the surface clearly visible.

Scratches appear bright in the dark regions. Unfortunately, also the border of the doorknob as well as the borders of the four inner squares appear bright. To distinguish the bright borders from the scratches, we first segment the bright border regions. We then subtract the segmented regions from the doorknob region and reduce the region of interest for the scratch detection to the difference region.

Because the edges are locally brighter than the background, we can segment the doorknob in the image with dynamic thresholding (see Section 3.4.2):

```
KnobSizeMin := 100
mean_image (Image, ImageMean, KnobSizeMin, KnobSizeMin)
dyn_threshold (Image, ImageMean, RegionBrightBorder, 50, 'light')
fill_up (RegionBrightBorder, RegionDoorknob)
```

To estimate the gray value of the local background, we use a mean filter. We set the size of the mean filter to the minimum size at which the doorknob appears in the image to ensure that all local structures are eliminated by the smoothing, and hence to better suppress noise in the segmentation. Then, all image structures that are brighter than the background by 50 gray values are segmented, yielding the region of the bright border. The result of applying the dynamic thresholding to the image of Figure 4.24(b) is shown in Figure 4.25(a). To obtain the complete region of the doorknob, holes in the border region are filled up (see Figure 4.25(b)).

In the next step, the planar surface that we would like to inspect is determined. For this, we have to eliminate the bright outer border of the doorknob as well as the bright

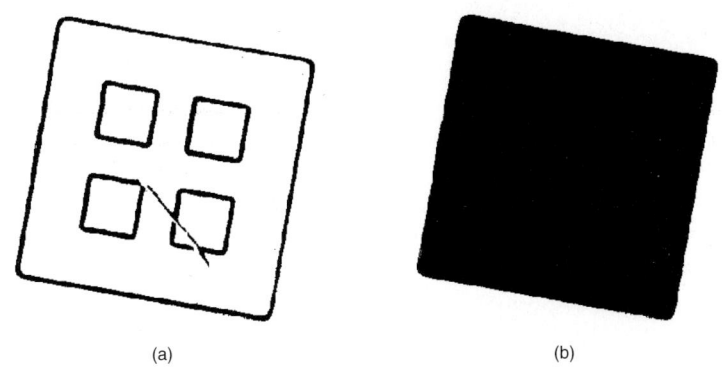

Fig. 4.25 (a) Result of the dynamic threshold operation. Note that for visualization purposes only a zoomed part enclosing the doorknob is shown. (b) Doorknob region that is obtained after filling up the holes in (a).

borders of the four small inner squares from the segmentation result. To do so, we first have to know the orientation and the size of the doorknob in the image:

```
erosion_circle (RegionDoorknob, RegionErosion, KnobSizeMin / 4)
smallest_rectangle2 (RegionErosion, Row, Column, Phi,
                     KnobLen1, KnobLen2)
KnobSize := KnobLen1 + KnobLen2 + KnobSizeMin/2
```

Because the doorknob has a square shape, its orientation corresponds to the orientation of the smallest enclosing rectangle of the segmented region (see Section 3.5.1). Since clutter or small protrusions of the segmented region would falsify the computation, they are eliminated in advance by applying an erosion to the doorknob region shown in Figure 4.25(b) (see Section 3.6.1). The radius of the circle that is used as the structuring element is set to a quarter of the minimum doorknob size. On the one hand, this allows relatively large protrusions to be eliminated. On the other hand, it ensures that the orientation can still be determined with sufficient accuracy. The size of the doorknob can be computed from the size of the smallest enclosing rectangle by adding the diameter of the circle that was used for the erosion.

Based on the orientation and the size of the doorknob, we segment the four inner squares by again using region morphology (see Section 3.6.1). First, small gaps in the previously segmented border of the inner squares are closed by performing two closing operations:

```
ScratchWidthMax := 11
gen_rectangle2 (StructElement1, 0, 0, Phi,
                ScratchWidthMax / 2, 1)
gen_rectangle2 (StructElement2, 0, 0, Phi+rad(90),
                ScratchWidthMax / 2, 1)
```

```
closing (Region, StructElement1, RegionClosing)
closing (RegionClosing, StructElement2, RegionClosing)
```

For this, two perpendicular rectangular structuring elements are generated in the appropriate orientation. The size of the rectangles is chosen such that gaps with a maximum width of ScratchWidthMax can be closed. Figure 4.26(a) shows a detailed view of the segmented region of Figure 4.25(a). Note the small gaps in the border of the inner square that are caused by the crossing scratch. The result of the closing operation is shown in Figure 4.26(b): the gaps could be successfully closed.

Fig. 4.26 (a) Detailed view of the segmented region of Figure 4.25(a). Note the gaps in the border region of the square. (b) Result of the closing visualized for the same part as in (a) with the gaps successfully closed.

Up to now, all scratches in the surface are still contained in our segmentation result of the bright border region. To be able to detect the scratches, we have to separate the scratches from the segmentation result. Because we know the appearance of the border region of the inner squares, we can get rid of the scratches by using opening operations with appropriate structuring elements. For this, we generate a structuring element that consists of two axis-parallel rectangles that represent two opposite borders of the inner squares:

```
InnerSquareSizeFraction = 0.205
InnerSquareSize := KnobSize * InnerSquareSizeFraction
gen_rectangle2 (Rectangle1, -InnerSquareSize / 2.0, 0, 0,
                InnerSquareSize / 4.0, 0)
gen_rectangle2 (Rectangle2, InnerSquareSize / 2.0, 0, 0,
                InnerSquareSize / 4.0, 0)
union2 (Rectangle1, Rectangle2, StructElementRef)
```

The size of the inner squares is computed as a predefined fraction of the size of the doorknob. The distance between the rectangles is set to the size of the inner squares. The structuring element that represents the upper and lower borders of the inner squares is obtained by rotating the generated region according to the determined orientation of the doorknob (see Section 3.3.1 for details about affine transformations). The structuring element that represents the left and right borders of the inner squares is obtained accordingly by adding a rotation angle of 90°:

```
hom_mat2d_rotate (HomMat2DIdentity, Phi, 0, 0, HomMat2DRotate)
affine_trans_region (StructElementRef, StructElement1,
                     HomMat2DRotate, 'false')
hom_mat2d_rotate (HomMat2DIdentity, Phi + rad(90), 0, 0, HomMat2DRotate)
affine_trans_region (StructElementRef, StructElement2,
                     HomMat2DRotate, 'false')
```

Figures 4.27(a) and (b) show the two generated structuring elements. Note that the rotation of the structuring elements could be omitted when creating the rectangles already in the appropriate orientations. However, this requires the centers of the rectangles to be transformed according to the orientation.

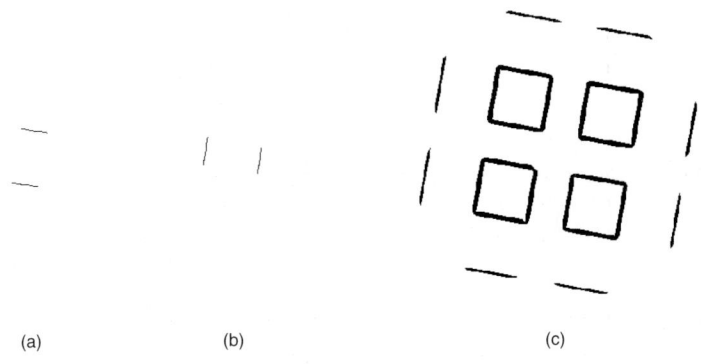

(a) (b) (c)

Fig. 4.27 (a) Structuring element that matches the upper and lower borders of the inner squares. (b) Structuring element that matches the left and right borders of the inner squares. (c) Result of the opening with the structuring elements of (a) and (b).

As we have seen in Section 3.6.1, the opening operation can be used as a template matching operation returning all points of the input region into which the structuring element fits:

```
opening (RegionClosing, StructElement1, RegionOpening1)
opening (RegionClosing, StructElement2, RegionOpening2)
union2 (RegionOpening1, RegionOpening2, RegionOpening)
```

Figure 4.27(c) shows the union of the results of both opening operations. As expected, the result contains the border of the inner squares. However, the result also contains parts of the outer border of the doorknob because the distance from the inner squares to the border of the doorknob is the same as the size of the inner squares. To exclude the part of the outer border, we intersect the result of the opening with an eroded version of the doorknob region:

```
erosion_circle (RegionDoorknob, RegionInner, InnerSquareSize / 2)
intersection (RegionInner, RegionOpening, RegionSquares)
```

The obtained region RegionSquares now only contains the border of the four inner squares. Finally, the region of the planar surface is the difference of the doorknob region and the border of the inner squares:

```
BorderWidth := 7
BorderTolerance := 3
erosion_circle (RegionDoorknob, RegionInner,
                BorderWidth + BorderTolerance)
dilation_circle (RegionSquares, RegionSquaresDilation,
                 BorderTolerance)
difference (RegionInner, RegionSquaresDilation, RegionSurface)
```

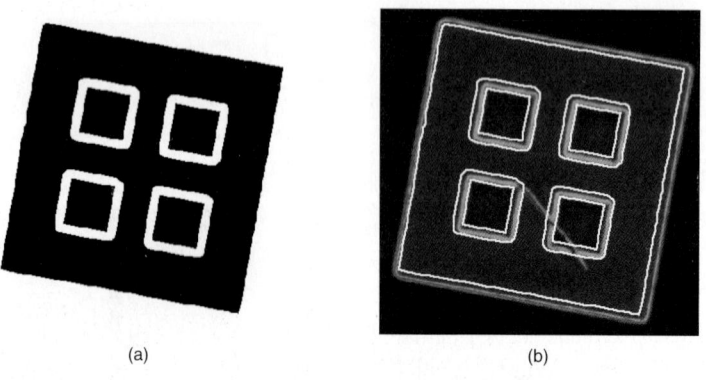

Fig. 4.28 (a) Region of interest (black) containing the planar surface of the doorknob. (b) Margin of the region of interest overlaid in white onto the original image. Note that the white border of the doorknob as well as the white borders of the inner squares are not contained. The contrast of the image has been reduced for visualization purposes.

Before computing the difference, the doorknob region is eroded using a circular structuring element to exclude the border from the surface inspection. The radius of the circle is the sum of BorderWidth and BorderTolerance, which are both predefined values. By adding BorderTolerance to the radius, pixels in the close vicinity of the border are also excluded from the inspection because their gray values are still influenced by the border region, and hence would be erroneously interpreted as defects. For the same reason, the region that represents the border of the inner squares is slightly dilated accordingly. The resulting region of interest RegionSurface that contains the planar surface of the doorknob is shown in Figure 4.28. Note that the white border of the doorknob as well as the white borders of the inner squares are not contained in the region.

Within the region of interest, we can now perform the defect detection:

```
ScratchGrayDiffMin := 15
reduce_domain (Image, RegionSurface, ImageReduced)
median_image (ImageReduced, ImageMedian, 'circle',
              ScratchWidthMax, 'mirrored')
dyn_threshold (ImageReduced, ImageMedian, RegionDeviation,
               ScratchGrayDiffMin, 'light')
```

The defects are detected by using the dynamic thresholding operation again. However, now we can use the median filter (see Section 3.2.3) to estimate the background. Based on the known maximum scratch width ScratchWidthMax, we can eliminate all scratches by passing ScratchWidthMax for the radius of the median filter. Because of the dark-field front light illumination, scratches appear as bright regions, which can easily be segmented by using the predefined threshold ScratchGrayDiffMin. In Figure 4.29(a), the result of the dynamic thresholding is shown. As can be seen, noise is included in the resulting region, which must be eliminated in a post-processing step:

```
connection (RegionDeviation, ConnectedRegions)
select_shape (ConnectedRegions, RegionDeviationNoNoise, 'area',
              'and', 4, 10000000)
```

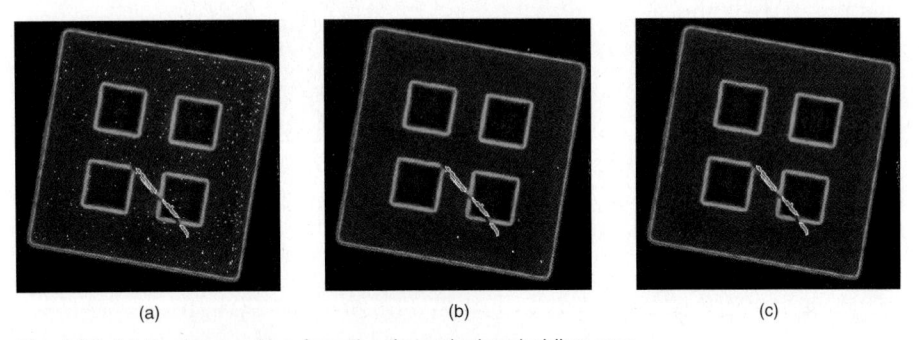

Fig. 4.29 (a) Region resulting from the dynamic thresholding overlaid in white onto the original image (with reduced contrast). Note that noise is contained in the region. (b) Result that is obtained after eliminating connected components of (a) that are smaller than 4 pixels. Note that not all the noise could be eliminated. (c) Result of the surface inspection with the detected scratch displayed in white.

First, connected components (see Section 3.4.2) that are smaller than 4 pixels are assumed to be noise in any case, and hence are eliminated. Unfortunately, this does not completely eliminate the noise from the segmentation, as can be seen from Figure 4.29(b). However, further increasing the threshold would also eliminate parts of interrupted defect regions, which is not desirable. In order to separate the interrupted defect regions from the noise, we assume that noise is equally distributed while defect regions that belong to the same scratch lie close to each other. Therefore, we can close small gaps in the defect regions by performing a dilation:

```
ScratchAreaMin := 20
dilation_circle (RegionDeviationNoNoise, RegionDilation, 2)
connection (RegionDilation, ConnectedRegions)
intersection (ConnectedRegions, RegionDeviationNoNoise,
              RegionIntersection)
select_shape (RegionIntersection, RegionErrors, 'area',
              'and', ScratchAreaMin, 10000000)
```

After the dilation, previously interrupted defect regions are merged. From the merged regions, the connected components are recomputed. To obtain the original shape of the defects, the connected components are intersected with the original non-dilated region. Note that the connectivity of the components is preserved during the intersection computation. Thus, with the dilation we can simply extend the neighborhood definition that is used in the computation of the connected components. Finally, we select all components with an area exceeding a predefined threshold for the minimum scratch area. The final result is shown in Figure 4.29(c).

In Figure 4.30(a), the zoomed part of a second example image is shown. Note the low contrast of the small scratch at the lower right corner of the doorknob. The surface inspection detects this defect, as shown in Figure 4.30(b).

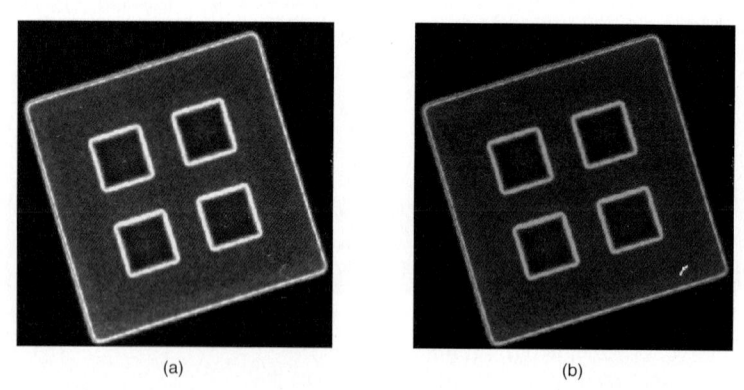

Fig. 4.30 (a) Zoomed part of a second example image. (b) Result of the surface inspection of the image in (a). The margin of the found defect is overlaid in white onto the original image (with reduced contrast).

Exercises

1. In the above application, defects can only be identified on the planar surface of the doorknob. Modify the program such that it is also possible to detect defects

in the bright border region of the doorknob and in the bright border regions of the inner squares.

2. The program described above uses region morphology to locate the doorknob and to segment its surface. Alternatively, the doorknob could be located by using a robust matching algorithm (see Section 3.11.5). Furthermore, the inspection could be performed by using an appropriate variation model (see Section 3.4.1). Use these two methods to perform the inspection of the surface and of the bright border regions simultaneously, without using region morphology.

4.7
Measuring of Spark Plugs

In shape and dimensional inspection applications, often distances between objects or object parts must be measured to decide whether they lie within the required tolerances.

In this application, we measure the size of spark plug gaps. The size of the gap between the center electrode and the side or ground electrode must lie within a certain tolerance range to ensure that the spark plug can be used within a specific engine type. We use a camera with a telecentric lens as well as a telecentric back light illumination for image acquisition. First, we calibrate the camera setup. Then, a robust matching algorithm is used to find the spark plug in an image. Based on the matching pose, the region of interest in which the measurement is performed is aligned. The actual measuring is performed with 1D edge extraction in a gray value profile across the spark gap.

The algorithms used in this application are:
- Camera calibration (Section 3.9)
- Robust template matching (Section 3.11.5)
- Region features (Section 3.5.1)
- Affine transformations (Section 3.3.1)
- 1D edge extraction (Section 3.7.2)

The corresponding example program is:
.../machine_vision_book/spark_plug_measuring/spark_plug_measuring.dev

Figures 4.31(a) and (b) show an image of a spark plug with a correct gap size and with a gap size that is too large, respectively. We use a telecentric lens to avoid perspective distortions in the image that would make an exact measurement difficult (see Section 2.2.4). With the telecentric illumination, very sharp edges at the silhouette of the spark plug are obtained (see Section 2.1.5). Furthermore, reflections at the spark plug on the camera side are avoided.

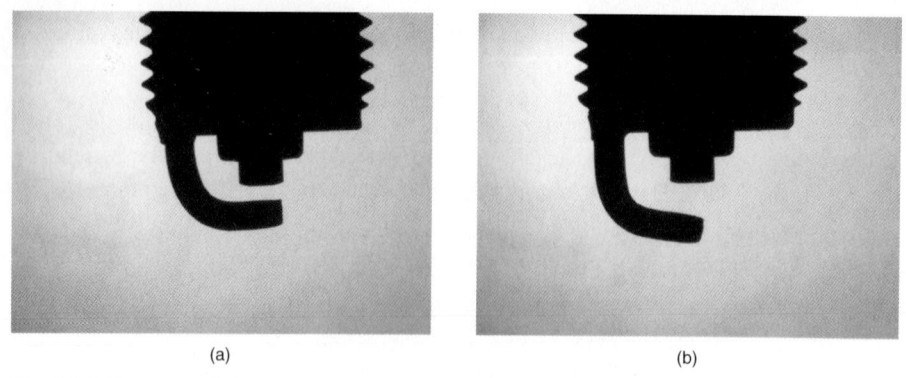

Fig. 4.31 Two example images of a spark plug. (a) Spark plug with a correct gap size. (b) Spark plug with a gap size that is too large.

In the first step, we calibrate the camera. In this application, camera calibration is essential for two reasons. First, we want to measure the spark gap accurately, and hence must consider possible lens distortions. Second, the size of the gap should be determined in metric units. To calibrate the camera, we acquire multiple images of a transparent calibration target in various poses. Figure 4.32 shows two of the 14 images that are used for the calibration.

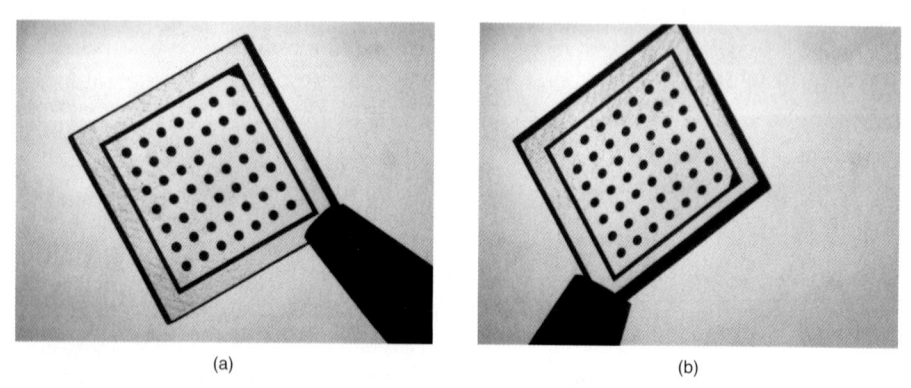

Fig. 4.32 Two out of a set of 14 images of a transparent calibration target in different poses. The images are used to calibrate the camera.

First, we specify the used calibration target, define initial values for the interior orientation of the camera, and initialize some arrays that are used to collect the image coordinates of the calibration marks as well as the initial poses of the calibration targets:

```
CaltabName    := 'caltab_10mm.descr'
StartCamParam := [0,0,39e-6,39e-6,320,240,640,480]
StartPoses    := []
Rows          := []
Cols          := []
```

Since we use a telecentric camera, which does not have a focal length, the focal length is set to zero. The initial value of the radial distortion is also set to zero, the pixel size in world coordinates is 39 μm, and the principal point is assumed to be in the center of the image. These values are just initial values. They will be improved during the calibration process.

Then, within a loop over all calibration images, the subpixel-accurate positions of the circular calibration marks are determined as well as initial estimations for the poses of the calibration target with respect to the camera:

```
find_caltab (Image, Caltab, CaltabName, 3, 100, 5)
find_marks_and_pose (Image, Caltab, CaltabName, StartCamParam,
                     128, 10, 18, 0.6, 15, 100,
                     RCoord, CCoord, StartPose)
StartPoses := [StartPoses,StartPose]
Rows := [Rows,RCoord]
Cols := [Cols,CCoord]
```

The approximate position of the calibration target in the image is determined with the operator `find_caltab`, which returns a region that contains the calibration target. Within this region, the calibration marks are extracted. The initial poses are determined based on the calibration marks and the initial values for the interior orientations.

Now, we determine the exact interior orientation of the camera:

```
caltab_points (CaltabName, X, Y, Z)
camera_calibration (X, Y, Z, Rows, Cols, StartCamParam, StartPoses,
                    'all', CamParam, FinalPoses, Errors)
```

First, the world coordinates of the calibration marks are read from the description file of the used calibration target. Then, the calibration is carried out with the operator `camera_calibration`. In the online phase, the interior orientation is used to transform the image measurements into world units. Note that, although the exterior orientation of the camera is also determined, it is not used below because we do not need to know the position of the spark plug with respect to a world coordinate system.

The position of the spark plug in the image can be determined by using a robust matching algorithm. Here, we use the shape-based matching, which is described in Section 3.11.5. For this, we create a model representation of the spark plug from a template image:

```
gen_rectangle1 (ModelRegion, 120, 230, 220, 445)
reduce_domain (ModelImage, ModelRegion, TemplateImage)
create_shape_model (TemplateImage, 'auto', rad(-30), rad(60), 'auto',
                    'none', 'use_polarity', 'auto', 'auto', ModelID)
area_center (ModelRegion, Area, RefRow, RefCol)
```

The template image is shown in Figure 4.33(a). Because the thread rolling may be different for different spark plugs, we must not include it in the model representation. Furthermore, the side electrode may bend as the gap size changes, and hence also must not be used. Therefore, the model is created only from a small part of the spark plug that always appears in the same way, and hence permits a robust matching. In Figure 4.33(a), the corresponding image region is indicated by the gray rectangle. It only contains the center electrode, the insulator, and the top part of the thread. The contour representation of the resulting model is shown in Figure 4.33(b). Note that, because the orientation of the spark plugs is similar in all images, the model is created only within an angle range of $\pm 30°$. Finally, to be able to correctly align the measurement in the online phase, we must know the reference point of the model, which is the center of gravity of the model region. In Figure 4.33(a), the reference point is displayed as a gray cross. Relative to the reference point of the model, we define a rectangle in which the actual gap measuring is to be performed:

```
RectRelRow := 80
RectRelCol := -26
RectRelPhi := rad(90)
RectLen1  := 50
RectLen2  := 15
```

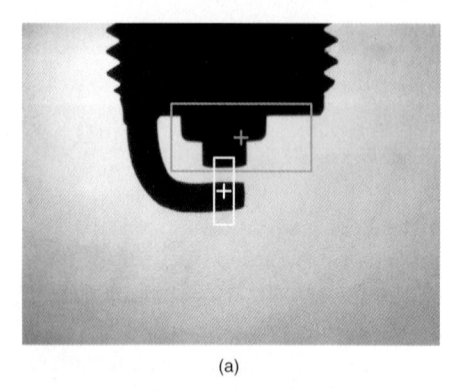

Fig. 4.33 (a) Template image from which the model representation is created. The model region is indicated by the gray rectangle. The reference point of the model is visualized as a gray cross. The rectangle within which the measuring of the gap size is performed and its center are displayed in white. (b) Contour representation of the created model.

The center of the rectangle is 80 pixels below and 26 pixels to the left of the model reference point. The rectangle's orientation is chosen such that its major axis points in the direction in which the measurement, i.e., the 1D edge extraction, is performed. Thus, edges that are perpendicular to the major axis can be measured. Here, the major axis is oriented vertically in the model image. During measuring in the online phase, the gray values are averaged perpendicular to the major axis to obtain a 1D edge profile. The range in which the averaging is performed is defined by the width of the rectangle. In Figure 4.33(a), the measurement rectangle and its center point are displayed in white.

All the steps described so far are performed offline. In the online phase, we use the shape model to find the spark plug in an image:

```
find_shape_model (Image, ModelID, rad(-30), rad(60), 0.7, 1, 0.5,
                 'least_squares', 0, 0.9, Row, Column, Angle, Score)
```

Based on the pose of the found spark plug, we create the measurement rectangle:

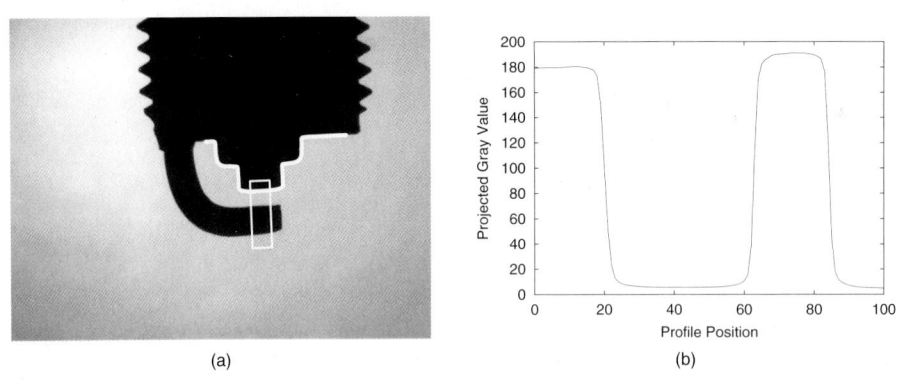

```
vector_angle_to_rigid (0, 0, 0, Row, Column, Angle, HomMat2D)
affine_trans_pixel (HomMat2D, RectRelRow, RectRelCol,
                    TransRow, TransCol)
gen_measure_rectangle2 (TransRow, TransCol, Angle+RectRelPhi,
                        RectLen1, RectLen2, Width, Height,
                        'bilinear', MeasureHandle)
```

Fig. 4.34 (a) Image of a spark plug to be measured. The transformed model contour at the pose of the found spark plug is overlaid as a bold white line. The transformed measurement rectangle is displayed as a thin white line. (b) Gray value profile of the measurement rectangle. The 1D edges are extracted by computing the local extrema of the first derivative of the gray value profile.

First, we transform the matching pose into a 2D rigid transformation, which aligns the model with the spark plug in the image. Then, with the obtained transformation, the relative position of the center of the measurement rectangle is transformed. Finally, the orientation of the rectangle is adapted according to the orientation of the spark plug. Figure 4.34(a) shows an image of the spark plug to be measured. The contour of the found spark plug is overlaid as a bold white line. The transformed measurement rectangle is displayed as a thin white line. With gen_measure_rectangle2, the gray value profile is computed by using bilinear interpolation. It is shown in Figure 4.34(b). The actual measuring is performed with the following operation:

```
measure_pairs (Image, MeasureHandle, 1, 30, 'positive_strongest',
               'all', Row1, Col1, Amplitude1, Row2, Col2,
               Amplitude2, IntraDistance, InterDistance)
```

The 1D edges are extracted by computing the local extrema of the first derivative of the gray value profile. Because we are only interested in the edge pair that corresponds to the gap, we use `measure_pairs` that automatically groups the edges to pairs. To ensure that the right edge pair is returned, we demand from the first edge of the pair to have a positive transition. With this, we can avoid obtaining the edge pair that corresponds to the lower and upper borders of the side electrode. Finally, the position of the resulting 1D edge pair is transformed back into the image and returned in (`Row1`,`Col1`) and (`Row2`,`Col2`).

To eliminate the effect of radial distortions and to obtain the result in metric units, we must transform the two edge points into the camera coordinate system. This can be done by constructing the two lines of sight that pass through the two edge points in the image:

```
get_line_of_sight ([Row1,Row2], [Col1,Col2], CamParam,
                   X, Y, Z, XH, YH, ZH)
DX := X[1]-X[0]
DY := Y[1]-Y[0]
GapSize := sqrt(DX*DX+DY*DY)
```

For both edge points, the line of sight is returned as two points on the line, given in world units. Because of the telecentric camera, all lines of sight are parallel, and hence the two points that define the line have the same X and Y coordinate. The size of the gap corresponds to the distance of the parallel lines that pass through the two edge points. In Figures 4.35(a)–(c), three examples are shown. The measured edge points are visualized by two lines that are perpendicular to the measurement direction. The measured gap size is overlaid as white text. Finally, based on a predefined range of valid gap sizes, we can decide whether the gap size is within the tolerances:

```
GapSizeMin := 0.78e-3
GapSizeMax := 0.88e-3
if (GapSize < GapSizeMin)
*    Gap size is too small.
else
  if (GapSize > GapSizeMax)
*      Gap size is too large.
  else
*      Gap size is within tolerances.
  endif
endif
```

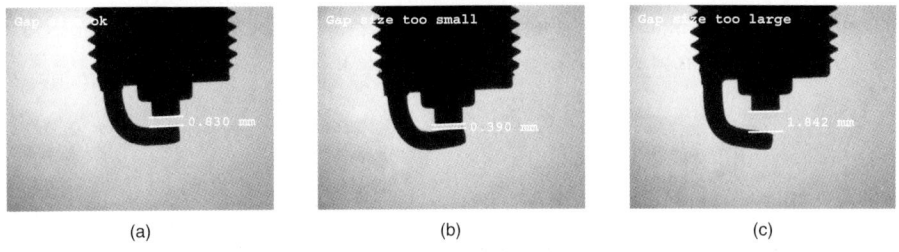

Fig. 4.35 Three example results of the spark plug measurement. The measured edge points are visualized by a line perpendicular to the measurement direction. (a) Spark plug with a correct gap size. (b) Spark plug with a gap size that is too small. (c) Spark plug with a gap size that is too large.

According to the above decision, an appropriate action can be performed, e.g., by rejecting the current spark plug.

Exercises

1. Alternatively to the shape-based matching, the normalized cross-correlation could be used for template matching (see Section 3.11.1). Adapt the above program such that the pose of the spark plug is determined with the normalized cross-correlation.

2. Overheating, oil leakage, or bad fuel quality may lead to the accumulation of deposits on the center electrode and on the insulator during the use of the spark plug. Therefore, in the automobile industry, especially in racing applications, used spark plugs are analyzed because they indicate conditions within the running engine. Extend the above program so that irregularities in the shape of the center electrode and of the insulator can be detected. Furthermore, measure the diameter of both components to detect heavy deposits and erosion.

4.8
Molding Flash Detection

Especially in the metal- or plastic-working industry, a frequent task is to detect flashes on castings or molded plastic parts. Flashes often cannot be avoided because the molds do not fit together tightly. Therefore, it is important to detect the flashes and either to remove them in a secondary operation or to reject the object.

In this application, we detect molding flashes on a circular plastic part that has been manufactured by injection molding. After the flashes are detected, their size and their position on the circular border are determined. For illustration purposes, this task is solved with two alternative approaches. While the first approach uses region morphology, the second approach is based on the processing of subpixel-precise contours.

The algorithms used in this application are:
- Thresholding (Section 3.4.1)
- Region morphology (Section 3.6.1)
- Circle fitting (Section 3.8.2)
- Contour features (Section 3.5.3)
- 2D edge extraction (Section 3.7.3)

The corresponding example program is:
.../machine_vision_book/molding_flash_detection/molding_flash_detection.dev

Figure 4.36(a) shows an image of the molded plastic we want to inspect. Note the flash at the upper right part of its border. The image was acquired with a diffuse bright-field back light illumination (see Section 2.1.5). With this kind of illumination, the object appears dark while the background appears bright, which simplifies the segmentation process significantly.

(a)   (b)

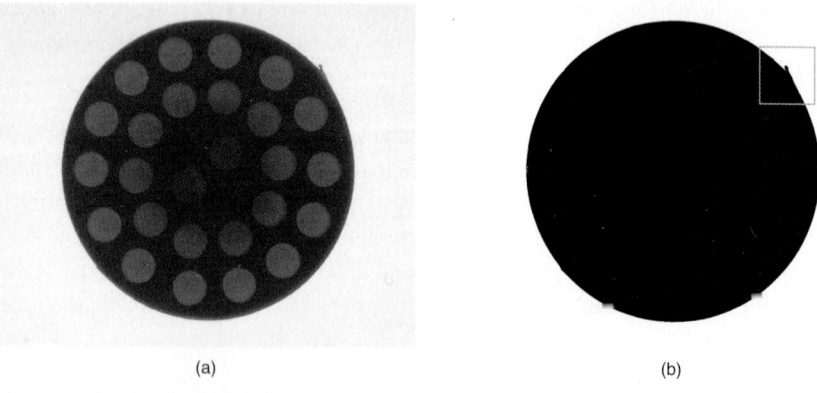

Fig. 4.36 (a) Circular molded plastic part with a flash at the upper right part of its border. (b) Result of thresholding the dark object. For visualization purposes, in the following, only a zoomed part of the image is displayed, which is indicated by the gray rectangle.

In the first step, which is identical for both presented approaches, the dark object is segmented by a thresholding operation:

```
threshold (Image, Object, 0, 180)
```

The result is shown in Figure 4.36(b).

Molding Flash Detection Using Region Morphology

The first approach, which uses region morphology, is described in the following. It is similar to the approach that is described in Section 3.6. However, here, an extended version is presented.

The flash appears as a protrusion of the object region (see Figure 4.36(b)). Therefore, the flash can be segmented by performing an opening on the object region and subtracting the opened region from the original segmentation:

```
opening_circle (Object, OpenedObject, 400.5)
difference (Object, OpenedObject, RegionDifference)
```

For the opening, a circle is used as the structuring element. The circle is chosen almost as large as the object to ensure that even large flashes can be eliminated and the circular shape of the molded plastic can be recovered. Note that the radius must be smaller than the radius of the molded plastic part. Otherwise, the opening would remove the object entirely. Figures 4.37(a) and (b) show the result of the closing and the difference operation, respectively, for a zoomed image part. Small components of the difference region that are caused by minor irregularities at the border must be eliminated. For this, the difference region is opened with a rectangular structuring element of size 5×5:

```
MinFlashSize := 5
opening_rectangle1 (RegionDifference, FlashRegion,
                   MinFlashSize, MinFlashSize)
```

The resulting region, which is shown in Figure 4.37(c), represents the molding flash.

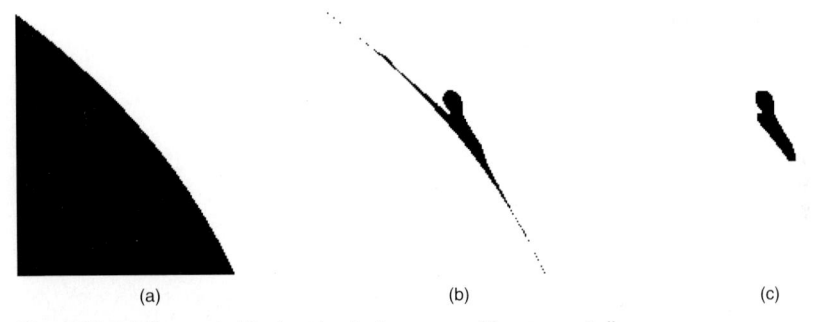

Fig. 4.37 (a) Opened object region in the zoomed image part displayed in Figure 4.36(b). (b) Difference between original and opened object region. (c) Segmented flash region obtained by opening the difference region with a rectangular structuring element of size 5×5.

An important quality feature in this application is the maximum distance of the flashes from the ideal circular shape of the object, which we will call the reference shape below. Here, the reference shape is represented by the opened object region. To determine the maximum distance, we compute the distance transform of the opened object region:

```
distance_transform (OpenedObject, DistanceImage,
                   'euclidean', 'false', Width, Height)
min_max_gray (FlashRegion, DistanceImage, 0,
              DistanceMin, DistanceMax, DistanceRange)
```

The resulting distance image, shown in Figure 4.38, contains the shortest distance to the reference shape for each point in the background region. The latter is the complement of the reference shape. Thus, the maximum distance `DistanceMax` is obtained by searching for the maximum value in the distance image within the flash region.

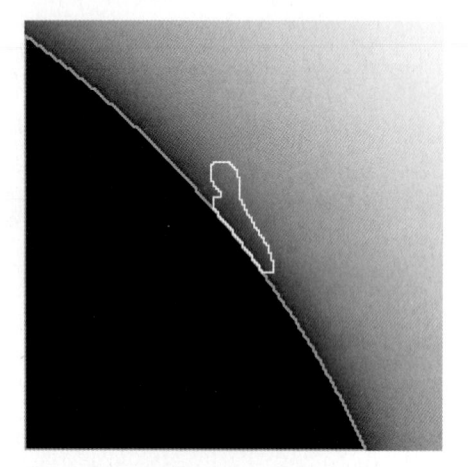

Fig. 4.38 Distance image that is obtained after applying a distance transform to the opened object region, whose margin is displayed in gray. The margin of the segmented flash region is displayed in white. Bright gray values in the distance image correspond to large distances. For better visualization, the distance image is displayed with a square root LUT.

To compute the angle range of the segmented flashes with respect to the object center, we need to know the start and end point of each flash on the border of the molded plastic part. These points can also be obtained with region morphology:

```
boundary (OpenedObject, RegionBoundary, 'outer')
connection (FlashRegion, FlashRegions)
intersection (FlashRegions, RegionBoundary, RegionCircleSeg)
junctions_skeleton (RegionCircleSeg, EndPoints, JuncPoints)
```

First, the one-pixel-wide boundary of the opened object region is computed (see Figure 4.39(a)). Note that the obtained boundary lies one pixel outside of the original region, and hence inside the flash region. Then, each connected component of the flash region is intersected with the boundary. Thus, for each flash a one-pixel-wide region at the boundary of the object is obtained. Finally, the region that only contains the end points of the intersected regions can be obtained with the operator `junctions_skeleton`. The intersected regions and the corresponding end points are visualized in Figure 4.39(b).

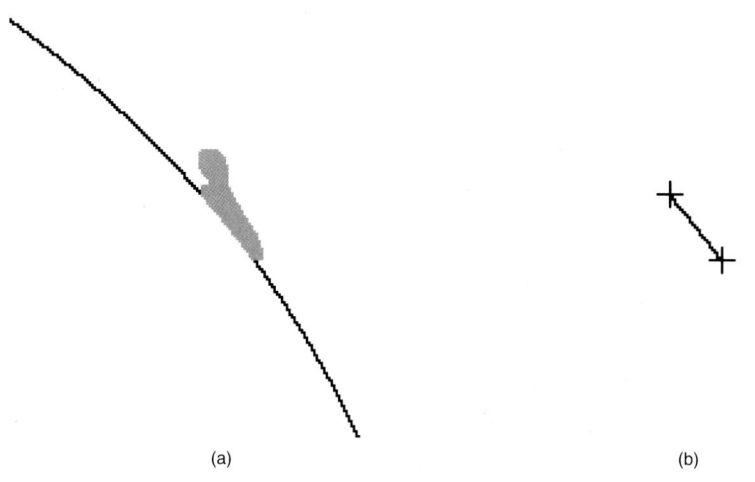

Fig. 4.39 (a) One-pixel-wide boundary region (black) of the opened object region of Figure 4.37(a) and the flash region (gray). (b) Intersection of the boundary region with the flash region. The computed end points of the intersection region are indicated by small crosses. They can be used to compute the angle range of the molding flash.

From the two end points of each segmented molding flash, the corresponding angle interval can be computed. For this, we have to know the center of the circular plastic part. The circle center is the center of gravity of the opened object region:

```
area_center (OpenedObject, Area, CenterRow, CenterCol)
```

Now, we are able to compute the angle range of each flash:

```
NumFlash := |EndPoints|
for Index := 1 to NumFlash by 1
  EndPointsSelected := EndPoints[Index]
  get_region_points (EndPointsSelected, SegRow, SegCol)
  Angle1 := atan2(CenterRow-SegRow[0],SegCol[0]-CenterCol)
  Angle2 := atan2(CenterRow-SegRow[1],SegCol[1]-CenterCol)
  AngleRange := Angle2-Angle1
endfor
```

Here `get_region_points` returns the row and column coordinates of the two end points in the two tuples SegRow and SegCol, respectively, each of which contains two elements, one for each end point. Finally, the corresponding angles are computed using the arc tangent function. The result is shown in Figure 4.40.

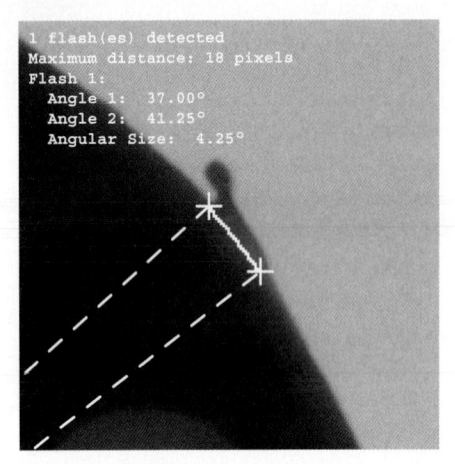

Fig. 4.40 Result of the molding flash detection with region morphology. The maximum distance, the angle of the start and end points of the detected flash, as well as the corresponding angle range are overlaid as white text in the upper left corner of the image. The associated circle sector (dashed), the enclosed segment of the boundary (solid), and its end points (crosses) are displayed in white. To provide a better visibility, the contrast of the image has been reduced.

Molding Flash Detection with Subpixel-Precise Contours

In the following, the second approach is described, which is based on the processing of subpixel-precise contours. The idea is to compute the distance of the contour that represents the boundary of the plastic part to the circular reference shape. The contour representation can be obtained by using subpixel-precise edge extraction (see Section 3.7.3). Because edge extraction is computationally expensive, we determine a region of interest first:

```
boundary (Object, RegionBorder, 'outer')
dilation_circle (RegionBorder, RegionDilation, 5.5)
reduce_domain (Image, RegionDilation, ImageReduced)
```

As in the first approach, we start with the segmented object region, which is shown in Figure 4.36(b). First, the one-pixel-wide boundary of the object region is computed. Using morphology, we broaden this border to get a band-shaped region of interest for the subpixel-precise extraction of the edges:

```
edges_sub_pix (ImageReduced, Edges, 'canny', 2.0, 20, 40)
union_adjacent_contours_xld (Edges, BorderContour, 20, 1, 'attr_keep')
```

Because of small irregularities of the object border, the obtained edges may be interrupted. To obtain one connected contour, adjacent edge segments are merged. The resulting contour is shown in Figure 4.41(a).

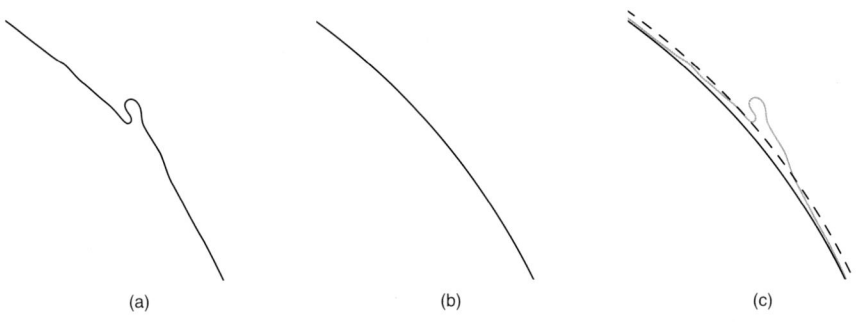

Fig. 4.41 (a) Result of the edge extraction in the zoomed image part displayed in Figure 4.36. Adjacent edge segments have been merged. (b) Circle that is robustly fitted to the contour of (a). (c) The extracted contour (gray line), the contour of the fitted circle (solid black line), and the contour of a circle with a radius enlarged by `MinFlashSize` (dashed black line). The intersection points of the extracted contour and the enlarged circle represent the desired start and end points of the flash.

In the next step, the reference shape must be extracted. This is done by fitting a circle to the extracted contour. The contour points that lie on the flash would falsify the result of the fitting, and hence should be treated as outliers. Therefore, we use a robust fitting algorithm as described in Section 3.8.1. Because we want to completely disregard the outliers from the computation, we use the Tukey weight function:

```
fit_circle_contour_xld (BorderContour, 'geotukey', -1, 0, 0, 3, 2,
                       CenterRow, CenterCol, Radius,
                       StartPhi, EndPhi, PointOrder)
```

The contour of the obtained circle is shown in Figure 4.41(b). Now, we can compute the distances of all contours points to the fitted circle:

```
dist_ellipse_contour_points_xld (BorderContour, 'unsigned', 0,
                                 CenterRow, CenterCol, 0,
                                 Radius, Radius, Distances)
```

In this approach, the maximum distance of the flashes from the ideal circular shape is obtained easily:

```
DistanceMax := max(Distances)
```

In the next step, we compute the start and end points of each contour interval whose points all have a distance exceeding the predefined threshold `MinFlashSize`. For this, we define a function whose argument is the index of the contour points. The function values are the distance values corrected by the threshold value. The function is shown in Figure 4.42. Thus, the indices of the start and end points correspond to the zero crossings of this function:

```
DistancesOffset := Distances-MinFlashSize
create_funct_1d_array (DistancesOffset, Function)
zero_crossings_funct_1d (Function, ZeroCrossings)
```

The principle is illustrated in Figure 4.41(c). Effectively, we enlarge the radius of the fitted circle by the threshold `MinFlashSize` and compute the intersection points of the contour with the enlarged circle. The obtained points are the start and end points of the intervals that represent the detected flashes. If the flash crosses the end of the contour, we have to take the 360° wrap-around into account. Note that, because of the closed contour, the first and last function values are identical. The flash crosses the end of the contour if the first (or the last) function value is positive. In this case, we move the last zero crossing to the first position in the tuple:

```
if (DistancesOffset[0] > 0)
  Num := |ZeroCrossings|
  ZeroCrossings := [ZeroCrossings[Num-1],ZeroCrossings[0:Num-2]]
endif
```

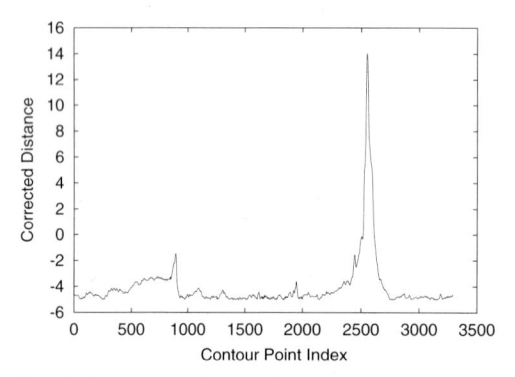

Fig. 4.42 Function that holds the corrected distance values `DistancesOffset` for all contour points. Intervals of contour points with positive function values represent the molding flashes.

Now, we can group two consecutive zero crossings to one interval. The corresponding angle range can be computed from the circle center and the two contour points at the position of the zero crossings:

```
get_contour_xld (BorderContour, ContRow, ContCol)
for Index := 0 to |ZeroCrossings|-1 by 2
  Start := round(ZeroCrossings[Index])
  End := round(ZeroCrossings[Index+1])
  Angle1 := atan2(CenterRow-ContRow[Start],ContCol[Start]-CenterCol)
  Angle2 := atan2(CenterRow-ContRow[End],ContCol[End]-CenterCol)
  AngleRange := Angle2-Angle1
endfor
```

Note that the zero crossings are extracted with subpixel precision. Therefore, we must round them to the nearest integer value before using them as indices to the array of contour points.

The result of the region-based molding flash detection is shown in Figure 4.43.

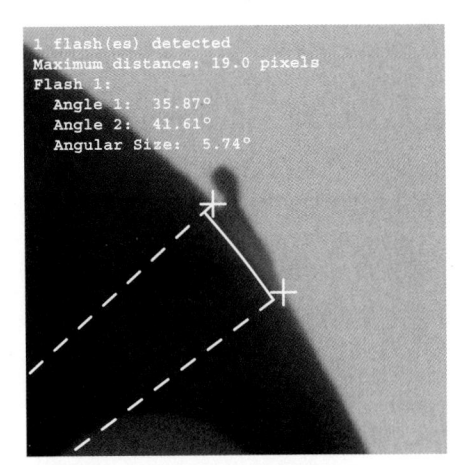

Fig. 4.43 Result of the molding flash detection with subpixel-precise contours. The maximum distance, the angle of the start and end points of the detected flash, as well as the corresponding angle range, are overlaid as white text in the upper left corner of the image. The associated sector of the fitted circle (dashed), the enclosed circle segment (solid), and the end points of the flash (crosses) are displayed in white. To provide a better visibility, the contrast of the image has been reduced.

Exercise

1. The described example program is able to detect flashes, i.e., protrusions, on the boundary of the molded part. Another molding defect that frequently must be detected are voids, i.e., indentations at the object boundary. Modify the program so that protrusions as well as indentations can be detected.

4.9
Inspection of Punched Sheets

A punched sheet is sheet metal that has been cut by using a punch. The punched sheets must be inspected to guarantee predefined tolerances.

In this application, we check the size of the holes in a punched sheet. First, we extract the boundaries of the holes in the sheet metal. Then the size of the holes is calculated from the extracted contours.

The algorithms used in this application are:

- Thresholding (Section 3.4.1)
- Region morphology (Section 3.6.1)
- 2D edge extraction (Section 3.7.3)
- Segmentation of contours into lines and circles (Section 3.8.4)
- Circle fitting (Section 3.8.2)

The corresponding example program is:
.../machine_vision_book/punching_sheet_inspection/
punching_sheet_inspection.dev

To inspect the outline of thin planar objects, the images are acquired with a diffuse bright-field back light illumination (see Section 2.1.5). With this kind of illumination, the object appears dark, while the surrounding areas are bright (see Figure 4.44(a)). This simplifies the segmentation process significantly. To avoid perspective distortions, the image plane of the camera must be parallel to the sheet metal. If this camera setup cannot be realized for the image acquisition, or if the lenses produce heavy radial distortions, the camera must be calibrated and the images must be rectified (see Section 3.9). If the radial distortions can be neglected, the rectification can also be done with a projective transformation (see Section 3.3.2).

(a)   (b)

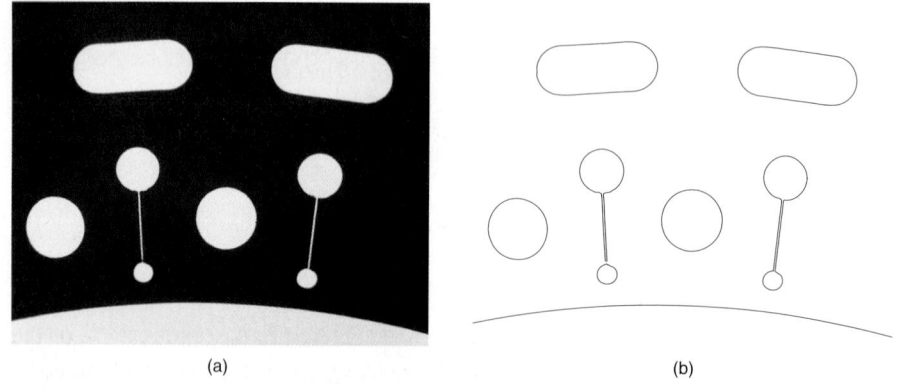

Fig. 4.44 (a) Image of a punched sheet that contains circular and oval holes. (b) Edges extracted from the image in (a).

4.9 Inspection of Punched Sheets

First, the boundaries of the punched sheet are extracted in the image. The resulting contours are then segmented into lines and circular arcs. Finally, circles are fitted into the circular arcs to obtain the radii of the holes in the sheet metal.

We use subpixel-precise edge extraction (see Section 3.7.3) for the determination of the boundary. Because edge extraction is computationally expensive, we determine a region of interest first:

```
threshold (Image, Region, 128, 255)
boundary (Region, RegionBorder, 'inner')
dilation_circle (RegionBorder, RegionBorderDilation, 3.5)
reduce_domain (Image, RegionBorderDilation, ImageReduced)
```

Because of the diffuse bright-field back light illumination, it is easy to find an appropriate threshold range for the segmentation of the bright background, e.g., 128 to 255. With this, we can be sure that the edges to be extracted lie in the vicinity of the border of the segmented region. Using morphology, we broaden this border to get a band-shaped region of interest for the subpixel-precise extraction of the edges.

```
edges_sub_pix (ImageReduced, Boundary, 'canny', 2.0, 20, 40)
```

The extracted edges are shown in Figure 4.44(b). In the next step, these edges are segmented into lines and circular arcs:

```
segment_contours_xld (Boundary, ContoursSplit, 'lines_circles',
                      5, 10, 5)
```

Now, we collect all circular arcs. The type of the split contours can be determined by reading out the global contour attribute 'cont_approx', which has the value 1 for contours that are best approximated by circular arcs. For contours that can be approximated by lines, the attribute has the value −1.

```
CircularArcs := []
Number := |ContoursSplit|
for i := 1 to Number by 1
  ObjectSelected := ContoursSplit[i]
  get_contour_global_attrib_xld (ObjectSelected, 'cont_approx', Attrib)
  if (Attrib = 1)
    CircularArcs := [CircularArcs,ObjectSelected]
  endif
endfor
```

The contours may have been oversegmented during the segmentation into lines and circular arcs. To obtain one contour for each circular border of the punched sheet, we merge cocircular arcs:

```
union_cocircular_contours_xld (CircularArcs, UnionContours,
                               0.1, 0.2, 0.1, 5, 10, 10, 'true', 1)
```

The resulting circular arcs are displayed in Figure 4.45(a). Figure 4.45(b) shows a detail of the image from Figure 4.44(a) along with the circular arc.

(a)   (b)

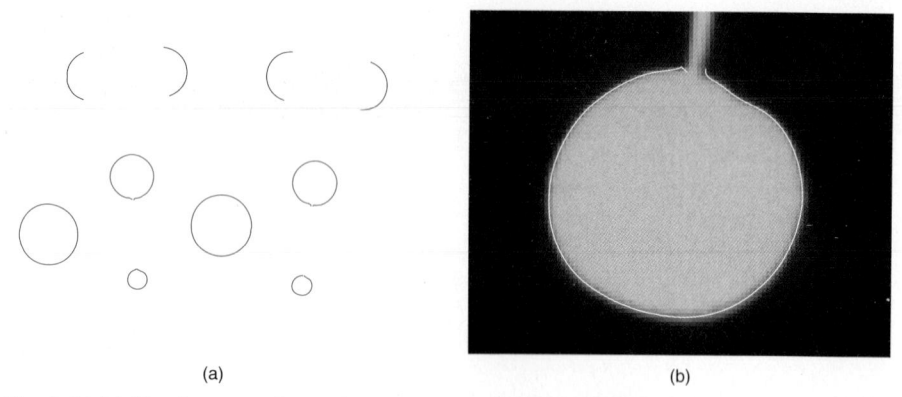

Fig. 4.45 (a) Circular arcs selected from the edges in Figure 4.44(b). (b) Detail of Figure 4.44(a) along with the selected edges. Note that the contrast of the image has been reduced for visualization purposes.

Now we determine the parameters of the best-fitting circle for each circular arc and select those circles that have a radius smaller than a relatively large value, e.g., 500 pixels, to get rid of the large arc at the lower image border.

```
fit_circle_contour_xld (UnionContours, 'geotukey',
                        -1, 0, 0, 3, 1, Row, Column,
                        Radius, StartPhi, EndPhi, PointOrder)
CircleIds := find(sgn(Radius-500),-1)
Circles := UnionContours[CircleIds+1]
```

To determine the indices of the circles that have a radius below 500 pixels, we use the following operations. First, we subtract 500 from the tuple that holds the radii and determine the sign of the resulting values. If a particular radius is smaller than 500, `sgn(Radius-500)` will be -1. Using the function `find`, we collect all indices of circles with a radius smaller than 500. The radii of these circles are displayed in Figure 4.46(a). Figure 4.46(b) shows a detail of the image from Figure 4.44(a) along with the fitted circle.

Finally, we determine the distance between the centers of the two circles of the two oval holes at the top of the image. For simplicity, we assume that the oval holes are the uppermost holes in the image. Hence we can select the four respective circle centers simply by sorting them appropriately:

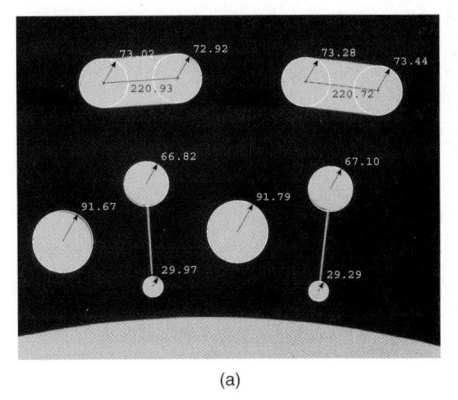

Fig. 4.46 (a) The overall inspection result. The fitted circles are displayed in white together with their radii. For the two oval holes, the distance between the centers of the two circles is given. (b) Detail of Figure 4.44(a) along with the fitted circle. Note that the shape of the hole is not perfectly circular. Note also that the contrast of the image has been reduced for visualization purposes.

```
IndicesUppermost := (sort_index(Row))[0:3]
RowOval := subset(Row,IndicesUppermost)
ColOval := subset(Column,IndicesUppermost)
```

First, we sort the row coordinates in ascending order and take the indices of the first four circle centers. By using these indices, we can select the four uppermost circles.

Now, we sort these circles according to their column position:

```
Indices := sort_index(ColOval)
RowOvalSorted := subset(RowOval,Indices)
ColOvalSorted := subset(ColOval,Indices)
```

The sorted indices are used to get the centers of the four circles arranged from left to right. With this, it is easy to determine the distance between the two circle centers of each oval hole:

```
for CircleNo := 0 to |Indices|-1 by 2
   distance_pp (RowOvalSorted[CircleNo], ColOvalSorted[CircleNo],
               RowOvalSorted[CircleNo+1], ColOvalSorted[CircleNo+1],
               Distance)
endfor
```

Because we have sorted the circle centers from left to right, we can simply step through the sorted array by using a step width of 2 and select two consecutive circle centers each time. This ensures that both circle centers belong to the same oval hole. The resulting distances are displayed in Figure 4.46(a).

Exercises

1. For the detection of the two oval holes, we made the assumption that the punched sheet appears in roughly the same orientation in each image. Write a procedure that determines the pairs of circles that belong to the same oval hole even if the punched sheet is arbitrarily orientated. Tip: You can test if the connection between two circle centers lies completely within a hole by means of set operations on regions (see Section 3.6.1).

2. As can be seen in Figure 4.46(b), the actual boundary of the hole in the sheet metal deviates from a perfect circle. Extend the above program such that the deviations of the extracted boundary from the fitted circle are determined. Visually highlight the parts of the boundary that deviate from the circle by more than some given threshold.

4.10
3D Plane Reconstruction with Stereo

A binocular stereo system consists of two cameras looking at the same object. If the interior orientations and the relative orientation of the two cameras are known, the 3D surface of the object can be reconstructed from the images of the two cameras.

In this application, we determine the angle between two planes on an intake manifold using a stereo camera setup. First, we calibrate the stereo camera setup. Then, from a stereo image pair of the intake manifold, a distance map is determined that describes the 3D surface of the object. We determine the two planes that form the indentation in the cylindrical surface. Finally, we calculate the indentation angle between the two planar faces of the indentation.

The algorithms used in this application are:

- Camera calibration (Section 3.9)
- Stereo reconstruction (Section 3.10)
- 2D edge extraction (Section 3.7.3)
- Region morphology (Section 3.6.1)

The corresponding example program is:
.../machine_vision_book/3d_plane_reconstruction_with_stereo/
3d_plane_reconstruction_with_stereo.dev

To be able to reconstruct the 3D surface of the object, we must calibrate the stereo system. For this, we acquire stereo image pairs of the calibration target. Figure 4.47 shows two stereo image pairs of a calibration target. In Figures 4.47(a) and (c), the

4.10 3D Plane Reconstruction with Stereo

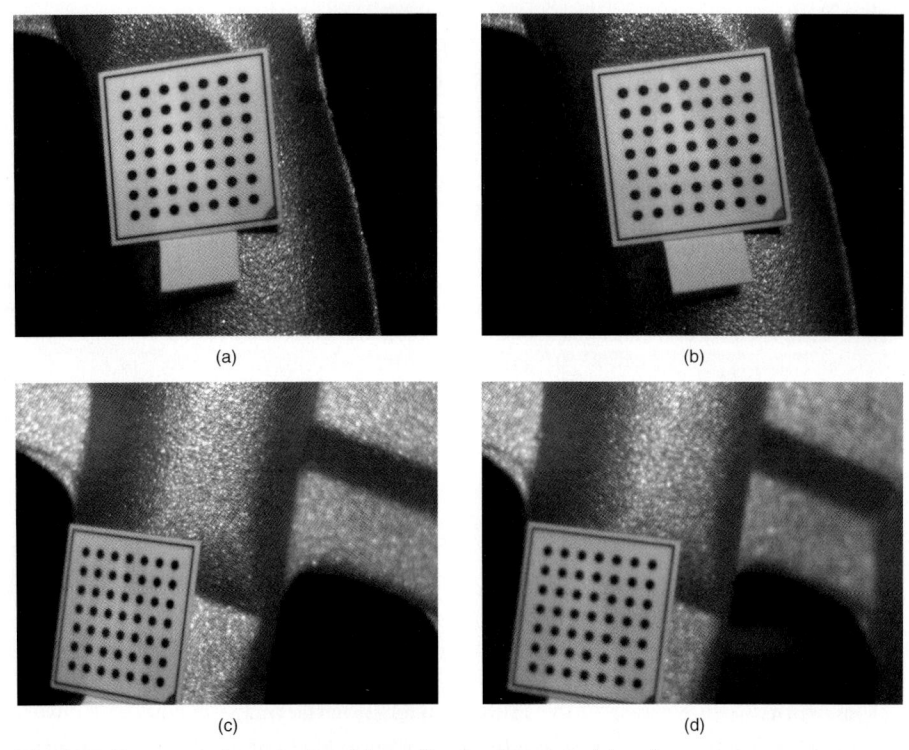

Fig. 4.47 Two stereo image pairs of the calibration target. In (a) and (c), the images of the first camera are shown; while (b) and (d) show the respective images of the second camera.

images of the first camera are shown, while Figures 4.47(b) and (d) show the respective images of the second camera.

First, we specify the used calibration target, define initial values for the interior orientation of both cameras, and initialize some arrays that are used to collect the image coordinates of the calibration marks as well as the initial poses of the calibration targets:

```
CaltabName := 'caltab_30mm.descr'
StartCamParam1 := [0.025, 0, 7.5e-6, 7.5e-6, Width1/2.0, Height1/2.0,
                   Width1,Height1]
StartCamParam2 := [0.025, 0, 7.5e-6, 7.5e-6, Width2/2.0, Height2/2.0,
                   Width2,Height2]
Rows1 := []
Cols1 := []
StartPoses1 := []
Rows2 := []
Cols2 := []
StartPoses2 := []
```

Both cameras have a focal length of approximately 25 mm and the pixel size is 7.5 μm. The radial distortion is set to zero and the principal point is assumed to be in the center of the image. These values are just initial values. They will be improved during the calibration process.

Then, within a loop over all available calibration image pairs, the subpixel-accurate positions of the calibration marks as well as initial estimates for the poses of the calibration target with respect to the two cameras are determined:

```
find_caltab (Image1, Caltab, CaltabName, 5, 85, 10)
find_marks_and_pose (Image1, Caltab, CaltabName, StartCamParam1,
                    128, 10, 18, 0.5, 15, 100,
                    RCoord, CCoord, StartPose)
Rows1 := [Rows1,RCoord]
Cols1 := [Cols1,CCoord]
StartPoses1 := [StartPoses1,StartPose]
find_caltab (Image2, Caltab, CaltabName, 5, 85, 10)
find_marks_and_pose (Image2, Caltab, CaltabName, StartCamParam2,
                    128, 10, 18, 0.5, 15, 100,
                    RCoord, CCoord, StartPose)
Rows2 := [Rows2,RCoord]
Cols2 := [Cols2,CCoord]
StartPoses2 := [StartPoses2,StartPose]
```

The approximate position of the calibration target in the image is determined with the operator `find_caltab`, which returns a region that contains the calibration target. Within this region, the calibration marks are extracted. The initial poses are determined based on the calibration marks and the initial values for the interior orientations.

Now, we determine the exact interior and exterior orientations as well as the relative pose of the two cameras:

```
caltab_points (CaltabName, X, Y, Z)
binocular_calibration (X, Y, Z, Rows1, Cols1, Rows2, Cols2,
                      StartCamParam1, StartCamParam2,
                      StartPoses1, StartPoses2, 'all',
                      CamParam1, CamParam2, NFinalPose1, NFinalPose2,
                      RelPose, Errors)
```

First, the world coordinates of the calibration marks are read from the description file of the used calibration target. Then, the calibration is carried out with the operator `binocular_calibration`.

For the calibrated stereo system, we determine two rectification maps that are used to rectify the stereo image pairs to the epipolar standard geometry:

```
gen_binocular_rectification_map (Map1, Map2, CamParam1, CamParam2,
                                RelPose, 1, 'geometric', 'bilinear',
                                CamParamRect1, CamParamRect2,
                                CamPoseRect1, CamPoseRect2,
                                RelPoseRect)
```

4.10 3D Plane Reconstruction with Stereo

The rectification maps define a mapping from the original stereo images to the rectified stereo images in which corresponding points, i.e., points that belong to the same object point, have identical row coordinates.

All steps described so far are performed offline. In the online phase, we use the calibrated stereo system to reconstruct the 3D surface of the intake manifold. For this, we need a stereo image pair of the object. Figure 4.48 shows one stereo image pair of an intake manifold acquired with the calibrated stereo system.

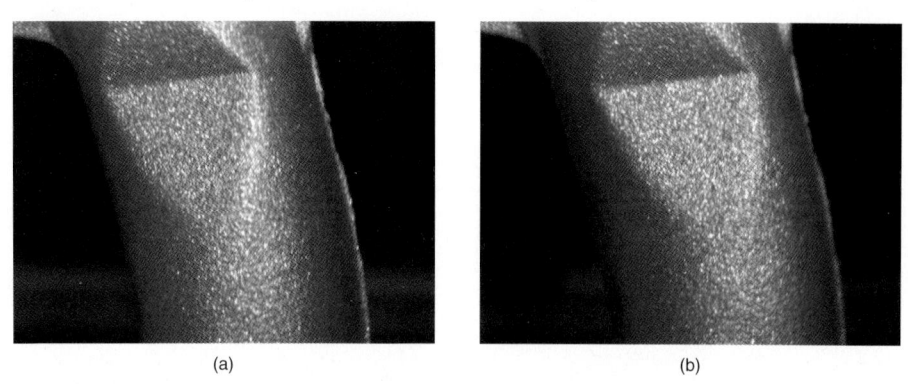

Fig. 4.48 Stereo image pair of an intake manifold acquired with the stereo system. In (a), the image taken by the first camera is shown, while (b) displays the image from the second camera.

First, the images are rectified to the epipolar standard geometry with the rectification maps created above:

```
map_image (Image1, Map1, ImageRect1)
map_image (Image2, Map2, ImageRect2)
```

The rectified images of the two cameras are shown in Figure 4.49.

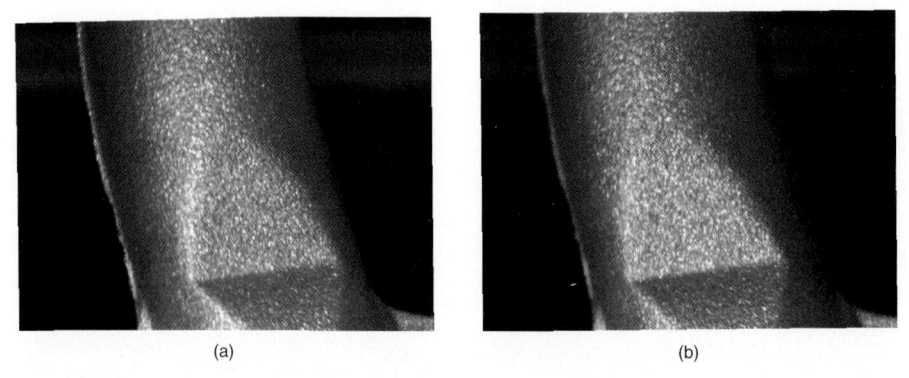

Fig. 4.49 The images of the first (a) and second (b) camera rectified to the epipolar standard geometry.

Now, we determine the distance of each point on the object surface from the stereo system:

```
binocular_distance (ImageRect1, ImageRect2, Distance, Score,
            CamParamRect1, CamParamRect2, RelPoseRect,
            'ncc', 21, 21, 5, MinDisparity, MaxDisparity,
            4, 0.1, 'left_right_check', 'interpolation')
```

The distance of the object surface from the stereo system is shown in Figure 4.50(a). Figure 4.50(b) shows the quality of the matches between the two rectified images. Bright values indicate good matches.

Fig. 4.50 (a) The distance of the object surface from the stereo system. Bright values indicate a larger distance than dark values. For better visibility, $d_{r,c}^{1/2}$ is displayed. (b) The quality of the match between the two rectified images. Bright values indicate good matches.

Now, we determine the two planes that form the indentation. First we determine the direction of the gradients in the distance image. Then, we segment the image into regions that have homogeneous gradient directions. Finally, we test how well these regions represent planes. We select the best two regions and determine the normal vectors of the respective planes.

To determine the gradient direction of the distance image, the following operations are used:

```
get_domain (Distance, Domain)
min_max_gray (Domain, Distance, 0, Min, Max, Range)
scale_image (Distance, DistanceScaled,
            pow(2,16)/Range, -Min*pow(2,16)/Range)
convert_image_type (DistanceScaled, DistanceUInt2, 'uint2')
edges_image (DistanceUInt2, ImaAmp, ImaDir, 'canny',
            1.5, 'none', 20, 40)
```

Because the distance image is a real-valued image and this data type is not supported by the edge extraction operator, we have to convert the image such that the distance values are represented by integer values. To preserve the accuracy of the distance values, the distance image is converted into an image with a gray value depth of 16 bits. The original (real) gray values are scaled such that they fully exploit the gray value range of the 16-bit image. Then, the gradient directions are determined with the operator edges_image. In the resulting direction image, regions that have homogeneous directions are determined.

In the direction image, the gradient directions are stored in 2-degree steps, i.e., a direction of x degrees with respect to the horizontal axis is stored as $x/2$ in the direction image. Points with edge amplitude 0 are assigned the edge direction 255, which indicates an undefined direction. Hence, only the gray values in the range $[0, 179]$ are of interest for the following analysis. Because we search for planes, there must exist regions in the image that have a homogeneous gradient direction. For this reason, the histogram of the direction image must contain peaks, which represent the mean directions of the regions with homogeneous gradient direction. The histogram is obtained with the following operation:

```
gray_histo (Domain, ImaDir, AbsoluteHisto, RelativeHisto)
```

Figure 4.51(a) shows the histogram of the direction image. It has four major peaks, one of them at the wrap-around at 360 degrees. Two of the peaks belong to the planes we want to extract. The two other peaks represent areas on the cylindrical surface of the object. To illustrate this, Figure 4.51(b) displays an image of the intake manifold along with the mean gradient directions in the regions that correspond to the four major peaks of the histogram of the direction image.

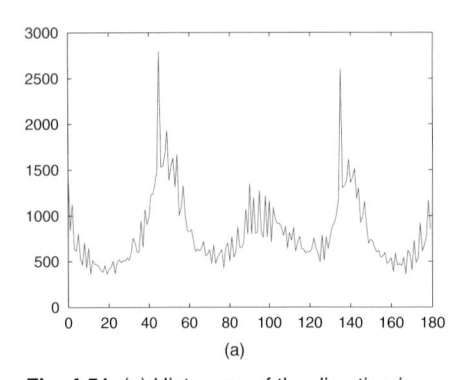

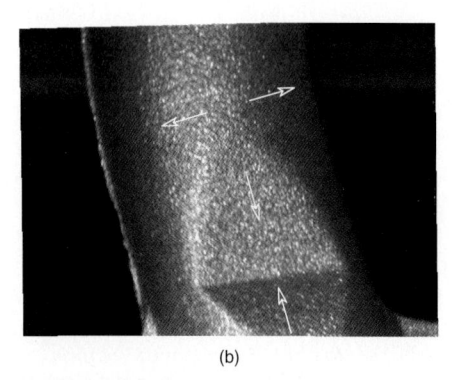

Fig. 4.51 (a) Histogram of the direction image. Note that the gradient directions are stored in 2-degree steps in the direction image. (b) Image of the intake manifold along with four arrows that indicate the mean gradient direction in the regions that correspond to the four major peaks of the histogram displayed in (a).

The appropriate thresholds for the segmentation of the image into regions with homogeneous gradient directions are the minima in the histogram of the direction image. To determine the minima in a stable manner, the histogram must be smoothed. Because of the cyclic nature of the direction image, we must take care to ensure a correct smoothing around the position 0 degrees. This is done by creating a function that consists of two consecutive copies of the histogram. Now we smooth this function. From the local minima of the smoothed function, we have to select a sequence that covers the range of 360 degrees, while we should not use the first or the last minimum because their positions might be disturbed if they lie close to 0 degrees or 360 degrees, respectively.

```
create_funct_1d_array ([AbsoluteHisto[0:179],AbsoluteHisto[0:179]],
                       HistoFunction)
smooth_funct_1d_gauss (HistoFunction, 4, SmoothedHistoFunction)
local_min_max_funct_1d (SmoothedHistoFunction, 'strict_min_max', 'true',
                       Minima, Maxima)
MinThreshShifted := Minima[1:(find(sgn(Minima-(Minima[1]+179)),1))[0]-1]
MinThresh := sort(fmod(MinThreshShifted,180))
```

We can now segment the distance image using the thresholds determined above. Because of the cyclic nature of the direction image from which the thresholds were determined, we must merge the region that starts at 0 degrees with the one that ends at 360 degrees if there is no minimum in the direction histogram at 0 degrees:

```
threshold (ImaDir, Region, [0,int(MinThresh)+1], [int(MinThresh),179])
Number := |Region|
FirstRegion := Region[1]
LastRegion := Region[Number]
union2 (FirstRegion, LastRegion, RegionUnion)
copy_obj (Region, ObjectsSelected, 2, Number-2)
Regions := [RegionUnion,ObjectsSelected]
```

The resulting regions all have a homogeneous gradient direction. To eliminate clutter, we select the largest connected component from each region and apply an opening operation to it. Then, we select the two regions that represent planes:

```
Number := |Regions|
for Index := 1 to Number by 1
  ObjectSelected := Regions[Index]
  connection (ObjectSelected, ConnectedRegions)
  area_center (ConnectedRegions, Area, _, _)
  LargestRegion := ConnectedRegions[(sort_index(-Area))[0]+1]
  opening_circle (LargestRegion, PlaneRegion, 3.5)
  area_center (PlaneRegion, _, Row, Col)
  fit_surface_first_order (PlaneRegion, Distance, 'regression', 5, 2,
                       Alpha, Beta, Gamma)
```

```
    gen_image_surface_first_order (ImageSurface, 'real',
                                   Alpha, Beta, Gamma, Row, Col,
                                   Width, Height)
    reduce_domain (ImageSurface, PlaneRegion, ImageReduced)
    sub_image (Distance, ImageReduced, DeviationFromPlane, 1, 0)
    intensity (PlaneRegion, DeviationFromPlane, Mean, Deviation)
endfor
```

Similar to the fitting of lines described in Section 3.8.1, we fit planes into the regions of the distance image that were described above. Then, we determine the deviation of the respective parts of the distance image from these planes. The two regions for which the distance image deviates least from the respective planes are selected.

We determine the normal vectors for these regions from the plane parameters and the pixel size of the rectified image:

```
Nx     := -Sy*Alpha
Ny     := -Sx*Beta
Nz     := Sx*Sy
Length := sqrt(Nx*Nx+Ny*Ny+Nz*Nz)
Nx     := Nx/Length
Ny     := Ny/Length
Nz     := Nz/Length
```

From the two normal vectors, the angle between the two planes can be determined easily.

Figure 4.52 shows the final results. In Figure 4.52(a), the rectified image of the intake manifold is shown along with the outlines of the two planes. Figure 4.52(b) shows a 3D plot of the surface of the intake manifold.

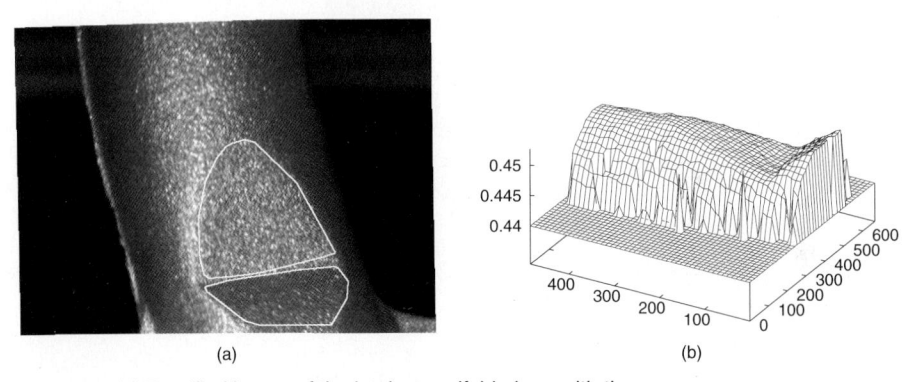

Fig. 4.52 (a) Rectified image of the intake manifold along with the outlines of the two planes. (b) A 3D plot of the surface of the intake manifold.

4.11
Pose Verification of Resistors

Typically, electronic components are mounted on a printed circuit board (PCB) by a pick-and-place machine. Before the completed boards are tested, they are visually inspected for missing or misaligned components. Furthermore, in some applications it is necessary to verify whether the correct type of electronic component has been mounted at the intended place.

In this application, we verify the pose and the type of different resistors that have been mounted on a PCB. First, the pose of the resistor on the board is determined and missing resistors are detected. Then, the type of the resistor (if present) is extracted and compared to the known reference type.

The algorithms used in this application are:

- Robust template matching (Section 3.11.5)
- Affine transformations (Section 3.3.1)
- Image transformations (Section 3.3.3)

The corresponding example program is:
.../machine_vision_book/resistor_verification/resistor_verification.dev

To prevent specular reflections in the images, we use a diffuse bright-field front light illumination for the acquisition of the PCBs (see Section 2.1.5). Figures 4.53(a) and (b) show the two types of resistors that we want to verify: $33\,\Omega$ and $1.1\,\Omega$, respectively.

Fig. 4.53 The two different types of resistors that must be verified: (a) $33\,\Omega$ and (b) $1.1\,\Omega$.

The first task is to determine the pose of the resistor in the image. All resistors have a rectangular shape. However, the size of the resistors as well as the aspect ratio of their sides are not identical. Consequently, determining the pose of the resistor implies determining its position, its orientation, and two scaling factors, which represent the size as well as the aspect ratio of the resistor's sides. We determine the pose of the resistor by using HALCON's shape-based matching (see Section 3.11.5). First, we create an artificial template image of a generic resistor with average size and aspect ratio:

```
MeanModelHeight := 185
MeanModelWidth := 100
gen_image_const (Image, 'byte', Width, Height)
gen_rectangle2_contour_xld (Rectangle, Height/2, Width/2, 0,
                            MeanModelWidth/2.0, MeanModelHeight/2.0)
paint_xld (Rectangle, Image, ModelImageGeneric, 128)
```

For this, we generate a rectangular contour that represents the boundary of an average resistor. Then, the contour is painted into an empty image. Note that `paint_xld` paints the contour onto the background using anti-aliasing. Consequently, the gray value of a pixel depends on the fraction by which the pixel is covered by the rectangle. For example, if only half of the pixel is covered, the gray value is set to the mean gray value of the background and the rectangle. The resulting template image is shown in Figure 4.54(a). After this, a shape model is created from the template image by using the following operations:

```
AngleTol := rad(5)
ScaleTol := 0.1
create_aniso_shape_model (ModelImageGeneric, 3,
                          -AngleTol, 2.0*AngleTol, 'auto',
                          1.0-ScaleTol, 1.0+ScaleTol, 'auto',
                          1.0-ScaleTol, 1.0+ScaleTol, 'auto',
                          'auto', 'use_polarity', 'auto',
                          10, ModelIDGeneric)
```

In this application, it can be assumed that the PCB is aligned horizontally and that the resistors are mounted on the board with an angle tolerance of ±5°. Furthermore, the length of the resistor's sides may vary by ±10% with respect to the average values `MeanModelHeight` and `MeanModelWidth`. The shape model of the generic resistor is created accordingly by passing the corresponding tolerance values. The model can be used to determine the pose of the resistor in the online phase even if it appears rotated and anisotropically scaled.

The second task is to determine the type of the resistor. This task can be solved by means of the printed characters on top of the resistors. For this, we create two additional model representations, one for the print "330" on the 33 Ω resistor and one for the print "1R1" on the 1.1 Ω resistor. In the online phase, the best matching model is assumed to represent the present resistor type. If none of the two models can be found in the online phase, the resistor type is set to "unknown." In the following, the model creation is described for the 33 Ω resistor. The model creation for the 1.1 Ω resistor is performed in the same manner.

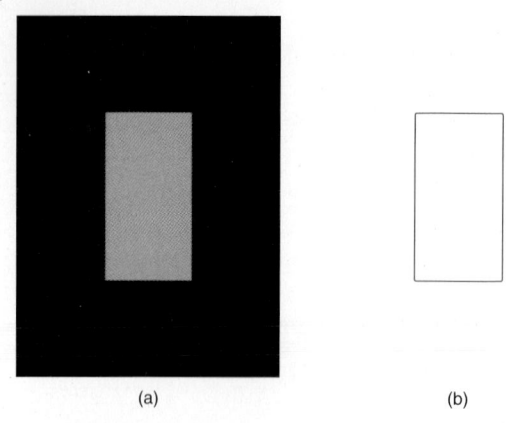

Fig. 4.54 (a) Artificial template image of a generic resistor. (b) Contour representation of the model that is used for matching.

The model is created from the print on the resistor shown in Figure 4.53(a). To ease the future use of new resistor types, the model generation process is done automatically. First, the pose of the resistor in the model image is determined by using the generic resistor model generated above:

```
find_aniso_shape_model (Image, ModelIDGeneric,
                       -AngleTol, 2.0*AngleTol,
                       1.0-ScaleTol, 1.0+ScaleTol,
                       1.0-ScaleTol, 1.0+ScaleTol,
                       0.7, 1, 0.5, 'least_squares', 0, 0.7,
                       Row, Column, Angle, ScaleR, ScaleC, Score)
```

The resulting pose parameters refer to the reference point of the generic model. The reference point is the center of gravity of the domain of the model image. Since the domain of the model image comprises the whole image, the reference point is simply the center of the model image, and hence the center of the resistor. Consequently, the pose of the generic model can be used to generate a rectangular region of interest that contains the print on the resistor. For this, we assume that the print is contained within a rectangle that has the same position and orientation as the resistor but only half of its side lengths:

```
PrintFraction := 0.5
gen_rectangle2 (Rectangle, Row, Column, Angle,
                PrintFraction*ScaleC*0.5*MeanModelWidth,
                PrintFraction*ScaleR*0.5*MeanModelHeight)
reduce_domain (Image, Rectangle, ModelImage)
create_shape_model (ModelImage, 'auto', rad(-90), rad(360),
                    'auto', 'auto', 'use_polarity',
                    ['auto_contrast_hyst',20], 'auto', ModelID330)
```

The domain of the model image is reduced to the generated rectangle. Figure 4.55(a) shows the model image, the border of the found resistor, and the generated rectangle. Because the size and the aspect ratio of the print are constant in all images, this time it is sufficient to create a model that allows only rotated instances of the print to be found. Note, however, that the model is created within the full angle range of 360°. The reason for this is that the resistors might be mounted not only in the reference orientation shown in Figure 4.55 but also rotated by 180°. The contour representation of the resulting shape model is shown in Figure 4.55(b). The model image and the resulting shape model of the print "1R1" are shown in Figures 4.55(c) and (d), respectively. Note that the print on the resistor in both model images should appear at an average orientation. If this cannot be guaranteed, either the model image must be rectified before creating the model, or the angle range that is used to search the print in the online phase must be adapted to the orientation difference of the print with respect to the resistor.

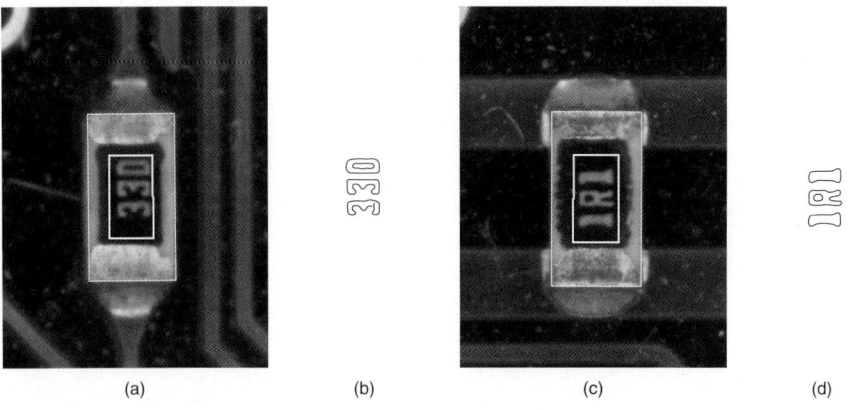

Fig. 4.55 (a), (c) Images that are used to create the models of the prints "330" and "1R1." The found resistor is indicated by the outer white rectangle. The region of interest for the model creation is indicated by the inner white rectangle. (b), (d) Contour representation of the created models.

All the steps described so far can be performed offline. In the online phase, we use the generated models to verify the pose and the type of the resistors. To verify the pose, the generic resistor model is used. It is searched in the same pose range in which it was created:

```
find_aniso_shape_model (ImageOnline, ModelIDGeneric,
                       -AngleTol, 2.0*AngleTol,
                       1.0-ScaleTol, 1.0+ScaleTol,
                       1.0-ScaleTol, 1.0+ScaleTol,
                       0.7, 1, 0.5, 'least_squares', 0, 0.7,
                       Row, Column, Angle, ScaleR, ScaleC, Score)
```

If the model cannot be found, we assume that the resistor is missing on the board. Otherwise, we proceed with the type verification by using the two models that represent the prints on the resistors. Since we know the pose of the resistor, we can restrict the search for the print to an appropriate image region. This region must contain the reference point of the model. Because the position of the print does not vary much with respect to the outline of the resistor, a relatively small region is sufficient. In this example, the region is set to a rectangle the side lengths of which are a quarter of the resistor's side lengths:

```
PrintRefFraction := 0.25
gen_rectangle2 (Rectangle, Row, Column, Angle,
                PrintRefFraction*ScaleC*0.5*MeanModelWidth,
                PrintRefFraction*ScaleR*0.5*MeanModelHeight)
reduce_domain (ImageOnline, Rectangle, ImageReduced)
```

Finally, the models of the prints are searched in the reduced image domain:

```
find_shape_models (ImageReduced,
                   [ModelID330,ModelID330,ModelID1R1,ModelID1R1],
                   [0,rad(180),0,rad(180)]-AngleTol, 2*AngleTol,
                   0.5, 1, 0.5, 'least_squares', 0, 0.9,
                   PrintRow, PrintColumn, PrintAngle,
                   PrintScore, PrintModel)
```

As mentioned above, the resistors may be mounted in the two orientations $0°$ and $180°$ (plus the tolerances). Consequently, we do not need to search the prints in the full range of $360°$. In contrast, each model is only searched in the two angle ranges $[-\text{AngleTol}, \text{AngleTol}]$ and $[\pi-\text{AngleTol}, \pi+\text{AngleTol}]$. With find_shape_models multiple models can be searched simultaneously. Therefore, altogether four model handles are passed to find_shape_models: each of the two models is passed twice, once for each angle range. Alternatively, find_shape_model could be called four times, once for each model and each angle range. However, this would be computationally more expensive because some computations must be performed multiple times. Furthermore, the best match would have to be determined in an additional post-processing step. The index of the model that yielded the best match is returned in PrintModel. This index refers to the tuple of shape model handles that was passed as input, and hence is in the range $[0\ldots 3]$. If no match was found, we assume that the resistor type is "unknown." Otherwise, the resistor type can be computed based on the returned index:

```
ResistorTypes := ['330','1R1']
Model        := PrintModel/2
ResistorType := ResistorTypes[Model]
```

Finally, if the determined resistor type does not correspond to the expected one, an appropriate action can be triggered. In Figures 4.56(a) and (b), for each of the two resistor types, an example of the verification result is shown.

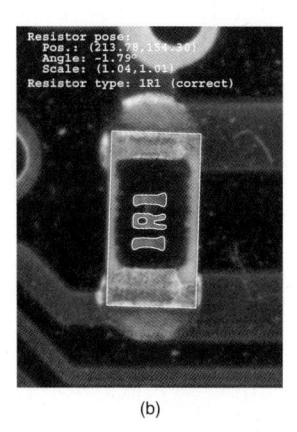

Fig. 4.56 Verification result for an example image that shows a 33 Ω (a) and a 1.1 Ω (b) resistor. The border of the found resistor is overlaid as a white rectangle. The pose of the found print is indicated by the overlaid white contours. The pose of the resistor as well as its type are displayed in the upper left corner of the image.

Exercises

1. In some applications, the resistors additionally appear in an orientation of ±90°. Modify the program appropriately so that resistors with orientations 0°, 90°, 180°, and 270° can be verified. Tip: To avoid problems with the wrap-around at 360°, create four models for each print.

2. Alternatively to the shape-based matching, the normalized cross-correlation could be used for template matching (see Section 3.11.1). Adapt the above program so that the pose and type of the resistor are determined with the normalized cross-correlation. Tip: The generic model should be extended to include more details of the resistor. It should contain the dark rectangular center part, the gray border regions at the sides, and the bright regions of the contacts at the upper and lower parts of the resistor. Note that the normalized cross-correlation is more robust against small deformations than the shape-based matching. Therefore, no (anisotropic) scaling of the resistor model needs to be taken into account.

3. In the above application, the type of the resistor is determined by using template matching. Alternatively, the print could be segmented by using region morphology (see Section 3.6.1) and read by using optical character recognition (see Section 3.12). Modify the above program accordingly.

4.12
Classification of Non-Woven Fabrics

Through classification, a sample can be assigned to one of a set of predefined categories. The goal of texture classification is to assign an unknown image to one of a set of known texture classes.

In this application, we classify images of different types of non-woven fabrics using a neural net classifier in the form of a multilayer perceptron (MLP).

The algorithms used in this application are:

- 2D edge extraction (Section 3.7.3)
- Gray value features (Section 3.5.2)
- Classification (Section 3.12.3)

The corresponding example program is:
.../machine_vision_book/classification_of_nonwoven_fabrics/
classification_of_nonwoven_fabrics.dev

The process of texture classification involves two phases: the training phase and the recognition phase. In the training phase, a classifier is created for the texture content of each texture class that is present in the training data. The training data consists of images with known class labels. The texture content of the images is captured by a set of texture features. These features characterize the texture properties of the images.

It is advantageous to separate the training data into a set of training images and a set of independent test images. The former are used to train the classifier as well as to carry out a first test of the classifier by reclassifying the training images. The latter are used to test the classifier and to improve it by adding those test images to the set of training images that could not be classified correctly. In real applications, to rate the performance of the final classifier, another set of independent test images must be available.

In the recognition phase, the texture content of the images is described by the same texture features that were used in the training phase. Then, the classifier assigns each image to the best-matching class based on the texture features.

Figure 4.57 shows three samples out of a set of 22 different types of non-woven fabrics. The images are taken with diffuse bright-field front light illumination. In the current application, only six training images and four independent test images are available for each of the 22 classes. Note that for real applications much more training data must be available to achieve a reliable classifier. It is not unusual that several thousands of training samples are used for each class.

First, we initialize the classifier:

```
NumHidden := 10
create_class_mlp (NumFeatures, NumHidden, NumClasses, 'softmax',
                 'normalization', -1, 42, MLPHandle)
```

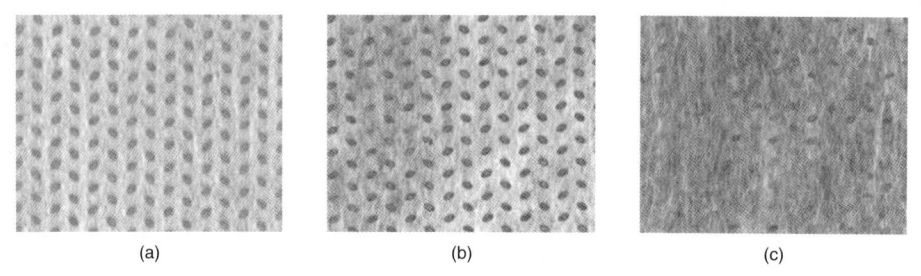

Fig. 4.57 Three samples out of a set of 22 different types of non-woven fabrics.

The number of hidden neurons defines the size of the neural net. Generally, it should be in the range of the number of features and the number of classes. If too few hidden neurons are chosen, the reclassification of the training images will give a large number of misclassified images. If the number of hidden neurons is chosen too large, the MLP may overfit the training data, which typically leads to bad generalization properties, i.e., the MLP learns the training data very well, but does not return very good results on unknown data.

Now we add the texture features of the training images to the classifier. We use features that characterize the number and strength of edges in the image as well as features that measure the distribution of the gray values across the image:

```
Features := []
gen_gauss_pyramid (Image, ImagePyramid, 'constant', 0.5)
for Index := 1 to 3 by 1
  ImageRR := ImagePyramid[Index]
  sobel_amp (ImageRR, EdgeAmplitude, 'sum_abs', 3)
  gray_histo_abs (EdgeAmplitude, EdgeAmplitude, 8, AbsoluteHisto)
  SobelFeatures := real(AbsoluteHisto)/sum(AbsoluteHisto)
  Features := [Features,SobelFeatures]
endfor
```

First, edges are extracted from the image. Then, the relative histogram of the edge amplitudes is derived. These features are calculated for the original image as well as for resolution-reduced copies of the image.

The distribution of the gray values across the image is measured by the entropy and the anisotropy, again calculated for the original image as well as for resolution-reduced copies of the image:

```
for Index := 1 to 3 by 1
  ImageRR := ImagePyramid[Index]
  entropy_gray (ImageRR, ImageRR, Entropy, Anisotropy)
  Features := [Features,Entropy,Anisotropy]
endfor
```

All features are calculated within the procedure gen_features. It is called for each training image, and the resulting features are added to the classifier:

```
gen_features (Image, Features)
add_sample_class_mlp (MLPHandle, Features, Class)
```

The internal weights of the classifier are determined based on the training data:

```
train_class_mlp (MLPHandle, 200, 0.1, 0.001, Error, ErrorLog)
```

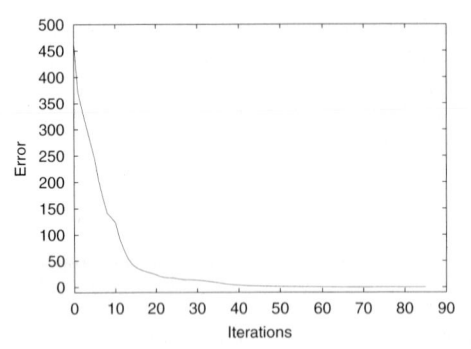

Fig. 4.58 The progression of the error during the training of the classifier. Note that initially the error drops off steeply while leveling out very flatly at the end.

This is an iterative process that terminates when the internal weights as well as the error of the MLP on the training samples become stable. To judge the training phase, the progression of the error can be plotted as a function over the number of iterations (see Figure 4.58). Initially, the error should drop off steeply while leveling out very flatly at the end.

With the classifier created above, we can reclassify all the training images. If all the images can be classified correctly, we can test the classifier by classifying the independent test images. Typically, some of the test images are classified incorrectly. Figure 4.59 shows two pairs of images. In the upper row, misclassified test images are displayed, while the lower row shows one image out of the class to which the image displayed above has been assigned erroneously.

If the number of misclassifications is very large, either the number of training samples must be increased significantly, or the number of hidden neurons must be reduced. Furthermore, it should be checked whether the used features are suitable for separating the different classes. If the number of misclassifications is moderate, the classifier is already reasonably good. The wrongly classified images should be added to the training images and the classifier should be trained anew, using the extended training data. This creates an improved classifier.

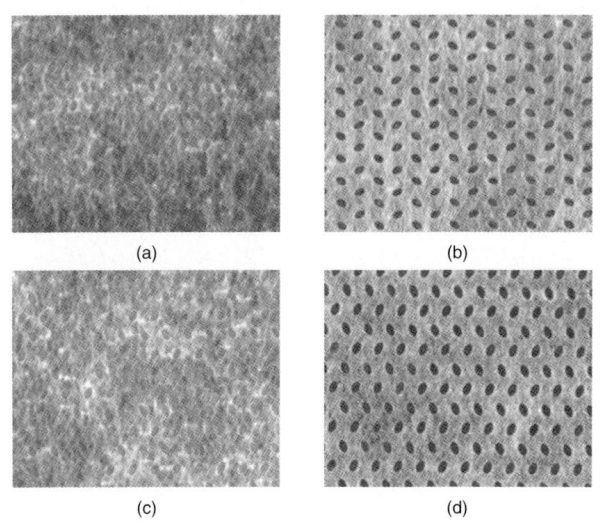

Fig. 4.59 Wrongly classified images. In the upper row, the misclassified images are displayed, while the lower row shows one image out of the class to which the image displayed above has been assigned erroneously.

If much training data is available, the classification of independent test images and the adding of the misclassified test images to the training images can be repeated several times. This procedure has two main advantages over feeding all the available training data into the classifier in one step. First, the training is faster because fewer training images are used to train the classifier. Second, the performance of the classifier can be judged already in an early stage of the training phase.

If the training phase is completed, the classifier can be used to classify images of unknown content:

```
gen_features (Image, Features)
classify_class_mlp (MLPHandleImproved, Features, 1,
                    ClassifiedClass, Confidence)
```

Figure 4.60 shows one sample out of each of the 22 classes of non-woven fabrics. The improved classifier is able to classify all images correctly.

Exercise

1. Explore the behavior of the classifier for different texture features and different sizes of the neural net.

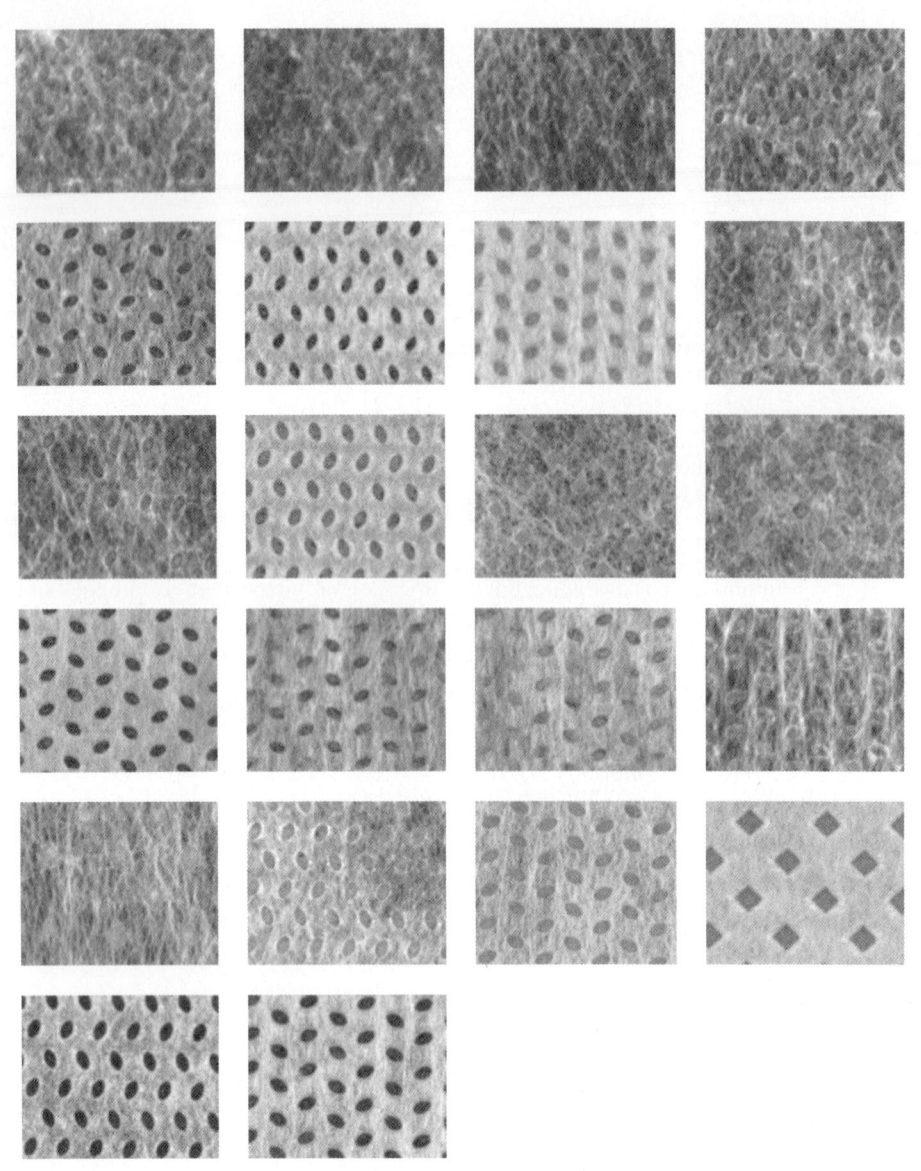

Fig. 4.60 One sample of each of the 22 classes of non-woven fabrics.

References

1 FRAUNHOFER ALLIANZ VISION. *Guideline for Industrial Image Processing*. Fraunhofer-Gesellschaft zur Förderung der angewandten Forschung, Erlangen, 2003.

2 D. CARO. *Automation Network Selection*. ISA – The Instrumentation, Systems, and Automation Society, Research Triangle Park, NC, 2003.

3 J. BERGE. *Fieldbuses for Process Control: Engineering, Operation, and Maintenance*. ISA – The Instrumentation, Systems, and Automation Society, Research Triangle Park, NC, 2004.

4 N. P. MAHALIK, EDITOR. *Fieldbus Technology: Industrial Network Standards for Real-Time Distributed Control*. Springer-Verlag, Berlin, 2003.

5 DIN 5031 TEIL 7. *Strahlungsphysik im optischen Bereich und Lichttechnik; Benennung der Wellenlängenbereiche*, January 1984.

6 H.-C. LEE. *Introduction to Color Imaging Science*. Cambridge University Press, Cambridge, 2005.

7 M. PLANCK. Ueber das Gesetz der Energieverteilung im Normalspectrum. *Annalen der Physik*, 309(3): 553–563, 1901.

8 G. WYSZECKI, W. S. STILES. *Color Science: Concepts and Methods, Quantitative Data and Formulae*. John Wiley & Sons, New York, 2nd edition, 1982.

9 M. BORN, E. WOLF. *Principles of Optics*. Cambridge University Press, Cambridge, 7th edition, 1999.

10 K. LENHARDT. Optical systems in machine vision. In A. HORNBERG, EDITOR, *Handbook of Machine Vision*. Wiley-VCH, Weinheim, 2006.

11 ISO 517:1996. *Photography – Apertures and Related Properties Pertaining to Photographic Lenses – Designations and Measurements*, 1996.

12 V. N. MAHAJAN. *Optical Imaging and Aberrations – Part I: Ray Geometrical Optics*. SPIE Press, Bellingham, WA, 1998.

13 ISO 9039:1994. *Optics and Optical Instruments – Quality Evaluation of Optical Systems – Determination of Distortion*, 1994.

14 ISO 15705:2002. *Optics and Optical Instruments – Quality Evaluation of Optical Systems – Assessing the Image Quality Degradation due to Chromatic Aberrations*, 2002.

15 A. EL GAMAL, H. ELTOUKHY. CMOS image sensors. *IEEE Circuits and Devices Magazine*, 21(3): 6–20, May/June 2005.

16 G. C. HOLST. *CCD Arrays, Cameras, and Displays*. SPIE Press, Bellingham, WA, 2nd edition, 1998.

17 O. YADID-PECHT, R. ETIENNE-CUMMINGS, EDITORS. *CMOS Imagers: From Phototransduction to Image Processing*. Kluwer Academic, Dordrecht, 2004.

18 M. WÄNY, G. P. ISRAEL. CMOS image sensor with NMOS-only global shutter and enhanced responsivity. *IEEE Transactions on Electron Devices*, 50(1): 57–62, January 2003.

19 S. KAVADIAS, B. DIERICKX, D. SCHEFFER, A. ALAERTS, D. UWAERTS, J. BOGAERTS. A logarithmic response CMOS image sensor with on-chip calibration. *IEEE Journal of Solid-State Circuits*, 35(8): 1146–1152, August 2000.

20 M. LOOSE, K. MEIER, J. SCHEMMEL. A self-calibrating single-chip CMOS camera

with logarithmic response. *IEEE Journal of Solid-State Circuits*, 36(4): 586–596, April 2001.

21 M. WILLEMIN, N. BLANC, G. K. LANG, S. LAUXTERMANN, P. SCHWIDER, P. SEITZ, M. WÄNY. Optical characterization methods for solid-state image sensors. *Optics and Lasers in Engineering*, 36(2): 185–194, 2001.

22 A. HAMASAKI, M. TERAUCHI, K. HORII. Novel operation scheme for realizing combined linear-logarithmic response in photodiode-type active pixel sensor cells. *Japanese Journal of Applied Physics*, 45(4B): 3326–3329, 2006.

23 G. STORM, R. HENDERSON, J. E. D. HURWITZ, D. RENSHAW, K. FINDLATER, M. PURCELL. Extended dynamic range from a combined linear-logarithmic CMOS image sensor. *IEEE Journal of Solid-State Circuits*, 41(9): 2095–2106, September 2006.

24 B. E. BAYER. Color imaging array. US Patent, 3 971 065, July 1976.

25 D. ALLEYSSON, S. SÜSSTRUNK, J. HÉRAULT. Linear demosaicing inspired by the human visual system. *IEEE Transactions on Image Processing*, 14(4): 439–449, April 2005.

26 K. HIRAKAWA, T. W. PARKS. Adaptive homogeneity-directed demosaicing algorithm. *IEEE Transactions on Image Processing*, 14(3): 360–369, March 2005.

27 EMVA. *Standard 1288: Standard for Characterization and Presentation of Specification Data for Image Sensors and Cameras*, Release A1.03, February 2007.

28 F. DIERKS. *Sensitivity and Image Quality of Digital Cameras*. Technical report, Basler AG, October 2004.

29 ISO 15739:2003. *Photography – Electronic Still-Picture Imaging – Noise Measurements*, 2003.

30 ITU-R BT.470-6. *Conventional Television Systems*, 1998.

31 AIA. *Camera Link: Specifications of the Camera Link Interface Standard for Digital Cameras and Frame Grabbers*, Version 1.1, January 2004.

32 IEEE STD 1394-1995. *IEEE Standard for a High Performance Serial Bus*, December 1995.

33 IEEE STD 1394A-2000. *IEEE Standard for a High Performance Serial Bus – Amendment 1*, March 2000.

34 IEEE STD 1394B-2002. *IEEE Standard for a High Performance Serial Bus – Amendment 2*, December 2002.

35 IEC 61883. *Consumer Audio/Video Equipment – Digital Interface. Part 1: General*, 1998, 2nd edition, 2003; *Part 2: SD-DVCR Data Transmission*, 1998, 2nd edition, 2004; *Part 3: HD-DVCR Data Transmission*, 1998, 2nd edition, 2004; *Part 4: MPEG2-TS Data Transmission*, 1998, 2nd edition, 2004; *Part 5: SDL-DVCR Data Transmission*, 1998, 2nd edition, 2004.

36 1394 TRADE ASSOCIATION. *IIDC 1394-Based Digital Camera Specification*, Version 1.31, February 2004.

37 USB IMPLEMENTERS FORUM. *Universal Serial Bus Specification*, Revision 2.0, April 2000.

38 USB IMPLEMENTERS FORUM. *Universal Serial Bus Device Class Definition for Video Media Transport Terminal*, Revision 1.1, June 2005.

39 R. M. METCALFE, D. R. BOGGS. Ethernet: Distributed packet switching for local computer networks. *Communications of the ACM*, 19(7): 395–404, 1976.

40 IEEE STD 802.3-2005. *IEEE Standard for Information Technology – Telecommunications and Information Exchange Between Systems – Local and Metropolitan Area Networks – Specific Requirements: Part 3: Carrier Sense Multiple Access with Collision Detection (CSMA/CD) Access Method and Physical Layer Specifications*, December 2005.

41 J. POSTEL. *Internet Protocol*. RFC 791, September 1981.

42 J. POSTEL. *Transmission Control Protocol*. RFC 761, January 1980.

43 J. POSTEL. *User Datagram Protocol*. RFC 768, August 1980.

44 R. T. FIELDING, J. GETTYS, J. C. MOGUL, H. FRYSTYK NIELSEN, L. MASINTER, P. J. LEACH, T. BERNERS-LEE. *Hypertext Transfer Protocol – HTTP/1.1*. RFC 2616, June 1999.

45 AIA. *GigE Vision: Camera Interface Standard for Machine Vision*, Version 1.0, May 2006.

46 R. DROMS. *Dynamic Host Configuration Protocol*. RFC 2131, March 1997.

47 S. Cheshire, B. Aboba, E. Guttman. *Dynamic Configuration of IPv4 Link-Local Addresses.* RFC 3927, May 2005.

48 EMVA. *GenICam Standard: Generic Interface for Cameras, Version 1.0,* June 2006.

49 D. M. McKeown Jr., S. D. Cochran, S. J. Ford, J. C. McGlone, J. A. Shufelt, D. A. Yocum. Fusion of HYDICE hyperspectral data with panchromatic imagery for cartographic feature extraction. *IEEE Transactions on Geoscience and Remote Sensing,* 3: 1261–1277, May 1999.

50 ISO 14524:1999. *Photography – Electronic Still-Picture Cameras – Methods for Measuring Opto-Electronic Conversion Functions (OECFs),* 1999.

51 S. Mann, R. Mann. Quantigraphic imaging: Estimating the camera response and exposures from differently exposed images. In *Computer Vision and Pattern Recognition,* Vol. I, pp. 842–849, 2001.

52 T. Mitsunaga, S. K. Nayar. Radiometric self calibration. In *Computer Vision and Pattern Recognition,* Vol. I, pp. 374–380, 1999.

53 A. Papoulis. *Probability, Random Variables, and Stochastic Processes.* McGraw-Hill, New York, 3rd edition, 1991.

54 R. Deriche. Fast algorithms for low-level vision. *IEEE Transactions on Pattern Analysis and Machine Intelligence,* 12(1): 78–87, January 1990.

55 T. Lindeberg. *Scale-Space Theory in Computer Vision.* Kluwer Academic, Dordrecht, 1994.

56 A. P. Witkin. Scale-space filtering. In *Eighth International Joint Conference on Artificial Intelligence,* Vol. 2, pp. 1019–1022, 1983.

57 J. Babaud, A. P. Witkin, M. Baudin, R. O. Duda. Uniqueness of the Gaussian kernel for scale-space filtering. *IEEE Transactions on Pattern Analysis and Machine Intelligence,* 8(1): 26–33, January 1986.

58 L. M. J. Florack, B. M. ter Haar Romeny, J. J. Koenderink, M. A. Viergever. Scale and the differential structure of images. *Image and Vision Computing,* 10(6): 376–388, July 1992.

59 R. Deriche. *Recursively Implementing the Gaussian and Its Derivatives.* Rapport de Recherche 1893, INRIA, Sophia Antipolis, April 1993.

60 I. T. Young, L. J. van Vliet. Recursive implementation of the Gaussian filter. *Signal Processing,* 44: 139–151, 1995.

61 T. S. Huang, G. J. Yang, G. Y. Tang. A fast two-dimensional median filtering algorithm. *IEEE Transactions on Acoustics, Speech, and Signal Processing,* 27(1): 13–18, February 1979.

62 M. Van Droogenbroeck, H. Talbot. Fast computation of morphological operations with arbitrary structuring elements. *Pattern Recognition Letters,* 17(14): 1451–1460, 1996.

63 E. O. Brigham. *The Fast Fourier Transform and Its Applications.* Prentice-Hall, Upper Saddle River, NJ, 1988.

64 W. H. Press, S. A. Teukolsky, W. T. Vetterling, B. P. Flannery. *Numerical Recipes in C: The Art of Scientific Computing.* Cambridge University Press, Cambridge, 2nd edition, 1992.

65 M. Frigo, S. G. Johnson. The design and implementation of FFTW3. *Proceedings of the IEEE,* 93(2): 216–231, February 2005.

66 R. Hartley, A. Zisserman. *Multiple View Geometry in Computer Vision.* Cambridge University Press, Cambridge, 2nd edition, 2003.

67 O. Faugeras, Q.-T. Luong. *The Geometry of Multiple Images: The Laws That Govern the Formation of Multiple Images of a Scene and Some of Their Applications.* MIT Press, Cambridge, MA, 2001.

68 R. M. Haralick, L. G. Shapiro. *Computer and Robot Vision,* Vol. I. Addison-Wesley, Reading, MA, 1992.

69 R. Jain, R. Kasturi, B. G. Schunck. *Machine Vision.* McGraw-Hill, New York, 1995.

70 R. Sedgewick. *Algorithms in C.* Addison-Wesley, Reading, MA, 1990.

71 M.-K. Hu. Visual pattern recognition by moment invariants. *IRE Transactions on Information Theory,* 8: 179–187, February 1962.

72 J. Flusser, T. Suk. Pattern recognition by affine moment invariants. *Pattern Recognition,* 26(1): 167–174, 1993.

73 A. G. Mamistvalov. n-dimensional moment invariants and conceptual mathematical theory of recognition *n*-dimensional solids. *IEEE Transactions on Pattern Analysis and Machine Intelligence*, 20(8): 819–831, August 1998.

74 G. TOUSSAINT. Solving geometric problems with the rotating calipers. In *Proceedings of IEEE MELECON '83*, Los Alamitos, CA, pp. A10.02/1–4. IEEE Press, New York, 1983.

75 E. WELZL. Smallest enclosing disks (balls and ellipsoids). In H. MAURER, EDITOR, *New Results and Trends in Computer Science*, Lecture Notes in Computer Science, Vol. 555, pp. 359–370. Springer-Verlag, Berlin, 1991.

76 M. DE BERG, M. VAN KREVELD, M. OVERMARS, O. SCHWARZKOPF. *Computational Geometry: Algorithms and Applications*. Springer-Verlag, Berlin, 2nd edition, 2000.

77 J. O'ROURKE. *Computational Geometry in C*. Cambridge University Press, Cambridge, 2nd edition, 1998.

78 J. M. MENDEL. Fuzzy logic systems for engineering: A tutorial. *Proceedings of the IEEE*, 83(3): 345–377, March 1995.

79 C. STEGER. *On the Calculation of Arbitrary Moments of Polygons*. Technical Report FGBV–96–05, Forschungsgruppe Bildverstehen (FG BV), Informatik IX, Technische Universität München, September 1996.

80 P. SOILLE. *Morphological Image Analysis*. Springer-Verlag, Berlin, 2nd edition, 2003.

81 D. S. BLOOMBERG. Implementation efficiency of binary morphology. In *International Symposium on Mathematical Morphology VI*, pp. 209–218, 2002.

82 L. LAM, S.-W. LEE, C. Y. SUEN. Thinning methodologies – a comprehensive survey. *IEEE Transactions on Pattern Analysis and Machine Intelligence*, 14(9): 869–885, September 1992.

83 U. ECKHARDT, G. MADERLECHNER. Invariant thinning. *International Journal of Pattern Recognition and Artificial Intelligence*, 7(5): 1115–1144, 1993.

84 G. BORGEFORS. Distance transformation in arbitrary dimensions. *Computer Vision, Graphics, and Image Processing*, 27: 321–345, 1984.

85 P.-E. DANIELSSON. Euclidean distance mapping. *Computer Graphics and Image Processing*, 14: 227–248, 1980.

86 M. MOGANTI, F. ERCAL, C. H. DAGLI, S. TSUNEKAWA. Automatic PCB inspection algorithms: A survey. *Computer Vision and Image Understanding*, 63(2): 287–313, March 1996.

87 J. GIL, R. KIMMEL. Efficient dilation, erosion, opening, and closing algorithms. *IEEE Transactions on Pattern Analysis and Machine Intelligence*, 24(12): 1606–1617, December 2002.

88 J. CANNY. A computational approach to edge detection. *IEEE Transactions on Pattern Analysis and Machine Intelligence*, 8(6): 679–698, June 1986.

89 R. DERICHE. Using Canny's criteria to derive a recursively implemented optimal edge detector. *International Journal of Computer Vision*, 1: 167–187, 1987.

90 C. STEGER. *Unbiased Extraction of Curvilinear Structures from 2D and 3D Images*. PhD thesis, Fakultät für Informatik, Technische Universität München. Herbert Utz Verlag, München, 1998.

91 S. ANDO. Consistent gradient operators. *IEEE Transactions on Pattern Analysis and Machine Intelligence*, 22(3): 252–265, March 2000.

92 S. LANSER, W. ECKSTEIN. A modification of Deriche's approach to edge detection. In *11th International Conference on Pattern Recognition*, Vol. III, pp. 633–637, 1992.

93 C. STEGER. Subpixel-precise extraction of lines and edges. In *International Archives of Photogrammetry and Remote Sensing*, Vol. XXXIII, part B3, pp. 141–156, 2000.

94 R. M. HARALICK, L. T. WATSON, T. J. LAFFEY. The topographic primal sketch. *International Journal of Robotics Research*, 2(1): 50–72, 1983.

95 R. M. HARALICK, L. G. SHAPIRO. *Computer and Robot Vision*, Vol. II. Addison-Wesley, Reading, MA, 1993.

96 ISO. *International Vocabulary of Basic and General Terms in Metrology*, 1993.

97 C. STEGER. Analytical and empirical performance evaluation of subpixel line and edge detection. In K. J. BOWYER, P. J. PHILLIPS, EDITORS, *Empirical Evaluation Methods in Computer Vision*, pp. 188–210, 1998.

98 V. BERZINS. Accuracy of Laplacian edge detectors. *Computer Vision, Graphics, and Image Processing*, 27: 195–210, 1984.

99 R. LENZ, D. FRITSCH. Accuracy of videometry with CCD sensors. *ISPRS Journal of Photogrammetry and Remote Sensing*, 45(2): 90–110, 1990.

100 S. LANSER. *Modellbasierte Lokalisation gestützt auf monokulare Videobilder*. PhD thesis, Forschungs- und Lehreinheit Informatik IX, Technische Universität München. Shaker Verlag, Aachen, 1997.

101 P. J. HUBER. *Robust Statistics*. John Wiley & Sons, New York, 1981.

102 F. MOSTELLER, J. W. TUKEY. *Data Analysis and Regression*. Addison-Wesley, Reading, MA, 1977.

103 M. A. FISCHLER, R. C. BOLLES. Random sample consensus: A paradigm for model fitting with applications to image analysis and automated cartography. *Communications of the ACM*, 24(6): 381–395, June 1981.

104 S. H. JOSEPH. Unbiased least squares fitting of circular arcs. *Computer Vision, Graphics, and Image Processing: Graphical Models and Image Processing*, 56(5): 424–432, 1994.

105 S. J. AHN, W. RAUH, H.-J. WARNECKE. Least-squares orthogonal distances fitting of circle, sphere, ellipse, hyperbola, and parabola. *Pattern Recognition*, 34(12): 2283–2303, 2001.

106 A. FITZGIBBON, M. PILU, R. B. FISHER. Direct least square fitting of ellipses. *IEEE Transactions on Pattern Analysis and Machine Intelligence*, 21(5): 476–480, May 1999.

107 S. LANSER, C. ZIERL, R. BEUTLHAUSER. Multibildkalibrierung einer CCD-Kamera. In G. SAGERER, S. POSCH, F. KUMMERT, EDITORS, *Mustererkennung, Informatik aktuell*, pp. 481–491. Springer-Verlag, Berlin, 1995.

108 J. HEIKKILÄ. Geometric camera calibration using circular control points. *IEEE Transactions on Pattern Analysis and Machine Intelligence*, 22(10): 1066–1077, October 2000.

109 P. L. ROSIN. Techniques for assessing polygonal approximations of curves. *IEEE Transactions on Pattern Analysis and Machine Intelligence*, 19(6): 659–666, June 1997.

110 P. L. ROSIN. Assessing the behaviour of polygonal approximation algorithms. *Pattern Recognition*, 36(2): 508–518, 2003.

111 U. RAMER. An iterative procedure for the polygonal approximation of plane curves. *Computer Graphics and Image Processing*, 1: 244–256, 1972.

112 D. M. WUESCHER, K. L. BOYER. Robust contour decomposition using a constant curvature criterion. *IEEE Transactions on Pattern Analysis and Machine Intelligence*, 13(1): 41–51, January 1991.

113 H.-T. SHEU, W.-C. HU. Multiprimitive segmentation of planar curves – a two-level breakpoint classification and tuning approach. *IEEE Transactions on Pattern Analysis and Machine Intelligence*, 21(8): 791–797, August 1999.

114 J.-M. CHEN, J. A. VENTURA, C.-H. WU. Segmentation of planar curves into circular arcs and line segments. *Image and Vision Computing*, 14(1): 71–83, 1996.

115 P. L. ROSIN, G. A. W. WEST. Nonparametric segmentation of curves into various representations. *IEEE Transactions on Pattern Analysis and Machine Intelligence*, 17(12): 1140–1153, December 1995.

116 A. GRUEN, T. S. HUANG, EDITORS. *Calibration and Orientation of Cameras in Computer Vision*. Springer-Verlag, Berlin, 2001.

117 R. GUPTA, R. I. HARTLEY. Linear pushbroom cameras. *IEEE Transactions on Pattern Analysis and Machine Intelligence*, 19(9): 963–975, September 1997.

118 S. J. AHN, H.-J. WARNECKE, R. KOTOWSKI. Systematic geometric image measurement errors of circular object targets: Mathematical formulation and correction. *Photogrammetric Record*, 16(93): 485–502, April 1999.

119 O. FAUGERAS. *Three-Dimensional Computer Vision: A Geometric Viewpoint*. MIT Press, Cambridge, MA, 1993.

120 D. SCHARSTEIN, R. SZELISKI. A taxonomy and evaluation of dense two-frame stereo correspondence algorithms. *International Journal of Computer Vision*, 47(1–3): 7–42, 2002.

121 H. HIRSCHMÜLLER, P. R. INNOCENT, J. GARIBALDI. Real-time correlation-based stereo vision with reduced border errors.

International Journal of Computer Vision, 47(1–3): 229–246, 2002.

122 K. MÜHLMANN, D. MAIER, J. HESSER, R. MÄNNER. Calculating dense disparity maps from color stereo images, an efficient implementation. *International Journal of Computer Vision*, 47(1–3): 79–88, 2002.

123 O. FAUGERAS, B. HOTZ, H. MATHIEU, T. VIÉVILLE, Z. ZHANG, P. FUA, E. THÉRON, L. MOLL, G. BERRY, J. VUILLEMIN, P. BERTIN, C. PROY. *Real Time Correlation-Based Stereo: Algorithm, Implementations and Applications*. Rapport de Recherche 2013, INRIA, Sophia-Antipolis, August 1993.

124 L. GOTTESFELD BROWN. A survey of image registration techniques. *ACM Computing Surveys*, 24(4): 325–376, December 1992.

125 L. DI STEFANO, S. MATTOCCIA, M. MOLA. An efficient algorithm for exhaustive template matching based on normalized cross correlation. In *12th International Conference on Image Analysis and Processing*, pp. 322–327, 2003.

126 M. GHARAVI-ALKHANSARI. A fast globally optimal algorithm for template matching using low-resolution pruning. *IEEE Transactions on Image Processing*, 10(4): 526–533, April 2001.

127 Y. HEL-OR, H. HEL-OR. Real time pattern matching using projection kernels. In *9th International Conference on Computer Vision*, Vol. 2, pp. 1486–1493, 2003.

128 S. L. TANIMOTO. Template matching in pyramids. *Computer Graphics and Image Processing*, 16: 356–369, 1981.

129 F. GLAZER, G. REYNOLDS, P. ANANDAN. Scene matching by hierarchical correlation. In *Computer Vision and Pattern Recognition*, pp. 432–441, 1983.

130 Q. TIAN, M. N. HUHNS. Algorithms for subpixel registration. *Computer Vision, Graphics, and Image Processing*, 35: 220–233, 1986.

131 S.-H. LAI, M. FANG. Accurate and fast pattern localization algorithm for automated visual inspection. *Real-Time Imaging*, 5(1): 3–14, 1999.

132 V. A. ANISIMOV, N. D. GORSKY. Fast hierarchical matching of an arbitrarily oriented template. *Pattern Recognition Letters*, 14(2): 95–101, 1993.

133 W. FÖRSTNER. A framework for low level feature extraction. In J.-O. EKLUNDH, EDITOR, *Third European Conference on Computer Vision*, Lecture Notes in Computer Science, Vol. 801, pp. 383–394. Springer-Verlag, Berlin, 1994.

134 C. SCHMID, R. MOHR, C. BAUCKHAGE. Evaluation of interest point detectors. *International Journal of Computer Vision*, 37(2): 151–172, 2000.

135 G. BORGEFORS. Hierarchical chamfer matching: A parametric edge matching algorithm. *IEEE Transactions on Pattern Analysis and Machine Intelligence*, 10(6): 849–865, November 1988.

136 W. J. RUCKLIDGE. Efficiently locating objects using the Hausdorff distance. *International Journal of Computer Vision*, 24(3): 251–270, 1997.

137 O.-K. KWON, D.-G. SIM, R.-H. PARK. Robust Hausdorff distance matching algorithms using pyramidal structures. *Pattern Recognition*, 34: 2005–2013, 2001.

138 C. F. OLSON, D. P. HUTTENLOCHER. Automatic target recognition by matching oriented edge pixels. *IEEE Transactions on Image Processing*, 6(1): 103–113, January 1997.

139 D. H. BALLARD. Generalizing the Hough transform to detect arbitrary shapes. *Pattern Recognition*, 13(2): 111–122, 1981.

140 P. V. C. HOUGH. Method and means for recognizing complex patterns. US Patent, 3 069 654, December 1962.

141 R. O. DUDA, P. E. HART. Use of the Hough transformation to detect lines and curves in pictures. *Communications of the ACM*, 15(1): 11–15, January 1972.

142 M. ULRICH, C. STEGER, A. BAUMGARTNER. Real-time object recognition using a modified generalized Hough transform. *Pattern Recognition*, 36(11): 2557–2570, 2003.

143 Y. LAMDAN, J. T. SCHWARTZ, H. J. WOLFSON. Affine invariant model-based object recognition. *IEEE Transactions on Robotics and Automation*, 6(5): 578–589, October 1990.

144 N. AYACHE, O. D. FAUGERAS. HYPER: A new approach for the recognition and positioning of two-dimensional objects.

IEEE Transactions on Pattern Analysis and Machine Intelligence, 8(1): 44–54, January 1986.

145 W. E. L. GRIMSON, T. LOZANO-PÉREZ. Localizing overlapping parts by searching the interpretation tree. *IEEE Transactions on Pattern Analysis and Machine Intelligence*, 9(4): 469–482, July 1987.

146 M. W. KOCH, R. L. KASHYAP. Using polygons to recognize and locate partially occluded objects. *IEEE Transactions on Pattern Analysis and Machine Intelligence*, 9(4): 483–494, July 1987.

147 J. A. VENTURA, W. WAN. Accurate matching of two-dimensional shapes using the minimal tolerance error zone. *Image and Vision Computing*, 15: 889–899, 1997.

148 M. S. COSTA, L. G. SHAPIRO. 3D object recognition and pose with relational indexing. *Computer Vision and Image Understanding*, 79(3): 364–407, September 2000.

149 S. H. JOSEPH. Analysing and reducing the cost of exhaustive correspondence search. *Image and Vision Computing*, 17: 815–830, 1999.

150 C. STEGER. Similarity measures for occlusion, clutter, and illumination invariant object recognition. In B. RADIG, S. FLORCZYK, EDITORS, *Pattern Recognition*, Lecture Notes in Computer Science, Vol. 2191, pp. 148–154. Springer-Verlag, Berlin, 2001.

151 C. STEGER. Occlusion, clutter, and illumination invariant object recognition. In *International Archives of Photogrammetry and Remote Sensing*, Vol. XXXIV, part 3A, pp. 345–350, 2002.

152 C. STEGER. System and method for object recognition. *European Patent*, 1 193 642, March 2005.

153 C. STEGER. System and method for object recognition. *Japanese Patent*, 3 776 340, May 2006.

154 C. STEGER. System and method for object recognition. *US Patent*, 7 062 093, June 2006.

155 M. ULRICH, C. STEGER. Empirical performance evaluation of object recognition methods. In H. I. CHRISTENSEN, P. J. PHILLIPS, EDITORS, *Empirical Evaluation Methods in Computer Vision*, pp. 62–76. IEEE Computer Society Press, Los Alamitos, CA, 2001.

156 M. ULRICH, C. STEGER. Performance comparison of 2d object recognition techniques. In *International Archives of Photogrammetry and Remote Sensing*, Vol. XXXIV, part 3A, pp. 368–374, 2002.

157 R. G. CASEY, E. LECOLINET. A survey of methods and strategies in character segmentation. *IEEE Transactions on Pattern Analysis and Machine Intelligence*, 18(7): 690–706, July 1996.

158 S. THEODORIDIS, K. KOUTROUMBAS. *Pattern Recognition*. Academic Press, San Diego, CA, 1999.

159 A. WEBB. *Statistical Pattern Recognition*. Arnold Publishers, London, 1999.

160 M. A. T. FIGUEIREDO, A. K. JAIN. Unsupervised learning of finite mixture models. *IEEE Transactions on Pattern Analysis and Machine Intelligence*, 24(3): 381–396, March 2002.

161 H. X. WANG, B. LUO, Q. B. ZHANG, S. WEI. Estimation for the number of components in a mixture model using stepwise split-and-merge EM algorithm. *Pattern Recognition Letters*, 25(16): 1799–1809, December 2004.

162 C. M. BISHOP. *Neural Networks for Pattern Recognition*. Oxford University Press, Oxford, 1995.

163 B. SCHÖLKOPF, A. J. SMOLA. *Learning with Kernels – Support Vector Machines, Regularization, Optimization, and Beyond*. MIT Press, Cambridge, MA, 2002.

164 N. CHRISTIANINI, J. SHAWE-TAYLOR. *An Introduction to Support Vector Machines and Other Kernel-Based Learning Methods*. Cambridge University Press, Cambridge, 2000.

Index

3D reconstruction 198–210, 320–327

a

a posteriori probability 246
a priori probability 246, 249
aberrations *see* lens aberrations
absolute sensitivity threshold 48
absolute sum of normalized dot products 238
accumulator array 227, 229
accuracy 162–163
– camera parameters 196–198
– contour moments 125
– edges 73, 164–168
– gray value features 73
– gray value moments 121–123, 125
– hardware requirements 168
– region moments 121–123
– subpixel-precise threshold 73
achromatic lens 33
active pixel sensor 40
ADC 36, 40, 52
affine transformation 94–95, 119, 212, 230, 237, 301–307, 328–334
Airy disk 27
algebraic distance 176
algebraic error 176
aliasing 37, 38, 41, 90–91, 98–99, 217
alignment 94, 107, 108
amplifier noise 46
analog-to-digital converter *see* camera, ADC
Ando filter 156
anisometry 118, 121, 123, 244, 269–276
anti-bloom drain 39
aperture stop 22, 30, 31, 33, 74
apochromatic lens 33
APS *see* camera, active pixel sensor
area 116, 121–123, 125, 269–276, 293–301, 307–316
area of interest 41
area sensor 37–41, 180
aspherical lens 31
astigmatism 31, 167

asynchronous reset 39, 42, 61, 62

b

back light 13
back porch 51
Bayes decision rule 246, 247
Bayes' theorem 246
bilateral telecentric lens 29
binary image 67, 113, 116, 126, 128, 132, 138
binocular stereo reconstruction 198–210, 320–327
black body 5–6
blooming 39
boundary 113, 120, 132–133
bounding box 119, 124, 244, 269–280, 293–301
bright-field illumination 13

c

calibration
– geometric 167, 168, 180–199, 301–307
– – accuracy of interior orientation 196–198
– – binocular stereo calibration 199–200, 320–327
– – calibration target 188–190
– – camera constant *see* calibration, geometric, principal distance
– – camera coordinate system 182–183, 185–186, 199
– – camera motion vector 185
– – distortion coefficient 183–184, 187, 196–199, 202
– – exterior orientation 168, 190–193, 320–327
– – focal length 181–182, 199
– – image coordinate system 184–185, 187
– – image plane coordinate system 183, 186–187
– – interior orientation 168, 190–193, 199, 320–327
– – pixel size 184–185, 187, 199

Machine Vision Algorithms and Applications. Carsten Steger, Markus Ulrich, and Christian Wiedemann
Copyright © 2008 WILEY-VCH Verlag GmbH & Co. KGaA, Weinheim
ISBN: 978-3-527-40734-7

– – principal distance   18, 24, 181–183, 186–187, 196–199
– – principal point   184–185, 187, 196–199
– – projection center   18, 23, 181, 199, 202, 204
– – relative orientation   199
– – world coordinate system   182–183, 186
– radiometric   73–77, 167
– – calibration target   73
– – chart-based   73–74
– – chart-less   74–77
– – defining equation for chart-less calibration   75
– – discretization of inverse response function   75–76
– – gamma response function   74, 77
– – inverse response function   75
– – normalization of inverse response function   76
– – polynomial inverse response function   77
– – response function   74, 77
– – smoothness constraint   76–77
camera   65–66
– absolute sensitivity threshold   48
– active pixel sensor   40
– ADC   36, 40, 52
– analog-to-digital converter   see camera, ADC
– area scan   37–41
– asynchronous reset   39, 42, 61, 62
– calibration
– – image rectification   195
– – world coordinates from single image   167–168, 193–195
– – world coordinates from stereo reconstruction   198–210
– CCD   36–40
– – anti-bloom drain   39
– – blooming   39
– – frame transfer sensor   37–38
– – full frame sensor   37
– – interlaced scan   40, 50
– – interline transfer sensor   38–39
– – lateral overflow drain   39
– – line sensor   36–37
– – progressive scan   40, 51
– – vertical overflow drain   39
– CMOS   40–42
– – area of interest   41
– – global shutter   41
– – line sensor   41
– – rolling shutter   41
– color   43–45
– – single-chip   43–44
– – three-chip   44
– color filter array   43

– – Bayer   43
– – demosaicking   44
– digital pixel sensor   41
– dynamic range   48
– exposure time   26, 37, 39, 74
– fill factor   37, 38, 41, 123, 165, 168
– gamma response function   74, 77
– gray value response   42, 73
– – linear   42, 73, 167, 168
– – linear–logarithmic   42
– – logarithmic   42
– – nonlinear   42, 73, 166
– inverse response function   75
– line scan   36–37, 41, 181, 185–188, 195
– – encoder   37
– noise
– – amplifier noise   46
– – dark current noise   46
– – dark noise   46
– – dark signal non-uniformity   47
– – fixed pattern noise   47
– – gain noise   47
– – noise floor   46
– – offset noise   47
– – overall system gain   47
– – pattern noise   47
– – photon noise   46
– – photoresponse non-uniformity   47
– – quantization noise   47
– – reset noise   46
– – signal-to-noise ratio   47
– – spatial noise   47
– – temporal noise   47
– performance   45–48
– pinhole   18, 23–24, 180–185, 194–195
– quantum efficiency   46
– response function   74, 77
– sensor size   45
– spectral response   42–45
– telecentric   180–185, 193–194, 301–307
– trigger   37, 39, 62–63
Camera Link   54–55
Canny filter   151–152, 157, 163, 281–285, 307–316
– edge accuracy   164, 165
– edge precision   163
cardinal elements   21
CCD   see camera, CCD
CCIR   49
center of gravity   116–118, 121, 122, 125, 170, 172, 269–276, 301, 307
central moments   117–118, 121, 124, 170
CFA   see camera, color filter array
CGH,   see computer-generated hologram
chamfer-3-4 distance   140
characteristic function   67, 121, 140

charge-coupled device *see* camera, CCD
chessboard distance 139
chief ray *see* principal ray
chromatic aberration 32–33, 167
circle fitting 174–175, 269–276, 307–320
– outlier suppression 174–175
– robust 174–175
circle of confusion 24, 30
city-block distance 138, 139
classification 243, 246–261, 269–276, 334–337
– a posteriori probability 246
– a priori probability 246, 249
– Bayes classifier 248–251
– classification accuracy 259
– classifier types 248
– curse of dimensionality 249, 256
– decision theory 246–248
– – Bayes decision rule 246, 247
– – Bayes' theorem 246
– error rate 248
– expectation maximization algorithm 251
– features 243, 269–276
– Gaussian mixture model classifier 250–251, 260
– generalized linear classifier 255
– *k* nearest-neighbor classifier 249–250
– linear classifier 251–252
– neural network 251–255, 258–260, 269–276, 334–337
– – hyperbolic tangent activation function 254
– – logistic activation function 254
– – multilayer perceptron 252–255, 258–260, 269–276, 334–337
– – multilayer perceptron training 254–255
– – sigmoid activation function 253–254
– – single-layer perceptron 251–252
– – softmax activation function 254
– – threshold activation function 251, 253
– – universal approximator 253, 254
– nonlinear classifier 252–260
– polynomial classifier 255
– rejection 259–260
– support vector machine 255–260
– – Gaussian radial basis function kernel 258
– – homogeneous polynomial kernel 258
– – inhomogeneous polynomial kernel 258
– – kernel 257
– – margin 256
– – separating hyperplane 256–257
– – sigmoid kernel 258
– – universal approximator 258
– test set 248
– training set 248, 255
– training speed 259

closing 135–137, 142–144, 293–301
clutter 223, 226, 230, 237
CMOS *see* camera, CMOS
color filter 11, 44
color filter array *see* camera, color filter array
coma 31, 167
compactness 120, 244, 269–276
complement 126
complementary metal–oxide–semiconductor *see* camera, CMOS
completeness checking 1
component labeling 113
composite video 51
connected components 111–113, 241, 269–276, 285–301
connectivity 111–113, 126, 133, 138
contour 69
contour feature *see* features, contour
contour length 120
contour segmentation 177–180, 223, 276–280, 316–320
– lines 177–178
– lines and circles 178–180, 276–280, 316–320
– lines and ellipses 178–180
contrast enhancement 70–73
contrast normalization 70–73
– robust 71–73, 245
convex hull 119–120, 124
convexity 119–120, 269–276
convolution 83–84, 90
– kernel 83
coordinates
– homogeneous 94, 95
– inhomogeneous 95
– polar 100
correlation 90, *see also* normalized cross-correlation, 263–269
cumulative histogram 72, 121
curvature of field 31

d

dark current noise 46
dark noise 46
dark signal non-uniformity 47
dark-field illumination 13
data structures
– images 65–66
– regions 66–69
– subpixel-precise contours 69
DCS *see* distributed control system
decision theory 246–248
– a posteriori probability 246
– a priori probability 246, 249
– Bayes decision rule 246, 247
– Bayes' theorem 246

demosaicking   44
depth of field   24–27, 29
depth-first search   112
Deriche filter   152–153, 157, 164
– edge accuracy   164, 166
– edge precision   164
derivative
– directional   147
– first   145, 149
– gradient   147
– Laplacian   147
– partial   147, 155, 157
– second   146, 149
DFT   *see* discrete Fourier transform
diaphragm   *see* lens, iris diaphragm
difference   126, 293–301, 307–320
diffraction   26
diffuse bright-field back light illumination   15–16
diffuse bright-field front light illumination   13–14
diffuse illumination   12
digital input/output   3
digital pixel sensor   41
digital signal processor   3
dilation   127–129, 132–133, 141–142, 160, 281–285, 293–301, 316–320
dimensional inspection   1
direct memory access   52, 61
directed bright-field front light illumination   14–15
directed dark-field front light illumination   15
directed illumination   12
discrete Fourier transform   90–93, *see also* Fourier transform
disparity   205–206
dispersion   19
distance
– chamfer-3-4   140
– chessboard   139
– city-block   138, 139
– Euclidean   139, 140, 307–316
distance transform   138–140, 224, 226, 307–316
distortion   31–32, 167, 183–184, 187
– barrel   31–32, 183–184, 187
– pincushion   31–32, 183–184, 187
distributed control system   3
DMA   *see* direct memory access
DSP   *see* digital signal processor
duality
– dilation–erosion   132, 142
– hit-or-miss transform   134
– opening–closing   135, 143
dynamic range   48

dynamic thresholding   104–107, 144, 241, 269–276, 293–301

e

edge
– amplitude   111, 147, 156
– definition
– – 1D   145–147
– – 2D   147–148
– gradient magnitude   111, 147, 156
– gradient vector   147
– Laplacian   147, 160–162
– non-maximum suppression   146, 153, 158
– polarity   147
edge extraction   144–168, 223, 316–320
– 1D   149–154, 301–307
– – Canny filter   151–152
– – Deriche filter   152–153
– – derivative   145, 146, 149
– – gray value profile   149–150, 301–307
– – non-maximum suppression   153
– – subpixel-accurate   153–154
– 2D   111, 155–162, 281–285, 307–327, 334–337
– – Ando filter   156
– – Canny filter   157, 163
– – Deriche filter   157, 164
– – Frei filter   155
– – gradient   147
– – hysteresis thresholding   158–159
– – Lanser filter   157, 164
– – Laplacian   147, 160–162
– – non-maximum suppression   158
– – Prewitt filter   155
– – Sobel filter   155
– – subpixel-accurate   159–162
edge filter   111, 151–153
– Ando   156
– Canny   151–152, 157, 163, 281–285, 307–316
– – edge accuracy   164, 165
– – edge precision   163
– Deriche   152–153, 157, 164
– – edge accuracy   164, 166
– – edge precision   164
– Frei   155
– Lanser   157, 164
– – edge accuracy   164
– – edge precision   164
– optimal   151–153, 157
– Prewitt   155
– Sobel   155
EIA-170   49
electromagnetic radiation   5–6
– black body   5–6
– spectrum   5

ellipse fitting   175–177
– algebraic error   176
– geometric error   176
– outlier suppression   176
– robust   176
ellipse parameters   117–119, 121, 123, 125, 170, 172, 176
enclosing circle   119, 124
enclosing rectangle   119, 124, 244, 269–280, 293–301
encoder   37
entrance pupil   22, 23, 29
epipolar image rectification   204–205, 320–327
epipolar line   201
epipolar plane   201
epipolar standard geometry   202–204, 320–327
epipole   201
erosion   130–133, 142, 293–301, 316–320
Euclidean distance   139, 140, 307–316
exit pupil   22, 23, 29
exposure time   26, 37, 39, 74
exterior orientation   168, 182–183, 186, 190–193, 301–307, 320–327
– world coordinate system   182–183, 186

f

f-number   25, 29, 31, 33, 75
facet model   160
fast Fourier transform   92, *see also* Fourier transform, 263–269
feature extraction   115–125, 285–293
features
– contour   115, 124–125, 276–280, 285–293
– – area   125
– – center of gravity   125
– – central moments   124
– – contour length   124, 276–280
– – ellipse parameters   125
– – major axis   125
– – minor axis   125
– – moments   124
– – normalized moments   124
– – orientation   125
– – smallest enclosing circle   124
– – smallest enclosing rectangle   124, 276–280
– gray value   115, 120–123, 334–337
– – α-quantile   121
– – anisometry   121, 123
– – area   121–123
– – center of gravity   121, 122
– – central moments   121
– – ellipse parameters   121, 123
– – major axis   121, 123

– – maximum   71, 120
– – mean   120
– – median   121
– – minimum   71, 120
– – minor axis   121, 123
– – moments   121–123
– – normalized moments   121
– – orientation   121
– – standard deviation   121
– – variance   120
– region   115–120, 269–276, 285–301, 307–316
– – anisometry   118, 244, 269–276
– – area   116, 122, 269–276, 293–301, 307–316
– – center of gravity   116–118, 122, 269–276, 301, 307
– – central moments   117–118
– – compactness   120, 244, 269–276
– – contour length   120
– – convexity   119–120, 269–276
– – ellipse parameters   117–119
– – major axis   117–118
– – minor axis   117–118
– – moments   116–119, 269–276
– – normalized moments   116–117
– – orientation   117–119
– – smallest enclosing circle   119
– – smallest enclosing rectangle   119, 244, 269–276, 293–301
FFT   *see* fast Fourier transform
field angle   23
field-programmable gate array   3
fieldbus   3
fill factor   37, 38, 41, 123, 165, 168
filter
– anisotropic   85, 157
– border treatment   81–82
– convolution   83–84, 90
– – kernel   83
– definition   83
– edge   151–153
– – Ando   156
– – Canny   151–152, 157, 163, 281–285, 307–316
– – Deriche   152–153, 157, 164
– – Frei   155
– – Lanser   157, 164
– – optimal   151–153, 157
– – Prewitt   155
– – Sobel   155
– Gaussian   85–87, 99, 103, 105, 106, 152, 157, 217
– – frequency response   86, 90
– isotropic   85, 86, 157
– linear   83–84, 149

– mask   83
– maximum   *see* morphology, gray value, dilation
– mean   80–83, 87, 99, 105, 106, 150, 217, 269–276, 293–301
– – frequency response   84–85, 90, 217
– median   87–89, 105, 106, 293–301
– minimum   *see* morphology, gray value, erosion
– nonlinear   87–89, 142
– optical
– – anti-aliasing   43
– – color   11, 44
– – infrared cut   11, 45
– – infrared pass   11, 42
– – polarizing   11
– rank   89, 142
– recursive   83, 84, 86, 142
– runtime complexity   82–83
– separable   82, 84, 86
– smoothing   78–89
– – optimal   85–86, 152, 157
– spatial averaging   80–83
– temporal averaging   78–80, 108
FireWire   *see* IEEE 1394
fitting
– circles   174–175, 269–276, 307–320
– – outlier suppression   174–175
– – robust   174–175
– ellipses   175–177
– – algebraic error   176
– – geometric error   176
– – outlier suppression   176
– – robust   176
– lines   170–174, 276–280
– – outlier suppression   171–174
– – robust   171–174
fixed pattern noise   47
fluorescent lamp   7–8
focal length   20, 181–182, 199
focal point   20
focusing plane   24
Fourier transform   84, 89–93, 263–269
– 1D   89
– – inverse   89
– 2D   89, 263–269
– – inverse   89, 263–269
– continuous   89–90
– convolution   90
– discrete   90–93
– – inverse   91
– fast   92, 263–269
– frequency domain   89
– Nyquist frequency   90
– real-valued   92, 263–269
– spatial domain   89

– texture removal   92–93
FPGA   *see* field-programmable gate array
frame grabber   49
– analog   52–53
– – line jitter   52–53, 164
– – pixel clock   52, 53
– digital   54
frame transfer sensor   37–38
Frei filter   155
frequency domain   89
front light   13
front porch   50
full frame sensor   37
fuzzy membership   121–123
fuzzy set   121–123

g

gain pattern noise   47
gamma response function   74, 77
Gaussian filter   85–87, 99, 103, 105, 106, 152, 157, 217
– frequency response   86, 90
Gaussian optics   18–29
generalized Hough transform   226–230
– accumulator array   227, 229
– R-table   229
geometric camera calibration   167, 168, 180–199, 301–307
– binocular stereo calibration   199–200, 320–327
– calibration target   188–190
– exterior orientation   168, 182–183, 186, 190–193, 301–307, 320–327
– – world coordinate system   182–183, 186
– interior orientation   168, 181–185, 190–193, 199, 301–307, 320–327
– – accuracy   196–198
– – camera constant   *see* geometric camera calibration, interior orientation, principal distance
– – camera coordinate system   182–183, 185–186, 199
– – camera motion vector   185
– – distortion coefficient   183–184, 187, 196–199, 202
– – focal length   181–182, 199
– – image coordinate system   184–185, 187
– – image plane coordinate system   183, 186–187
– – pixel size   184–185, 187, 199
– – principal distance   18, 24, 181–183, 186–187, 196–199
– – principal point   184–185, 187, 196–199
– – projection center   18, 23, 181, 199, 202, 204
– relative orientation   199

– – base   199
– – base line   202
geometric error   176
geometric hashing   230–232
geometric matching   230–240
Gigabit Ethernet   *see* GigE Vision
GigE Vision   59–60
– control channel   60
– GigE Vision Control Protocol   60
– GigE Vision Streaming Protocol   60
– GVCP   *see* GigE Vision, GigE Vision Control Protocol
– GVSP   *see* GigE Vision, GigE Vision Streaming Protocol
– message channel   60
– stream channel   60
global shutter   41
gradient   147
– amplitude   111, 147, 156
– angle   147, 229
– direction   147, 226, 229
– length   147
– magnitude   111, 147, 156
– morphological   144
gray value   66
– 1D histogram   71–73, 102–104, 121
– – cumulative   72, 121
– – maximum   102–104
– – minimum   102–104
– – peak   102–104
– 2D histogram   76
– α-quantile   121
– camera response   42, 73
– – linear   42, 73, 167, 168
– – linear–logarithmic   42
– – logarithmic   42
– – nonlinear   42, 73, 166
– feature   *see* features, gray value
– maximum   71, 120
– mean   120
– median   121
– minimum   71, 120
– normalization   70–73
– – robust   71–73, 245
– profile   150
– robust normalization   245
– scaling   70
– standard deviation   121
– transformation   70–73, 120, 123
– variance   120

h

Hausdorff distance   224–226
Hessian normal form   170
histogram
– 1D   71–73, 102–104, 121

– – cumulative   72, 121
– – maximum   102–104
– – minimum   102–104
– – peak   102–104
– 2D   76
hit-or-miss opening   135
hit-or-miss transform   133–134, 138
homogeneous coordinates   94, 95
horizontal blanking interval   50
horizontal synchronization pulse   50
Huber weight function   172
hypothesize-and-test paradigm   230
hysteresis Thresholding   158–159

i

identification   1
IEEE 1394   55–57
– asynchronous data transfer   56
– IIDC   56–57
– isochronous data transfer   56
IIDC   56–57
illumination   5–17
– back light   13
– bright-field   13
– dark-field   13
– diffuse   12
– diffuse bright-field back light illumination   15–16
– diffuse bright-field front light illumination   13–14
– directed   12
– directed bright-field front light illumination   14–15
– directed dark-field front light illumination   15
– front light   13
– light sources   7–8
– – fluorescent lamp   7–8
– – incandescent lamp   7
– – LED   *see* illumination, light sources, light-emitting diode
– – light-emitting diode   8
– – xenon lamp   7
– telecentric   13
– telecentric bright-field back light illumination   16–17
image   65–66
– binary   67, 113, 116, 126, 128, 132, 138
– bit depth   66
– complement   142
– domain   *see* region of interest
– enhancement   70–89
– function   66–67
– gray value   66
– gray value normalization   70–73
– – robust   71–73, 245

– gray value scaling   70
– gray value transformation   70–73
– label   67, 113
– multichannel   66
– noise   *see* noise
– pyramid   217–220, 239
– rectification   97, 99–100, 195, 204–205, 242, 320–327
– RGB   66
– segmentation   *see* segmentation
– single-channel   66
– smoothing   78–89
– spatial averaging   80–83
– temporal averaging   78–80, 108
– transformation   96–102, 107, 281–285, 328–334
image acquisition modes   61–63
– asynchronous acquisition   61–62
– continuous acquisition   63
– queued acquisition   63
– synchronous acquisition   61
– triggered acquisition   62–63
image distance   20, 24
image plane   18, 21, 24, 199, 202, 204, 205
incandescent lamp   7
infrared cut filter   11, 45
infrared pass filter   11, 42
inhomogeneous coordinates   95
interior orientation   168, 181–188, 190–193, 199, 301–307, 320–327
– accuracy   196–198
– camera constant   *see* interior orientation, camera constant
– camera coordinate system   182–183, 185–186, 199
– camera motion vector   185
– distortion coefficient   183–184, 187, 196–199, 202
– focal length   181–182, 199
– image plane coordinate system   183, 186–187
– image plane system   184–185, 187
– pixel size   184–185, 187, 199
– principal distance   18, 24, 181–183, 186–187, 196–199
– principal point   184–185, 187, 196–199
– projection center   18, 23, 181, 199, 202, 204
interlaced scan   40, 50
interline transfer sensor   38–39
interpolation
– bilinear   97–98, 113, 150
– nearest-neighbor   96–97, 150
intersection   126, 269–276, 293–301, 307–316
invariant moments   119
iris diaphragm   22, 25, 31

j
junction   69

k
kernel   *see* convolution, kernel *and* support vector machine, kernel

l
label image   67, 113
labeling   113
Lanser filter   157, 164
– edge accuracy   164
– edge precision   164
Laplacian   147, 160–162
lateral overflow drain   39
law of refraction   19
LED   *see* light-emitting diode
lens   18–29
– achromatic   33
– Airy disk   27
– aperture stop   22, 30, 31, 33, 74
– apochromatic   33
– aspherical   31
– cardinal elements   21
– chief ray   *see* lens, principal ray
– circle of confusion   24, 30
– depth of field   24–27, 29
– diffraction   26
– entrance pupil   22, 23, 29
– exit pupil   22, 23, 29
– f-number   25, 29, 31, 33, 75
– field angle   23
– focal length   20
– focal point   20
– focusing plane   24
– image distance   20, 24
– image plane   18, 21, 24
– iris diaphragm   22, 25, 31
– magnification factor   21
– nodal point   20
– numerical aperture   29
– object distance   20, 24
– optical axis   20
– principal plane   20
– principal ray   23
– pupil magnification   23
– sagittal focal surface   31
– sagittal image   31
– surface vertex   20
– system of lenses   21–22
– tangential focal surface   31
– tangential image   31
– telecentric   27–29
– thick   20–21
– vignetting   34–35, 74
lens aberrations   19, 30–34

– astigmatism 31, 167
– chromatic aberration 32–33, 167
– coma 31, 167
– curvature of field 31
– distortion 31–32, 167, 183–184, 187
– – barrel 31–32, 183–184, 187
– – pincushion 31–32, 183–184, 187
– spherical aberration 30–31
light
– absorption 10
– infrared 5, 42
– polarized 9
– reflection 8
– refraction 9, 19
– spectrum 5
– – black body 5–6
– ultraviolet 5, 42
– visible 5, 42
light sources 7–8
– fluorescent lamp 7–8
– incandescent lamp 7
– LED *see* light sources, light-emitting diode
– light-emitting diode 8
– xenon lamp 7
light-emitting diode 8
line
– Hessian normal form 170
line fitting 170–174, 276–280
– outlier suppression 171–174
– robust 171–174
line jitter 52–53, 164
line scan camera 36–37, 41, 181, 185–188, 195
line sensor 36–37, 41, 180
look-up table 70, 75
low-voltage differential signaling 54
LUT *see* look-up table
LVDS *see* low-voltage differential signaling

m

magnification factor 21
major axis 117–118, 121, 123, 125, 170, 172
maximum filter *see* morphology, gray value, dilation
maximum likelihood estimator 250
mean filter 80–83, 87, 99, 105, 106, 150, 217, 269–276, 293–301
– frequency response 84–85, 90, 217
mean squared edge distance 224
median filter 87–89, 105, 106, 293–301
minimum filter *see* morphology, gray value, erosion
Minkowski addition 127–128, 132, 140–141
Minkowski subtraction 129–130, 132, 142
minor axis 117–118, 121, 123, 125, 170
moments 116–119, 121–123, 269–276

– invariant 119
morphology 125–144, 242, 293–301, 307–320
– duality
– – dilation–erosion 132, 142
– – hit-or-miss transform 134
– – opening–closing 135, 143
– gray value
– – closing 142–144
– – complement 142
– – dilation 141–142, 281–285
– – erosion 142
– – gradient 144
– – Minkowski addition 140–141
– – Minkowski subtraction 142
– – opening 142–144
– – range 144
– region 125–140, 293–301, 307–327
– – boundary 132–133
– – closing 135–137, 293–301
– – complement 126
– – difference 126, 293–301, 307–320
– – dilation 127–129, 132–133, 160, 293–301, 316–320
– – distance transform 138–140, 224, 226, 307–316
– – erosion 130–133, 293–301, 316–320
– – hit-or-miss opening 135
– – hit-or-miss transform 133–134, 138
– – intersection 126, 269–276, 293–301, 307–316
– – Minkowski addition 127–128, 132
– – Minkowski subtraction 129–130, 132
– – opening 134–135, 293–301, 307–316
– – skeleton 137–138, 307–316
– – translation 126–127
– – transposition 127
– – union 125–126, 293–301
– structuring element 127, 133, 140, 141, 293–301

n

neighborhood 111–113, 133, 138
neural network 251–255, 258–260, 269–276, 334–337
– activation function
– – hyperbolic tangent 254
– – logistic 254
– – sigmoid 253–254
– – softmax 254
– – threshold 251, 253
– multilayer perceptron 252–255, 258–260, 269–276, 334–337
– – training 254–255
– single-layer perceptron 251–252
– universal approximator 253, 254

nodal point 20
noise 78
– amplifier noise 46
– dark current noise 46
– dark noise 46
– dark signal non-uniformity 47
– fixed pattern noise 47
– gain noise 47
– noise floor 46
– offset noise 47
– pattern noise 47
– photon noise 46
– photoresponse non-uniformity 47
– quantization noise 47
– reset noise 46
– signal-to-noise ratio 47
– spatial noise 47
– suppression 78–89
– temporal noise 47
– variance 78–79, 81, 86–87, 150, 163
non-maximum suppression 146, 153, 158
normal distribution 250
normalized cross-correlation 208, 214–216, 328–334
normalized moments 116–117, 121, 124, 170, 172
NTSC 49
numerical aperture 29
Nyquist frequency 90

o
object distance 20, 24
object identification 1
object-side telecentric lens 28–29
occlusion 222, 225, 230, 237
OCR *see* optical character recognition
offset noise 47
opening 134–135, 142–144, 293–301, 307–316
optical anti-aliasing filter 43
optical axis 20
optical character recognition 73, 94, 95, 115, 116, 241–261, 269–276, 328–334
– character segmentation 241–243, 269–276
– – touching characters 242–243
– classification *see* classification
– features 243–246, 269–276
– image rectification 97, 99–100, 242
orientation 117–119, 121, 125
– exterior 168, 182–183, 186, 190–193, 301–307, 320–327
– – world coordinate system 182–183, 186
– interior 168, 181–188, 190–193, 199, 301–307, 320–327
– – accuracy 196–198

– – camera constant *see* orientation, interior, principal distance
– – camera coordinate system 182–183, 185–186, 199
– – camera motion vector 185
– – distortion coefficient 183–184, 187, 196–199, 202
– – focal length 181–182, 199
– – image plane coordinate system 183, 186–187
– – image plane system 184–185, 187
– – pixel size 184–185, 187, 199
– – principal distance 18, 24, 181–183, 186–187, 196–199
– – principal point 184–185, 187, 196–199
– – projection center 18, 23, 181, 199, 202, 204
– relative 199
– – base 199
– – base line 202
outlier 169, 171
outlier suppression 171–176
– Huber weight function 172
– random sample consensus 173
– RANSAC 173
– Tukey weight function 172
overall system gain 47

p
PAL 49
paraxial approximation 19
pattern noise 47
perspective transformation 95–96, 99–100, 242
photon noise 46
photoresponse non-uniformity 47
pinhole camera 18, 23–24, 180–185, 194–195
– projection center 18, 23, 181, 199, 202, 204
pixel 65–66
pixel clock 52–54
PLC *see* programmable logic controller
polar coordinates 100
polar transformation 100–102, 242, 269–276
polarization 9
polarizing filter 11
polygonal approximation 177–178
– Ramer algorithm 177–178
pose 94, 182–183, 186, 211, 301–307
position detection 1
precision 162–163
– edge angle 229
– edges 163–164
– hardware requirements 164
Prewitt filter 155
principal plane 20
principal point 184–185, 187, 196–199

principal ray   23
print inspection   107–111
programmable logic controller   2
progressive scan   40, 51
projection center   18, 23, 181, 199, 202, 204
projective transformation   95–96, 99–100, 212, 242
pupil magnification   23

q
quantization noise   47
quantum efficiency   46

r
radiometric camera calibration   73–77, 167
– calibration target   73
– chart-based   73–74
– chart-less   74–77
– – defining equation   75
– gamma response function   74, 77
– inverse response function   75
– – discretization   75–76
– – normalization   76
– – polynomial   77
– – smoothness constraint   76–77
– response function   74, 77
Ramer algorithm   177–178
rank filter   89, 142
reflection   8
reflectivity   9
refraction   9, 19
refractive index   19
region   66–69
– as binary image   67, 113, 116, 126, 128, 132, 138
– boundary   120, 132–133
– characteristic function   67, 121, 140
– complement   126
– connected components   111–113, 241, 269–276, 285–301
– convex hull   119–120
– definition   66
– difference   126, 293–301, 307–320
– feature   see features, region
– intersection   126, 269–276, 293–301, 307–316
– run-length representation   68–69, 112, 116, 126, 128, 138
– translation   126–127
– transposition   127
– union   125–126, 293–301
region of interest   67, 94, 96, 129, 140, 160, 212
relative orientation   199
– base   199
– base line   202

repeatability   see precision
reset noise   46
RGB video   51
rigid transformation   182, 211, 222, 231, 301–307
ROI   see region of interest
rolling shutter   41
rotation   95, 97, 119, 211, 222, 242, 293–301
R-table   229
run-length encoding   68–69, 112, 116, 126, 128, 138

s
sagittal focal surface   31
sagittal image   31
scaling   95, 98–99, 119, 212, 222
segmentation   102–115, 285–293, 307–320
– connected components   111–113, 241, 269–276, 285–301
– dynamic thresholding   104–107, 144, 241, 269–276, 293–301
– hysteresis thresholding   158–159
– subpixel-precise thresholding   113–115, 161, 276–280, 285–293
– thresholding   102–104, 241, 281–293, 307–320
– – automatic threshold selection   102–104, 241
– variation model   107–111, 281–285, 293–301
sensor   65–66
serial interface   3
shape inspection   1
shape-based matching   236–240, 263–269, 301–307, 328–334
shutter
– electronic   39
– global   41
– mechanical   37, 38
– rolling   41
signal-to-noise ratio   46, 47, 152, 164, 229
similarity measure   212–217, 237–239
– absolute sum of normalized dot products   238
– normalized cross-correlation   208, 214–216, 328–334
– sum of absolute gray value differences   207, 212–216
– sum of absolute normalized dot products   238
– sum of normalized dot products   238
– sum of squared gray value differences   207, 212–214
– sum of unnormalized dot products   237–238
similarity transformation   212, 222, 231, 285–293

skeleton   137–138, 307–316
skew   95
slant   95
smallest enclosing circle   119, 124
smallest enclosing rectangle   119, 124, 244, 269–280, 293–301
smart camera   2
smoothing filter   78–89
– Gaussian   85–87, 99, 103, 105, 106, 152, 157, 217
– – frequency response   86, 90
– mean   80–83, 87, 99, 105, 106, 150, 217, 269–276, 293–301
– – frequency response   84–85, 90, 217
– median   87–89, 105, 106, 293–301
– optimal   85–86, 152, 157
– spatial averaging   80–83
– temporal averaging   78–80, 108
Sobel filter   155
spatial averaging   80–83
spatial domain   89
spatial noise   47
spectral response
– Gaussian filter   86, 90
– human visual system   42
– mean filter   84–85, 90, 217
– sensor   42–45, 65
spherical aberration   30–31
stereo geometry   199–206
– corresponding points   200
– disparity   205–206
– epipolar line   201
– epipolar plane   201
– epipolar standard geometry   202–204, 320–327
– epipole   201
– image rectification   204–205, 320–327
stereo matching   207–210, 320–327
– robust   210
– – disparity consistency check   210
– – excluding weakly textured areas   210
– similarity measure
– – normalized cross-correlation   208
– – sum of absolute gray value differences   207
– – sum of squared gray value differences   207
– subpixel-accurate   209
– window size   209–210
stereo reconstruction   198–210, 320–327
stochastic process   78, 81
– ergodic   81
– stationary   78
structuring element   127, 133, 140, 141, 293–301
subpixel-precise contour   69

– convex hull   124
– features   see features, contour
subpixel-precise thresholding   113–115, 161, 276–280, 285–293
sum of absolute gray value differences   207, 212–216
sum of absolute normalized dot products   238
sum of normalized dot products   238
sum of squared gray value differences   207, 212–214
sum of unnormalized dot products   237–238
support vector machine   255–260
– kernel   257
– – Gaussian radial basis function   258
– – homogeneous polynomial   258
– – inhomogeneous polynomial   258
– – sigmoid   258
– margin   256
– separating hyperplane   256–257
– universal approximator   258
surface inspection   1
surface vertex   20
S-Video   51
SVM   see support vector machine

t

tangential focal surface   31
tangential image   31
telecentric bright-field back light illumination   16–17
telecentric camera   180–185, 193–194, 301–307
telecentric illumination   13
telecentric lens   27–29
– bilateral   29
– object-side   28–29
template matching   107, 211–242
– clutter   223, 226, 230, 237
– erosion   131
– generalized Hough transform   226–230
– – accumulator array   227, 229
– – R-table   229
– geometric hashing   230–232
– geometric matching   230–240
– Hausdorff distance   224–226
– hierarchical search   218–220, 239
– hit-or-miss transform   133
– hypothesize-and-test paradigm   230
– image pyramid   217–220, 239
– linear illumination changes   214
– matching geometric primitives   232–234
– mean squared edge distance   224
– nonlinear illumination changes   223, 230, 237–238
– occlusion   222, 225, 230, 237
– opening   134, 293–301

– robust   222–241, 263–269, 281–285, 293–307, 328–334
– rotation   222
– scaling   222
– shape-based matching   236–240, 263–269, 301–307, 328–334
– similarity measure   212–217, 237–239
– – absolute sum of normalized dot products   238
– – normalized cross-correlation   208, 214–216, 328–334
– – sum of absolute gray value differences   207, 212–216
– – sum of absolute normalized dot products   238
– – sum of normalized dot products   238
– – sum of squared gray value differences   207, 212–214
– – sum of unnormalized dot products   237–238
– stopping criterion   215–217
– – normalized cross-correlation   216
– – sum of absolute gray value differences   215–216
– – sum of normalized dot products   239
– subpixel-accurate   221, 239–240
– translation   212–220
temporal averaging   78–80, 108
temporal noise   47
texture
– removal   92–93
thick lens   20–21
– cardinal elements   21
– focal length   20
– focal point   20
– image distance   20, 24
– magnification factor   21
– nodal point   20
– object distance   20, 24
– optical axis   20
– principal plane   20
– surface vertex   20
thresholding   102–104, 241, 281–293, 307–320
– automatic threshold selection   102–104, 241
– subpixel-precise   113–115, 161, 276–280, 285–293
transformation
– affine   94–95, 119, 212, 230, 237, 301–307, 328–334
– geometric   94–102, 242, 269–276, 285–307, 328–334
– gray value   70–73
– image   96–102, 107, 269–276, 281–285, 301–307, 328–334
– perspective   95–96, 99–100, 242

– polar   100–102, 242, 269–276
– projective   95–96, 99–100, 212, 242
– rigid   182, 211, 222, 231, 301–307
– rotation   95, 97, 119, 211, 222, 242, 293–301
– scaling   95, 98–99, 119, 212, 222
– similarity   212, 222, 231, 285–293
– skew   95
– slant   95
– translation   95, 119, 126–127, 211, 222
translation   95, 119, 126–127, 211, 222
transmittance   9
transposition   127
trigger   37, 39, 62–63
Tukey weight function   172

u

union   125–126, 293–301
universal serial bus   see USB
USB   57–58
– bulk data transfers   58
– control transfers   57
– interrupt data transfers   58
– isochronous data transfers   58

v

variation model   107–111, 281–285, 293–301
vertical blanking interval   51
vertical overflow drain   39
vertical synchronization pulse   51
video signal
– analog
– – back porch   51
– – CCIR   49
– – EIA-170   49
– – front porch   50
– – horizontal blanking interval   50
– – horizontal synchronization pulse   50
– – interlaced scan   50
– – NTSC   49
– – PAL   49
– – progressive scan   51
– – vertical blanking interval   51
– – vertical synchronization pulse   51
– color
– – composite video   51
– – RGB   51
– – S-Video   51
– – Y/C   51
– digital   54–60
– – Camera Link   54–55
– – FireWire   see video signal, digital, IEEE 1394
– – frame valid   54
– – Gigabit Ethernet   see video signal, digital, GigE Vision

– – GigE Vision   59–60
– – IEEE 1394   56–57
– – IIDC   56–57
– – line valid   54
– – low-voltage differential signaling   54
– – LVDS   *see* video signal, digital, low-voltage differential signaling
– – pixel clock   54
– – USB   57–58
vignetting   34–35, 74

w
weight function
– Huber   172
– Tukey   172

world coordinates
– from single image   167–168, 193–195
– – line scan camera   195
– – pinhole camera   194–195
– – telecentric camera   193–194
– from stereo reconstruction   198–210

x
xenon lamp   7

y
Y/C video   51

z
zero crossing   146, 148, 160